BELIEFS: A HIDDEN VARIABLE IN MATHEMATICS EDUCATION?

Mathematics Education Library

VOLUME 31

The titles published in this series are listed at the end of this volume.

BELIEFS: A HIDDEN VARIABLE IN MATHEMATICS EDUCATION?

Edited by

GILAH C. LEDER
La Trobe University, Australia

ERKKI PEHKONEN
University of Turku, Finland

and

GÜNTER TÖRNER
University of Duisburg, Germany

KLUWER ACADEMIC PUBLISHERS
DORDRECHT / BOSTON / LONDON

Library of Congress Cataloging-in-Publication Data

ISBN 1-4020-1057-5

Published by Kluwer Academic Publishers,
P.O. Box 17, 3300 AA Dordrecht, The Netherlands.

Sold and distributed in North, Central and South America
by Kluwer Academic Publishers,
101 Philip Drive, Norwell, MA 02061, U.S.A.

In all other countries, sold and distributed
by Kluwer Academic Publishers,
P.O. Box 322, 3300 AH Dordrecht, The Netherlands.

Printed on acid-free paper

Printed in the Netherlands.

TABLE OF CONTENTS

Tables and Figures vii
Acknowledgements xi
Contributors xiii

Chapter 1: Setting the Scene 1
Gilah C. Leder, Erkki Pehkonen, and Günter Törner

PART 1 BELIEFS: CONCEPTUALIZATION AND MEASUREMENT

Chapter 2: Framing Students' Mathematics-Related Beliefs. A Quest for Conceptual Clarity and a Comprehensive Categorization 13
Peter Op't Eynde, Erik de Corte, and Lieven Verschaffel

Chapter 3: Rethinking Characterizations of Beliefs 39
Fulvia Furinghetti and Erkki Pehkonen

Chapter 4: Affect, Meta-Affect, and Mathematical Belief Structures 59
Gerald A. Goldin

Chapter 5: Mathematical Beliefs – A Search for a Common Ground: Some Theoretical Considerations on Structuring Beliefs, Some Research Questions, and Some Phenomenological Observations 73
Günter Törner

Chapter 6: Measuring Mathematical Beliefs and Their Impact on the Learning of Mathematics: A New Approach 95
Gilah C. Leder and Helen J. Forgasz

Chapter 7: Synthesis - Beliefs and Mathematics Education: Implications For Learning, Teaching, and Research 115
Douglas B. McLeod and Susan H. McLeod

PART 2 TEACHERS' BELIEFS

Chapter 8: Mathematics Teacher Change and Development. The Role of Beliefs 127
Melvin (Skip) Wilson and Thomas J. Cooney

Chapter 9: Mathematics Teachers' Beliefs and Experiences with Innovative Curriculum Materials. The Role of Curriculum in Teacher Development 149
Gwendolyn M. Lloyd

Chapter 10: A Four Year Follow-up Study of Teachers' Beliefs after Participating in a Teacher Enhancement Project 161
Lynn C. Hart

Chapter 11: Belief Structure and Inservice High School Mathematics Teacher Growth 177
Olive Chapman

Chapter 12: Participation and Reification in Learning to Teach: The Role of Knowledge and Beliefs 195
Salvador Llinares

Chapter 13: A Study of the Mathematics Teaching Efficacy Beliefs of Primary Teachers 211
George Philippou and Constantinos Christou

Chapter 14: Situating Research on Mathematics Teachers' Beliefs and on Change 233
Stephen Lerman

PART 3 STUDENTS' BELIEFS

Chapter 15: Beliefs about Mathematics and Mathematics Learning in the Secondary School: Measurement and Implications for Motivation 247
Peter Kloosterman

Chapter 16: "The Answer is Really 4.5": Beliefs about Word Problems 271
Brian Greer, Lieven Verschaffel, and Erik De Corte

Chapter 17: Beliefs about the Nature of Mathematics in the Bridging of Everyday and School Mathematical Practices 293
Norma Presmeg

Chapter 18: Beliefs and Norms in the Mathematics Classroom 313
Erna Yackel and Chris Rasmussen

Chapter 19: Intuitive Beliefs, Formal Definitions and Undefined Operations: Cases of Division by Zero 331
Pessia Tsamir and Dina Tirosh

Chapter 20: Implications of Research on Students' Beliefs for Classroom Practice 345
Frank K. Lester, Jr.

Index 355

TABLES AND FIGURES

List of Tables

Chapter 3	Table 1: The nine characterizations of belief included in the questionnaire	47
	Table 2: Responses to the 10 items by individual respondents	49
	Table 3: Degree of agreement/disagreement of the respondents with the characterizations	49
Chapter 6	Table 1: Selected definitions of beliefs	96
	Table 2: Summary of selected methods for measuring attitudes/beliefs	98
	Table 3:Beliefs in recent mathematics education research	101
Chapter 10	Table 1: Percent of teachers responding by item on factors they believe influenced their change	169
Chapter 13	Table 1: Proportions of indicative items by institution and overall mean	222
	Table 2: A summary of responses in the interviews by scale dimension and preservice education	227
Chapter 14	Table 1: Teachers' positions within different pedagogic modes	239
Chapter 16	Table 1: Examples of P-items used in Greer (1993) and Verschaffel et al. (1994)	275
	Table 2: Percentages of Realistic Reactions (RRs) on selected P-items in various studies	276
	Table 3. Percentages of Realistic Reactions (RRs) on selected P-items in the study of Community College students by Mukhopadhyay and Greer (2000) (N = 13)	278
	Table 4. Percentages of Realistic Reactions (RRs) on selected P-items in the study of teachers in training by Verschaffel et al. (1997) (N = 332)	280
	Table 5. Percentages of scores for realistic and non-realistic answers on P-items for the first-year and third-year student teachers (Verschaffel et al., 1997)	280

Chapter 17	Table 1: Student beliefs concerning the nature of mathematics	296
	Table 2: Activities described by students	297
	Table 3: Other activities in which potential for math was seen by students	298
	Table 4: Beliefs of students about the nature of mathematics, Fall 1999	303
Chapter 18	Interpretive framework for analyzing individual and collective activity classrooms	315
Chapter 19	Table 1: Percentages of responses to a÷0 (a≠0) expressions by grade and level of mathematics achievement (in %)	337
	Table 2: Percentages of responses to 0÷0 expressions by grade and level of mathematics achievement (in %)	337

List of Figures

Chapter 2	Figure 1: Different categorizations of students' beliefs	19
	Figure 2: Constitutive dimensions of students' mathematics-related belief system	27
	Figure 3: A framework of students' mathematics-related beliefs	28
Chapter 5	Figure 1: Different belief structures	88
Chapter 6	Figure 1: Excerpts from the experience Sampling Form	106
	Figure 2: The spread of activities in which students in our study were engaged at the time they were signalled, overall and by gender	107
	Figure 3: Overview of Caitlin and Boyd's activities when beeped	108
	Figure 4: Results from selected ESFs for two case studies	109
Chapter 11	Figure 1: Elise's beliefs	187
	Figure 2: Mark's beliefs	188
Chapter 14	Figure 1: Fields and sub-fields in the production of positions of teachers	239
Chapter 16	Figure 1: Factors shaping beliefs about word problems	285

Chapter 17	Figure 1: Melanie'semiotic chain for combinations of various stitch designs	306
	Figure 2: Derek's semiotic chain for the symmetries of a tennis court	309

ACKNOWLEDGEMENTS

Planning for this book began at a meeting held late in 1999 in Germany at the unique venue of the *Mathematisches Forschungsinstitut Oberwolfach*, well known to mathematicians for its hosting of special research meetings, workshops, and other research programs. We gratefully acknowledge the splendid environment so conducive to productive work and invaluable support offered by this Institute.

Thanks are also due to Christiane Dohren and Thorsten Bahne (Germany), and to Helen Neville (Australia), for their help in preparing the final manuscript.

I would also like to extend my sincere appreciation to my coeditors, Erkki Pehkonen and Günter Törner, as well as to all the contributing authors, all co-travelers on the highs and lows of the journey from planning to realization of the manuscript.

Gilah Leder
Melbourne, Australia
August 2002

CONTRIBUTORS

Editors

Gilah Leder is Director of the Institute for Advanced Study at La Trobe University – Bundoora, Australia, and a Professor in the Institute for Education at the same institution. Her teaching and research interests embrace gender issues, affect, the interaction between teaching, learning and assessment of mathematics, and exceptionality. She has published widely in each of these areas. Gilah serves on various editorial boards and educational and scientific committees. She is Past President of the Mathematics Research Group of Australasia (MERGA), immediate Past President of the International Group for the Psychology of Mathematics Education (PME), and a Fellow of the Academy of the Social Sciences in Australia.
Email: G.Leder@latrobe.edu.au

Erkki Pehkonen is a Professor of Mathematics and Science Education at the University of Turku in Finland. He is interested in ways of improving mathematics instruction in the comprehensive school, especially via problem solving with a focus on motivating pupils, and in understanding pupils' and teachers' beliefs/conceptions about mathematics teaching and learning, and furthermore in conditions of teacher change.
Email: Erkki.Pehkonen@utu.fi

Günter Törner was appointed a full Professor at the University of Duisburg (Germany) in 1978. Since completing his PhD and Habilitation, he has worked as a mathematician as well as a mathematics educator. Because his research and lecturing is involved in both fields, he was elected as an Executive Member of the presidential board of the German Mathematical Society (DMV) as well as of the German Society for Mathematics Education (GDM). His research interests in mathematics cover algebra, geometry and discrete mathematics. In mathematics education he focuses on belief issues, problems of mental representations, teacher education and diverse subjects from secondary mathematics. He is also involved in the field of electronic publishing and a member of the initiative of Networked Digital Library of Theses and Dissertations (NDLTD) at Virginia Tec.
Email: toerner@math.uni-duisburg.de

Other contributors

Olive Chapman is Associate Professor of Mathematics Education at the Universit of Calgary, Canada. She works with preservice and inservice teachers at th undergraduate and graduate levels, respectively. Her research interests an publications focus on the learning and, in particular, teaching of mathematical wor

problems and problem solving; understanding the teaching of mathematics in classrooms through mathematics teachers' thinking/perspectives and classroom behaviors; preservice mathematics teacher education and inservice mathematics teacher professional development using metaphors and narratives; and the nature of mathematical thinking in classroom approaches to teaching/learning high school mathematics.
Email: Chapman@Ucalgary.Ca

Constantinos Christou is Associate Professor of Mathematics Education at the University of Cyprus. He studied at the Western College (majoring in mathematics), at the Antioch University (M.A. Mathematics Education) and the University of Toledo, OH (Ph.D. Mathematics Education). He has conducted research on problem solving, conceptions of students about mathematical knowledge, efficacy beliefs of mathematics teachers, and cognitive development of mathematical concepts. His published work has appeared in a variety of international journals.
Email: edchrist@ucy.ac.cy

Thomas J. Cooney is Professor Emeritus of Mathematics Education from the University of Georgia (USA). His research focuses on teachers' beliefs as they relate to teacher change and on issues related to mathematics teacher education. He is Founding Editor of the *Journal of Mathematics Teacher Education.* His current work also includes research and development activities involving teachers' use of open-ended assessment items.
Email: Proftcooney@aol.com

Erik De Corte is Professor of Educational Psychology and Director of the Center for Instructional Psychology and Technology at the University of Leuven, Belgium. His major research interest is to contribute to the development of theories of learning from instruction and the design of powerful learning environments, focusing thereby on learning, teaching, and assessment of thinking and problem solving, especially in mathematics. He was the founding editor of the journal *Learning and Instruction* (1990-1993), and from 1987 till 2002 associate editor of the *International Journal of Educational Research.* He co-edited (with F.E. Weinert) the International encyclopedia of developmental and instructional psychology (1996). He is currently President of the International Academy of Education (1998-2004). In March 2000 he was awarded a doctorate honoris causa of the Rand Afrikaans University, Johannesburg, South Africa.
(http://www.kuleuven.ac.be/~p1486000)

Helen Forgasz is a Senior Lecturer in Mathematics Education at Deakin University, Australia. Before embarking on an academic career, she was a secondary teacher of mathematics, physics and computing. Her current research interests include gender issues, the affective domain, and computers for the learning of mathematics. Helen has published widely on a range of mathematics education issues. She recently co-edited a book entitled *Sociocultural research on mathematics education: An*

international perspective. Helen is the Vice President (Research) of the Mathematics Education Research Group of Australasia.
Email: forgasz@deakin.edu.au

Fulvia Furinghetti teaches "elementary mathematics from an advanced standpoint" in the Department of Mathematics of University of Genoa (Italy). She is the director of a group of mathematics teachers and researchers working in the field of mathematics education. Her educational research concerns mathematical beliefs, the integration of the history in mathematics teaching, approaches to proof, teacher education and training. In 2000 she was appointed chairperson of the International Study Group on History and Pedagogy of Mathematics affiliated to ICMI. She is the author of papers and books in mathematics education and in the history of mathematics.
Email: Furinghe@dima.unige.it

Gerald Goldin received his B.A. in chemistry and physics from Harvard University, and his Ph.D. in theoretical physics from Princeton University. He taught at the University of Pennsylvania and Northern Illinois University, before joining the Rutgers University faculty in 1984. In addition to his work as a mathematical physicist, for which he received a Humboldt Research Award in 1998, he studies the psychology of mathematical learning and problem solving. Here he has focused on cognitive and affective systems of representation and their development. Gerald has served as Director of Rutgers' Center for Mathematics, Science, and Computer Education (1985-98), and as organizer and Principal Investigator of New Jersey's Statewide Systemic Initiative in mathematics, science, and technology education (1993-98). Currently he is the University Director for Science and Mathematics Partnerships, as well as Professor of Mathematics, Physics, and Education.
Email: gagoldin@dimacs.rutgers.edu

Brian Greer became interested in mathematics education after studying mathematics and psychology. Following early work with a group led by Zoltan Dienes, his orientation changed during a sabbatical year at the Shell Centre, where Alan Bell and Efraim Fischbein introduced him to the fascinating topic of multiplicative reasoning. He has enjoyed a long and fruitful collaboration with his Belgian colleagues, most recently in co-authoring the book *Making sense of word problems*". In 2000, he moved to the United States where he now works closely with his partner, Swapna Mukhopadhyay, attempting to elucidate the historical, cultural, and political embeddedness of mathematics education.
Email: ggreer@mail.sdsu.edu

Lynn Hart is a mathematics educator in the Department of Early Childhood Education, Georgia State University, Atlanta, Georgia, U.S.A. She conducts research in the area of teacher change in mathematics and the role of teacher beliefs in teacher change. She is one of the co-developers of the Reflective Teaching Model (RTM), a teacher development model for supporting teachers as they attempt to learn about and implement reform recommendations. She is also a director of the

Collaborative Masters Program in elementary education at GSU and she consults with schools and colleagues across the United States in the area of reform in mathematics teaching.
Email: ecelch@langate.gsu.edu

Peter Kloosterman is a Professor of Mathematics Education in the School of Education at Indiana University, Bloomington (USA). Although he has worked in the areas of gender issues, adult literacy, and teaching mathematics through application, his primary research focus has been students' attitudes, beliefs, and motivation in mathematics. In particular, he has applied general psychological theories of motivation to mathematics learning and has looked at reasons why students resist inquiry-based instruction in mathematics. Additional information, including a list of publications and copies of some of his conference papers, can be found on his web page (www.indiana.edu/~pwkwww/).
Email: klooster@indiana.edu

Stephen Lerman was a secondary school teacher of mathematics for many years. Whilst working on his PhD, which was a study of teachers' beliefs about mathematics and mathematics education, drawing on philosophy of mathematics, he began lecturing in mathematics education at the Institute of Education. He is now Professor of Mathematics Education at South Bank University in London. He is a former Chair of the British Society for Research in Learning Mathematics (BSRLM) and a former President of the International Group for the Psychology of Mathematics (PME). His research interests include classroom research, sociocultural theories, equity issues, and teacher education.
Email: lermans@sbu.ac.uk

Frank K. Lester, Jr. is Martha Lea and Bill Armstrong Professor in Teacher Education. He also is Professor of Mathematics Education and of Cognitive Science at Indiana University - Bloomington. His primary research interests lie in the areas of mathematical problem solving and metacognition, especially problem-solving instruction. From 1991 to 1996 he was the editor of the *Journal for Research in Mathematics Education*, a leading research journal in mathematics education, following a four-year term as editor of that journal's monograph series. He also serves as consulting editor for several other research journals. From 1999-2002, he served on the Board of Directors of the National Council of Teacher of Mathematics. In 2002, he began a four-year project involving the preparation of interpretive analyses of the past ten years of student mathematics achievement data gathered by the National Assessment of Educational Progress in the United States.
Email: lester@indiana.edu

Salvador LLinares graduated in mathematics from the University of Valencia (Spain) and gained his doctorate in education from the University of Seville (Spain) where he was a mathematics teacher educator in the Department of Mathematics Education for twenty years. He is currently Professor of Mathematics Education at the University of Alicante (Spain). His research has centered on processes of

learning to teach, mathematics teachers' professional knowledge and the practices of mathematics teachers. He developed different learning settings for teacher education. His current research focuses on the integration of ICTs in Teacher Education and on the uses of situated perspectives to analysis teacher learning and practice.
Email: sllinares@ua.es

Gwendolyn Lloyd is an Associate Professor in the Department of Mathematics at Virginia Tech. Dr. Lloyd is the Principal Investigator of an ongoing National Science Foundation-sponsored project that focuses on the mathematical and pedagogical conceptions of preservice elementary teachers. She is particularly interested in understanding changes to teachers' conceptions of mathematics curriculum. She also teaches mathematics and education courses for elementary and secondary education students, directs research seminars for graduate students, and organizes workshops for teachers.
Email: lloyd@calvin.math.vt.edu

Douglas B. McLeod is Professor of Mathematical Sciences at San Diego State University in the USA, where he has been a member of the faculty since 1972. He served for three years as Program Director for mathematics education at the National Science Foundation in Washington, DC, and for seven years as Professor of Mathematics and Education at Washington State University. He also taught at Haile Selassie I University in Addis Ababa, Ethiopia, and at Texas Southern University, Houston, Texas. His publications focus on affective issues in mathematics education and on curriculum reform in mathematics. His doctorate is from the University of Wisconsin.
Email: dmcleod@sciences.sdsu.edu

Susan H. McLeod is Professor and Director of the Writing Program at the University of California, Santa Barbara (USA). From 1986 to 2001, she was Professor of English at Washington State University, where she also served as Department Chair and Associate Dean of the College of Liberal Arts. She also taught at San Diego State University, Haile Selassie I University in Addis Ababa, Ethiopia, and at Texas Southern University, Houston, Texas. Her publications focus on writing across the curriculum, writing program administration, and affective issues in the teaching of writing. Her doctorate is from the University of Wisconsin.
Email: mcleod@mail.wsu.edu

Peter Op 't Eynde is a researcher at the Centre for Instructional Psychology and Technology at the University of Leuven. His major research interests are the role of motivational and emotional processes in learning and to contribute to the development of theories of learning and instruction that adequately address these conative and affective factors as well as the more (meta)cognitive ones. He is preparing a doctoral dissertation on the role of emotions in mathematical problem solving and is currently

coordinating the development and implementation of a Multimedia Interactive Learning Environment (MILE-Flanders) for mathematics education in teacher colleges.
Email: Peter.OptEynde@ped.kuleuven.ac.be

George N. Philippou is Professor of Mathematics Education at the University of Cyprus. He studied mathematics at the University of Athens, mathematics education at the American University of Beirut and received his Ph.D. from the University of Patras. He has taught mathematics in public schools and Colleges in Cyprus, and mathematics and mathematic education courses at the University of the Aegean. He has conducted research dealing with attitudes and beliefs, problem solving and mathematical evaluation. He has published work in Conference Proceedings (e.g., PME), and in a variety of international journals.
Email: edphilip@ucy.ac.cy

Norma Presmeg is a Professor in the Mathematics Department at Illinois State University in the USA. Her degrees include B.Sc. in mathematics and physics from Rhodes University in South Africa, B.Sc. Honours in mathematics and B.Ed. from University of Natal. Her M.Ed. dissertation, in the Department of Educational Psychology at the University of Natal, was an analysis of Albert Einstein's creativity with implications for mathematics education. After completing the Ph.D. degree in mathematics education at Cambridge University in England, she served on the faculty at the University of Durban-Westville in South Africa for five years before joining Florida State University in the USA in 1990. In Florida she was a faculty member in the Department of Curriculum and Instruction for ten years. She is currently an editor of *Educational Studies in Mathematics*, and serves on the Editorial Board of *Journal for Research in Mathematics Education*.
Email: npresmeg@msn.com

Chris Rasmussen is Associate Professor of Mathematics Education in the Department of Mathematics, Computer Science, and Statistics at Purdue University Calumet, Hammond, IN, USA. Formerly a mechanical engineer and high school mathematics teacher in the Peace Corps, his research focuses on how emerging analyses of student thinking, technology, context problems, and symbol-use can be profitably coordinated to promote student learning in undergraduate mathematics. Together with his colleagues, his integrated research and instructional design efforts in differential equations explore how theory-driven work at the elementary and secondary level can inform, guide, and sustain the learning and teaching of university mathematics.
Email: raz@calumet.purdue.edu

Dina Tirosh serves as the head of the Department of Mathematics and Science Education at Tel Aviv University. She was previously a mathematics teacher in high schools and in teachers colleges. Her main interests are in students' understanding of mathematical concepts, the role of intuitive rules in science and mathematics, teacher education and professional development of mathematics teachers.
Email: dina@post.tau.ac.il

Pessia Tsamir serves as the coordinator of the Mathematics Teacher Education Program at the Department of Mathematics and Science Education at Tel Aviv University. She was previously a mathematics teacher in high schools and in teachers colleges. Dr. Tsamir's research focuses on examining students' and teachers' intuitive, algorithmic and formal mathematical understanding and on the impact of different ways of teaching on students' solutions to mathematical tasks.
Email: pessia@post.tau.ac.il

Lieven Verschaffel is Professor and Director of the Section of Instructional Sciences at the Faculty of Psychology and Educational Sciences of the University of Leuven, Belgium. His major research interests are teaching and learning problem solving and learning skills, psychology of mathematics education, and computer-supported learning and instruction. From 1990 until 1993 he was assistant editor of *Learning and Instruction*. Since 1999 he is the editor of *Pedagogische Studiën*, the leading journal in educational and instructional sciences in The Netherlands and Flanders. Lieven Verschaffel was and is actively involved in the reform of the elementary school mathematics curriculum in Flanders, as a member of several programming committees.
Email: Lieven.Verschaffel@ped.kuleuven.ac.be

Melvin (Skip) Wilson is an Associate Professor at Virginia Tech in Blacksburg, Virginia (USA). He teaches mathematics and education courses for prospective and practicing secondary mathematics teachers. His research is related to how teachers' conceptions about mathematics and the teaching of mathematics are related to their abilities and tendencies to share responsibility with their students and involve them in exploratory instructional activities.
Email: skipw@vt.edu

Erna Yackel is Professor of Mathematics Education in the Department of Mathematics, Computer Science, and Statistics at Purdue University, Hammond, IN, USA. Her research involves developing theoretically grounded ways to investigate mathematics teaching and learning in the classroom, including both individual students' learning and the learning of the class as a collective. She is particularly interested in explanation, justification and argumentation.
Email: yackeleb@calumet.purdue.edu

CHAPTER 1

GILAH C. LEDER, ERKKI PEHKONEN, AND GÜNTER TÖRNER

SETTING THE SCENE

Abstract. Although the importance of engaging students both cognitively and affectively when they learn mathematics is now widely recognized, the place of beliefs in the teaching and learning of mathematics is not well researched. After a brief introduction in which some contextual issues are raised, the contents of the contributions that follow – each with a clear focus on beliefs in mathematics education - are described in this introductory chapter.

1. INTRODUCTION

Mathematics is widely recognized not only as a core component of the curriculum but also as a critical filter to many educational and career opportunities. Yet in recent years much concern has been expressed about students' reluctance to continue with the study of mathematics well beyond the compulsory years, a trend often described emotionally as the drift away from mathematics and from the physical sciences. The passage below, taken from a national Australian document, is representative of the sentiments voiced.

> Mathematics and science have a fundamental contribution to make both to understanding the world and to changing the world, particularly in the context of change and economic adjustment. The decline in interest in mathematics ... needs to be arrested. This is an urgent and complex matter related not only to education but to other issues. (Department of Employment, Education and Training, 1989, p. 14)

Lip service is certainly paid to the importance of engaging students of mathematics - of all ages - affectively as well as cognitively (see e.g., Jensen, Niss, & Wedege, 1998; the National Council of Teachers of Mathematics [NCTM], 2000). But (how) will that make a difference? What do we know about the interplay between beliefs and behaviors? Only a decade ago, McLeod (1992) noted: "(a)lthough affect is a central concern of students and teachers, research on affect in mathematics education continues to reside on the periphery of the field" (p. 575). Schoenfeld (1992) similarly argued that there was "a fairly extensive literature on" student beliefs, "a moderate but growing literature" about teacher beliefs, and as yet relatively little exploration of "general societal beliefs about doing mathematics" (p. 358). Research on mathematics and gender has been singled out as one area where "some aspects of beliefs about self have been researched quite thoroughly" (McLeod, 1992, p.580). A review of this work is beyond the scope of this

G. C. Leder, E. Pehkonen, & G. Törner (Eds.), Beliefs: A Hidden Variable in Mathematics Education? 1-10.

introductory chapter, but can be found in, for example, Fennema and Hart (1994), Forgasz and Leder (2001a, 2001b), Leder (1992), and Leder, Forgasz, and Solar (1996). At the same time, there are areas in research on beliefs, for example teacher change, where extensive research has been carried out over some two decades, but no consistent pattern has yet been identified for facilitating teacher change. Research reports about intervention programs commonly conclude that some teachers have changed, but others have not (e.g., Senger, 1999; Borko, Davinroy, Bliem, & Cumbo, 2000; Wood, 2001; Hart, this volume; Wilson & Cooney, this volume).

1.1. Genesis of the Book.

Each of the contributions in this book has a clear focus on beliefs about mathematics. Aspects of beliefs about mathematics, its teaching and learning, are examined broadly and from a variety of perspectives. Collectively, the contributions reflect the diverse approaches used in the conceptualization and study of beliefs. The genesis of the book can be traced to a specialist international meeting about mathematics-related beliefs held in the unique setting of the *Mathematisches Forschungsinstitut Oberwolfach* in November 1999 (Pehkonen & Törner, 1999). The majority of the authors of the book were invited participants to the Institute, and their chapters are based on the presentations they gave at the meeting. Thus the core ideas contained in the chapters, and the diverse ways in which beliefs in mathematics education can be explored, were discussed beforehand, during an international high-level forum. The presentations and discussions were subsequently formalized, elaborated into full papers, and subjected to extensive peer-review.

2. ABOUT BELIEFS

A careful reading of the early psychological literature reveals that beliefs and belief systems began to be explored in the beginning of this century, particularly by social psychologists (Thompson, 1992). However, as the more behavioral perspectives of learning attracted increased attention, research endeavors turned to the more readily observed parts of human behavior, with a consequent loss of interest in beliefs. With new developments in cognitive science in the 1970s, attention to beliefs and belief systems re-emerged (Abelson, 1979). Although work on beliefs can be found in areas as diverse as political science, history, psychology, sociology, and anthropology, it is "especially social psychologists, who have devoted much effort into studying the acquisition and change of beliefs, their structure, their contents, and their effects mainly on individuals' affect and behavior" (Bar-Tal, 1990).

Given the variety of perspectives and disciplines within which beliefs have been studied, it is not surprising that the field abounds with subtly different definitions and classifications of beliefs. A detailed overview of this work is again beyond the scope of this chapter, but can be found in later chapters in this volume (Furinghetti & Pehkonen; Op't Eynde, de Corte, & Verschaffel). Here it is convenient to reproduce some key features highlighted by Bar-Tal (1990):

- The study of beliefs can be classified into four areas: "(a) acquisition and change of beliefs, (b) structure of beliefs, (c) effects of beliefs, and (d) content of beliefs[1]" (p. 12)
- "Beliefs have been viewed by social psychologists as units of cognition. They constitute the totality of an individual's knowledge, including what people consider as facts, opinions, hypotheses, as well as faith" (p. 12). Descriptions such as this highlight the difficulty often shown in distinguishing between beliefs and knowledge[2].
- "Beliefs [can] be differentiated on the basis in which they are formed: (a) Descriptive beliefs are formed on the basis of direct experience.... (b) Inferential beliefs ... are based on rules of logic that allow inferences.... (c) Informational beliefs are formed on the basis of information provided by outside sources..." (based on the work of Bem (1970) and Fishbein & Ajzen (1975), as summarized by Bar-Tal, p. 12).
- "Psychologists have suggested different features and dimensions to characterize beliefs.... Krech and Crutchfield (1948) proposed the following seven characteristics to describe beliefs: kind, content, precision, specificity, strength, importance, and verifiability" (p. 15).
- It is useful to focus on four characteristics of beliefs: "Confidence, centrality, interrelationship, and functionality. Confidence differentiates beliefs on the basis of truth attributed to them; centrality characterizes the extent of beliefs in individuals' repertoire...; interrelationship indicates the extent to which the belief is related to other beliefs; functionality differentiates beliefs on the basis of the needs they fulfill" (p. 21).

Almost any one of these characterizations could have been used as an organizational theme for the various contributions in this volume. The grouping we finally selected is one we consider to be particularly constructive for the examination of beliefs about mathematics, its teaching and learning from a variety of different perspectives. We settled on a three clusters: contributions with a major focus on the concept of beliefs in mathematics education; on teachers' beliefs; and on students' beliefs.

3. ABOUT THE BOOK

As already indicated, the book is divided into three main sections, with different yet overlapping themes. The broad international mix of the contributing authors ensures a diversity of perspectives, as well as reference to relevant research beyond that published in English. Such coverage is less likely to be achieved with a culturally more homogeneous group of contributors. Thus the book offers a variety of different perspectives into the concept of beliefs, and into methods of investigating the place of beliefs in the teaching and learning of mathematics. A synthesis/critique chapter which, inter alia, highlights common and diverse themes, concludes each section.

3.1. Part 1: The Concept 'Belief'

The contributions in the first section are particularly concerned with examining what is meant by mathematical beliefs and how they differ from other, related concepts. The authors draw extensively on existing literature, highlight consensus and confusion in the ways various terms have been used in earlier work, and indicate directions for further research, without – collectively – offering a unified view on the main theoretical concepts explored: belief, conception, and knowledge. The overall thrust is rather on the diversity of different starting points which typically correspond to different emphases.

In the opening chapter, *Peter Op 't Eynde, Erik de Corte,* and *Lieven Verschaffel* (Belgium) draw on an extensive and methodically identified body of literature to delineate current understandings of the nature of beliefs and the functioning of belief systems. They trace the ways in which beliefs are thought to develop, and the influence of contextual factors on this development. Issues of overlap, and distinctions between beliefs and knowledge, are inevitably raised as part of this discussion. Finally these authors propose a theoretical model that integrates key components of major, and influential, models found in the literature and also allows a more elaborated definition of students' related mathematics beliefs.

Fulvia Furinghetti (Italy) and *Erkki Pehkonen* (Finland) examine, in the second chapter of this section, differences and overlap between beliefs, conceptions, and knowledge, in terms of the broad meanings they convey and the methods used to gauge them. Their review encompasses works within and beyond mathematics education and includes research carried out in a diversity of countries. In the second half of their chapter they describe the results of a study in which well-known mathematics educators were asked to delineate what "beliefs" conveyed to them. After identifying descriptors of beliefs with which the group generally agreed or disagreed, Furinghetti and Pehkonen recommend a more restrictive use of the term and contend that it is inappropriate to expect that one single definition of beliefs will be suitable for all the possible fields of application. "Contextualization and goal-orientation make the characterization an efficient one", they conclude.

In his chapter, *Gerald Goldin* (USA) argues that beliefs are multiply-encoded, useful, empowering to their holder, and consistent with evidence available at the time. By distinguishing between, for example, beliefs, belief structures, belief systems, knowledge, and values he is able to draw out subtle differences he considers to be critical for an informed exploration of what the concept of beliefs appropriately does and does not convey. The general discussion that takes up much of the chapter is given some specificity through a preliminary topology of types of mathematical beliefs that could be fruitfully considered. That device is also useful for highlighting the intersecting role of the individual, and social and cultural dimensions and their impact on beliefs. Implicit references to the heated debate, in California in particular, about what constitutes an optimum mathematics curriculum are a further, and timely, reminder that beliefs are subtly shaped by many contextual factors – including the political climate in which they are formed.

Various sources in the research literature are invoked by *Günter Törner* (Germany) to explain why a single definition of beliefs has proved so elusive.

Language specific nuances are introduced as another confounding factor. Specifically, Törner cautions in his chapter: "the word 'belief' cannot be translated into the German language without being open to interpretation". A critical appraisal of common definitions of beliefs will, he argues, yield a four-component definition of beliefs which has common features with definitions used in modern algebra: the belief object – which can be abstract or relatively concrete, the mental associations made by an individual with reference to the belief object, the threshold and potency of each association, and one or more evaluation maps – ranging from extremes of positive to negative. Törner argues that reliance on these dimensions allows commonalities and differences in the various definitions of beliefs to be captured effectively and should also prove a useful catalyst for further research on beliefs. His formulaic rather than purely narrative approach distinguishes the tone of this chapter.

Gilah Leder and *Helen Forgasz* (Australia) begin their chapter with a brief review of common definitions of beliefs and ways in which beliefs are frequently measured. This is done to provide a context for the study they report in the latter part of the chapter, rather than as an attempt to provide a detailed overview of the ways in which beliefs can best be defined and operationalized. Their eventual emphasis on the Experience Sampling Method [ESM] as a technique for tapping beliefs suggests that these authors, too, consider beliefs to be multi-dimensional. After describing the scope and method of administration of the ESM, they focus on data gathered from two students to illustrate the ways in which the measure captures students' beliefs and draws out students' values and beliefs about mathematics learning and about various daily activities more broadly.

A synthesis and critique of the contributions in this section, written by *Douglas McLeod* and *Susan McLeod* (USA) completes this part of the book. They place the chapters reviewed in a broader context, trace common themes running through the work in this section, and argue that, rather than striving for the elusive goal of reaching agreement on a single definition for the term "belief", it seems more productive to cluster the various definitions - e.g., in terms of intuitive definitions, exact definitions, and advanced definitions. Such an approach is deemed more consistent with the reality that beliefs are discussed for different purposes and with different audiences. Suggested pathways for future research conclude the chapter.

3.2. Part 2: Teachers' Beliefs

Each of the chapters in part 2 has a focus on teachers' beliefs; many as well on teacher change. The settings in which these are explored differ widely. Collectively these contributions give rich insights into the ways in which beliefs are shaped and reflected in the instructional strategies used in the mathematics classroom.

The chapter by *"Skip" Wilson* and *Tom Cooney* (USA) opens part 2. The authors begin with a discussion of common definitions of beliefs. Rather than concentrating on the nuances of overlap and difference between beliefs and knowledge, they focus on reports of teacher change that, they contend, can illuminate the interactive functions of knowledge and beliefs and the role played by beliefs in inhibiting or facilitating change. An extensive and thoughtful review of relevant articles published in recent years in three influential mathematics education journals, with a wide international readership, is used to sketch current research on mathematics teachers' beliefs. The review illustrates that a number of common themes are raised in many of the articles. These concern the importance of reflection, the relevance of the context in which teaching occurred, teachers' proficiencies in attending to students' understanding, and the failure – described in many of the articles – to attend to both content and pedagogy. "There does not appear to be a consensus", Wilson and Cooney conclude, "about what constitutes beliefs and whether they include or simply reflect behavior". The latter issue is tackled, in different ways, by other contributors to the book.

Much effort has been spent on producing new curriculum materials that reflect and support the principles about teaching and learning advocated by the Reform Movement in the USA. Their impact on teachers' and students' beliefs about mathematics, its learning, and teaching, are described in the chapter by *Gwendolyn Lloyd* (USA). Having to grapple with new materials, recognizing that they cover content areas not previously studied, and being exposed to a broader and new range of problem contexts and formulations are, Lloyd illustrates, important catalysts for change.

Lynn Hart (USA) draws on the experiences of teachers involved in a specific "teacher enhancement" project to explore what they themselves believed to be critical in helping them achieve change. The sequence used in the project – experience followed by reflection – persistently encouraged teachers to evaluate their own, and possibly changing, instructional practices. The high value attached to support from other colleagues and to collaboration illustrates the influence of the social context, and support of critical others, in achieving change. The impact of new curriculum materials *per se* was considered of lesser importance. The insights into their own behaviors and beliefs provided by the teachers who persisted in using strategies promoted by the project, and who exhibited long-term changes in their teaching behaviors, are, according to Hart, useful indicators for those who want to achieve instructional changes through pre- and inservice activities.

In her chapter, *Olive Chapman* (Canada) illustrates, through two case studies, how the tensions between teachers' beliefs about mathematics and their current teaching practices led to changes in the latter – though these changes were achieved neither easily nor readily. The descriptive accounts are set in a clear theoretical context. Evidence is presented of the centrality, psychological strength, and stability of the beliefs that underpinned the changes in instructional strategies gradually implemented by the teachers to achieve greater congruence between the ways in which mathematics was presented in the classroom and their beliefs about mathematics. In both cases discussed, behaviors were altered to accommodate deeply held beliefs. Implications drawn from the work presented for achieving change among the wider teaching force in the way in which mathematics is taught are also explored.

How the notions of situated learning and a community of practice, advanced in particular by Jean Lave and Etienne Wenger, can be used to deconstruct the attempts made by student teachers to understand the difficulties experienced by school students in solving set problems is the main theme of *Salvador Llinares'* (Spain) chapter. Excerpts of the student teachers' discussions illustrate their explicit attempts to draw out the knowledge needed to solve the problems. Implicit beliefs, which underpin the teacher-student role, seem rarely to be questioned. Llinares argues that a greater emphasis is needed in drawing out student teachers' beliefs about mathematics and mathematics teaching if they are to question, and possibly deviate from, their own school experiences and instead implement instructional strategies congruent with current emphases. Optimum strategies for implementing change, and making relevant beliefs explicit to those holding them, vary with the target group. Thus, Llinares claims, to be successful, preservice and inservice programs require different elements and organizations. In this chapter, as in others, there is an emphasis on the context and setting in which beliefs are formed and challenged.

Aspects of affect, teacher education, and mathematics education are the focus of the chapter written by *George Philippou* and *Constantinos Christou* (Cyprus). Prospective teachers, they argue, and particularly prospective elementary teachers, generally do not have positive attitudes to mathematics, or a strong belief in their own mathematical proficiency. When faced with classroom teaching, they rarely focus on the affective components of mathematics teaching and learning. To support their contention that affective competencies are both learnable and teachable, they describe the contents of a preservice teacher education program in Cyprus and its effectiveness in raising teachers' self-confidence about teaching mathematics and in turn making their students feel confident and, if necessary, to seek help from colleagues to achieve this. Reference is also made to interview data to enrich and support the questionnaire findings described in the chapter.

As for part 1, a summative and evaluative chapter – this one written by *Stephen Lerman* (UK) – completes this section. After a perceptive review which highlights common and divergent themes among the six chapters, Lerman interrogates the body of work from a sociological rather than psychological perspective. This allows him to put forward some alternative orientations for the study of teachers' beliefs and to explore briefly the influence, on teachers and educational systems, of the political climate in which change is promoted.

3.3. Part 3: Students' Beliefs

Students' beliefs are the focal point of the third, and last, section. There is considerable diversity in the ways in which these beliefs are explored, in the specificity of the setting, and in the samples used as the vehicle for these explorations.

Answers to an extensive set of interview questions aimed at elucidating students' beliefs about mathematics and themselves as learners of mathematics serve as the focus of *Peter Kloosterman*'s (USA) chapter, which opens this part. School and other influences on motivation, assessment practices, expectations of parents and teachers, and various related issues are also examined. By clustering student answers, Kloosterman is able to sketch the broad beliefs held by students about the nature of mathematics, school mathematics, and factors that facilitate mathematics learning. Extracts from individual interviews, which are included in the chapter, point to the limitations of assuming that group and individual beliefs necessarily coincide.

Brian Greer (USA), *Lieven Verschaffel,* and *Erik de Corte* (Belgium) provide far ranging evidence of the prevalence for both students and (preservice) teachers to believe that posing mathematics in a "real world" setting is of limited relevance and that the world depicted in mathematics word problems bears little resemblance to the real world of every day experiences. A note of optimism is introduced by means of experimental evidence that confirms that it is possible to change beliefs about word problems. Balanced against that is the authors' cautionary note that prevailing cultural beliefs and political goals may serve to reinforce the historical schism between the mathematics of the classroom and the reality experienced beyond school.

Barriers between the learning of mathematics and students' lived experiences is a theme also taken up by *Norma Presmeg* (USA). Drawing on experimental evidence, she argues that students' beliefs about the nature of mathematics can both enable and constrain "their ability to construct conceptual bridges between familiar everyday practices and mathematical concepts taught in school or university". Students' changing beliefs and perceptions about the interaction between cultural practices familiar to them and systematized mathematical concepts are captured through semiotic chains depicted by the students themselves.

In their chapter, *Erna Yackel* and *Chris Rasmussen* (USA) argue that changes in beliefs can be explained effectively by drawing on both sociological and psychological perspectives. In particular, they "illustrate the reflexive relationship between students' beliefs and classroom social and socio-mathematical norms" through data gathered in a semester long teaching experiment in a university level class on differential equations. The class requirement, that students submit an electronic journal each week, enabled students' changing beliefs and increased acceptance of the class social norms to be traced and also proved a rich source for other student reflections about the structure of the class.

The focus of the chapter by *Pessia Tsamir* and *Dina Tirosh* (Israel) is more limited in scope. These authors examine the extent to which secondary school students adhere to the "perform-the-operation belief in the cases of division by zero". Their quantitative summary is enriched by references to explanations provided by students for the answers they gave. These extracts illustrate the various ways in which students reach answers glibly coded as correct or incorrect.

Frank Lester's (USA) overview and critique of the chapters in this section completes part 3. From what he humbly calls an "amateur" perspective, Lester discusses what he considers inherent difficulties in research exploring people's beliefs and proposes some avenues for overcoming these problems. This broad sketch serves as a useful context for an incisive analysis of the chapters included in this section of the book. Lester's commitment to the book's broad theme is reinforced in his final sections in which he offers "some thoughts about why we should care about the beliefs our students hold about the nature of mathematics, how it is learned, and how the subject is taught".

4. NOTES

[1] Without wishing to draw the analogy too rigidly, "understandings" of what mathematics is can also vary. School students generally have a different images of mathematics from that of their (mathematics) teachers. Teachers, in turn, will vary in their views of mathematics. Those outside (mathematics) education are likely to give yet another description. Mathematics professors and educators also have their own particular view of mathematics - as is clearly demonstrated by research summarized in a later chapter in this volume. The diversity of answers to the question "what is mathematics" suggests that there are several different views of mathematics; that it cannot be assumed that there is only the one right view of mathematics and that the others are wrong. Philosophers of mathematics (e.g., Ernest, 1991; Hersh, 1997) have introduced several right views of mathematics and the way it is learnt - in terms which have been accepted among mathematics.

[2] An analogy with mathematics can again be drawn. For example, in the 1700s it was generally accepted among mathematicians that all infinite series, with the limit zero of the general term, were convergent. With the counter example published at the end of that century, this knowledge/belief was rejected. The understanding of π is another example, with a shift from an initial belief that π was equal to 3 to the current understanding of π as a transcendental number (e.g., Boyer, 1985).

5. REFERENCES

Abelson, R. (1979). Differences between belief systems and knowledge systems. *Cognitive Science, 3*, 355–366.

Bar-Tal, D. (1990). *Group beliefs. A conception for analyzing group structure, processes, and behavior.* New York: Springer-Verlag.

Bem, D. J. (1970). *Beliefs, attitudes, and human affairs.* Belmont, CA: Brooks/Cole.

Borko, H., Davinroy, K. H., Bliem, C. L., & Cumbo, K. B. (2000). Exploring and supporting teacher change: Two third-grade teachers' experiences in a mathematics and literacy staff development project. *The Elementary School Journal, 100*, 273-306.

Boyer, C. B. (1985). *A history of mathematics.* Princeton (NJ): Princeton University Press.

Department of Employment, Education and Training. (1989) *Discipline review of teacher education in mathematics and science.* Vol. 1. Canberra: Australian Government Publishing Service.

Ernest, P. (1991). *The Philosophy of Mathematics Education.* Hampshire (U.K.): Falmer Press.

Fennema, E., & Hart, L. E. (1994). Gender and the JRME. *Journal for Research in Mathematics Education, 25*(6), 648-659.

Fishbein, M., & Ajzen, I. (1975). *Belief, attitude, intention and behavior.* Reading, MA: Addison-Wesley

Forgasz, H. J., & Leder, G. C. (2001a). "A+ for girls, B for Boys": Changing perspectives on gender and equity. In B. Atweh, H. Forgasz, & B. Nebres (Eds.), *Sociocultural research on mathematics education. An international perspective* (pp. 347-366). Mahway, New Jersey: Lawrence Erlbaum Associates.

Forgasz, H. J., & Leder, G. C. (2001b). The Victorian Certificate of Education – a Gendered Affair? *Australian Educational Researcher, 28*(2), 53-66.

Hersh, R. (1997). *What is Mathematics, really?* New York: Oxford University Press.

Jensen, J. H., Niss, M., & Wedege, T. (1998). *Justification and enrolment problems in education involving mathematics or physics.* Roskilde, Denmark: Roskilde University Press.

Krech, D., & Crutchfield, R. S. (1948). *Theory and problems of social psychology.* New York: McGraw Hill.

Leder, G. C. (1992). Mathematics and gender: Changing perspectives. In D. A. Grouws (Ed.), *Handbook of research in mathematics teaching and learning* (pp. 597-622). New York: Macmillan.

Leder, G. C., Forgasz, H. J., & Solar, C. (1996). Research and intervention programs in mathematics education: A gendered issue. In A. Bishop, K. Clements, C. Keitel, J. Kilpatrick, & C. Laborde (Eds.), *International handbook of mathematics education, Part 2* (pp. 945-985). Dordrecht: Kluwer Academic Publishers.

McLeod, D. B. (1992). Research on affect in mathematics education: A reconceptualization. In D. A. Grouws (Ed.), *Handbook of research on mathematics teaching and learning* (pp. 575-596). New York: MacMillan.

National Council of Teachers of Mathematics (2000). *Principles and standards for school mathematics.* Reston, VA: NCTM.

Pehkonen, E., & Törner, G. (Eds.). (1999, November). *Mathematical Beliefs and their Impact on Teaching and Learning of Mathematics. Proceedings of the Workshop in Oberwolfach.* Schriftenreihe des Fachbereichs Mathematik, No. 457. Duisburg: Universität Duisburg.

Schoenfeld, A. H. (1992). Learning to think mathematically: problem solving, meta-cognition, and sense making in mathematics. In D. A. Grouws (Ed.), *Handbook of research on mathematics learning and teaching* (pp. 334–370). New York: Macmillan.

Senger, E. (1999). Reflective reform in mathematics: The recursive nature of teacher change. *Educational Studies in Mathematics,37*, 191-221.

Thompson, A. G. (1992). Teachers' Beliefs and Conceptions: A Synthesis of the Research. In D. A. Grouws (Ed.), *Handbook of research on mathematics learning and teaching* (pp. 127–146). New York: Macmillan.

Wood, T. (2000). Learning to teach mathematics differently: Reflection matters. In M. van den Heuvel-Panhuizen (Ed.), *Proceedings of the 25th annual PME conference* Vol. 4 (pp. 431-438).

PART 1

BELIEFS: CONCEPTUALIZATION AND MEASUREMENT

CHAPTER 2

PETER OP 'T EYNDE, ERIK DE CORTE, AND LIEVEN VERSCHAFFEL

FRAMING STUDENTS' MATHEMATICS-RELATED BELIEFS

A Quest For Conceptual Clarity And A Comprehensive Categorization

Abstract. Despite the general agreement among researchers today that students' beliefs have an important influence on mathematical problem solving there is still a lack of clarity from a conceptual viewpoint. In this chapter we present a literature review of available categorizations or models of students' beliefs related to mathematics learning and problem solving. These reveal that although they all cover a broad spectrum of relevant beliefs, there appears to be no consensus on the structure and the content of the relevant categories of students' beliefs. A philosophical and psychological analysis of the nature and the structure of beliefs enables us to come to a deeper understanding of the development and the functioning of students' beliefs and to clarify the relation between beliefs and knowledge. The insights developed through this analysis result in an elaborated and concrete definition of students' mathematics-related beliefs and allow us to develop a theoretical framework that coherently integrates the major components of prevalent models of students' beliefs. We differentiate between students' beliefs about mathematics education, students' beliefs about the self, and students' beliefs about the social context, i.e., the class context.

1. INTRODUCTION

Recent theories on cognition and learning (e.g., Greeno, Collins, & Resnick, 1996; Salomon & Perkins, 1998) point to the social-historical embeddedness and the constructive nature of thinking and problem solving. According to these theories, each form of knowing and thinking is constituted by the meanings and rules that function in the specific communities in which they are situated (e.g., the scientific community, the class, the group). Acquiring knowledge or learning, therefore, consists of getting acquainted with the concepts and rules that characterize the activities in the different contexts. As such, learning becomes fundamentally a social activity.

From such a perspective, learning is primarily defined as a form of engagement that implies the active use of certain cognitive and metacognitive knowledge and strategies, but cannot be reduced to it. Indeed, more and more researchers (e.g., Bereiter & Scardamalia, 1993) are convinced that referring only to cognitive and metacognitive factors does not capture the heart of learning. Several studies (e.g.,

G. C. Leder, E. Pehkonen, & G. Törner (Eds.), Beliefs: A Hidden Variable in Mathematics Education? 13-37.

Connell & Wellborn, 1990; Schiefele & Csikszentmihalyi, 1995) point to the key role conative and affective factors play as constituting elements of the learning process, as well as and in close interaction with (meta)cognitive factors. Motivation and volition (i.e., the conative factors) are no longer seen as just the fuel or the engine of the learning process, but are perceived as fundamentally determining the quality of learning. In a similar way, self-confidence and positive emotions (affective factors) are no longer considered as just positive side effects of learning, but become important constitutent elements of learning and problem solving.

Recent developments in the field of research on mathematical problem solving tend to illustrate this change in perspective. Studies on students' beliefs about mathematics (e.g., Garofalo, 1989; Kouba & McDonald, 1986; Schoenfeld, 1985a) and on their motivational beliefs (e.g., Pintrich & Schrauben, 1992; Seegers & Boekaerts, 1993), as well as research on the influence of emotions (Cobb, Yackel, & Wood, 1989; DeBellis, 1996) and on other affective factors such as "students' perceived confidence" (Vermeer, 1997) aim at unraveling the role of conative and affective factors in mathematical problem solving. On a conceptual level researchers try to capture the interrelated influence of (meta)cognitive, conative and affective factors on mathematical learning and problem solving in line with the notion of a "mathematical disposition". Such a disposition refers to the integrated mastery of five categories of aptitude (De Corte, Verschaffel, & Op 't Eynde, 2000):

1. A well-organized and flexibly accessible knowledge base involving the facts, symbols, algorithms, concepts, and rules that constitute the contents of mathematics as a subject-matter field.
2. Heuristics methods, i.e., search strategies for problem solving which do not guarantee, but significantly increase, the probability of finding the correct solution because they induce a systematic approach to the task.
3. Metaknowledge, which involves knowledge about one's cognitive functioning (metacognitive knowledge), on the one hand, and knowledge about one's motivation and emotions that can be used to deliberately improve volitional efficiency (metavolitional knowledge), on the other hand.
4. Mathematics-related beliefs, which include the implicitly and explicitly held subjective conceptions about mathematics education, the self as a mathematician, and the social context, i.e., the class-context.
5. Self-regulatory skills, which embrace skills relating to the self-regulation of one's cognitive processes (metacognitive skills or cognitive self-regulation), on the one hand, and of one's volitional processes (metavolitional skills or volitional self-regulation), on the other hand.

Acquiring such a mathematical disposition is necessary for students to become competent problem solvers, equipped to recognize and tackle mathematical problems in different contexts, and, as such, able to overcome the well-known phenomenon of inert knowledge (see the National Council of Teachers of Mathematics [NCTM], 2000). After all, according to Perkins (1995), the integrated mastery of these different kinds of knowledge (i.e., domain-specific, metacognitive, metavolitional), skills and beliefs results in a sensitivity to the occasions when it is appropriate to use them and an inclination to do so. This sensitivity to situations and contexts, and the inclination to follow through, are both determined by the concepts and beliefs a person holds. A person's beliefs about what counts as a mathematical context and what (s)he finds interesting or important will, as such, have a strong influence on the situations (s)he will be sensitive to, and whether or not (s)he will engage in them. The relevance of beliefs as a component of a mathematical disposition and their impact on mathematics learning is echoed in the *Curriculum and evaluation standards for school mathematics* (NCTM, 1989) in the U.S.A.: "These beliefs exert a powerful influence on students' evaluation of their own ability, on their willingness to engage in mathematical tasks, and on their ultimate mathematical disposition" (p. 233).

It is nowadays generally assumed that the impact of students' mathematics-related beliefs on their learning and problem-solving behavior is mediated through cognitive as well as conative and affective processes. First, several researchers have shown how students' "beliefs about mathematics" determine how they choose to approach a problem and which techniques and cognitive strategies will be used (e.g., Garofalo, 1989; Schoenfeld, 1985a). Secondly, others have pointed to the implications of students' "mathematically related beliefs" for their motivational decisions in mathematics learning and problem solving (e.g., Kloosterman, 1996). Finally, it is argued that "students' beliefs related to mathematics education" provide an important part of the context within which emotional responses to mathematics develop (e.g., Isoda & Nakagoshi, 2000; McLeod, 1992). Moreover, this relationship between beliefs and emotions seems to be reciprocal. It is explained that local emotional experiences over time provide the context for the development and strengthening of more stable "global" affect as attitudes and beliefs (Goldin, this volume).

Notwithstanding the general agreement among researchers that students' beliefs have an important influence on mathematical learning and problem solving, there is still a lack of clarity from a conceptual viewpoint (see also Furinghetti & Pehkonen, this volume). The diversity in the terms used to describe relevant beliefs, sometimes referring to the same, at other times to different beliefs, is symptomatic of the actual state of the research domain. Despite, or maybe precisely because of the attention paid to the multiple ways in which different student beliefs influence mathematical learning and problem solving, research on this topic has not yet resulted in a comprehensive model of, or theory on, students' mathematics-related beliefs. As a matter of fact, most of the studies are situated in, respectively, cognitive, motivational or affective research traditions and in many cases they operate in relative isolation from each other. This results in a conceptual and theoretical confusion that is not at all advantageous for the development of a comprehensive theory or model of mathematics-related beliefs and the impact of these beliefs on students' learning and problem solving. We are well aware that the state of the art of the research field does not allow the development of a comprehensive theory at the moment. Nevertheless an attempt will be made in this chapter to clarify the conceptual discussion and to introduce a framework for students' mathematics-related beliefs that might be a further step in that direction.

To direct our research activities we have chosen to start from the following very broad and to some extent intuitive, working definition of students' mathematics-related beliefs:

> *Students' mathematics-related beliefs are the implicitly or explicitly held subjective conceptions students hold to be true, that influence their mathematical learning and problem solving.*

This working definition is the result of an attempt to integrate partly differing definitions and/or defining characteristics of beliefs formulated in the work of other scholars (e.g., Pehkonen & Törner, 1996; Richardson, 1996; Thompson, 1992). We have tried to come to a coherent synthesis that could serve as a starting point for further study. From that perspective it was acceptable to us to work with a definition that leaves a lot of questions unanswered. For instance, first, what are the different categories of beliefs that influence mathematical learning and problem solving? Or, second and more fundamentally, if some beliefs are explicitly held conceptions, how do they then differ from knowledge? Although we might not be able to answer all the questions that arise when clarifying this definition, in this chapter we aim to investigate critically the conceptualization of students' mathematics-related beliefs in an attempt to clarify at least the two issues just mentioned.

In the first part of the chapter we present a review of the most representative categorizations or models of students' beliefs related to mathematics learning and problem solving currently available. Although they all cover a broad spectrum of relevant student beliefs, there appears to be no consensus about the grounds on which different categories of beliefs can be identified. The introduction of different categories, without investigating the underlying assumptions and the mutual relations, creates a conceptual divergence that is difficult to reconcile. Therefore, in the second part of the chapter, we analyze the nature and the structure of beliefs in an attempt to come to a deeper understanding of the development and the functioning of students' beliefs. In addition, we further clarify the relation between beliefs and knowledge. Thirdly, starting from this analysis of the nature and the structure of beliefs, we introduce a theoretical framework and a more elaborated definition of students' mathematics-related beliefs that thoughtfully tries to integrate the major components of the models reviewed in the first part. Some final comments focusing on directions for future research conclude the chapter.

2. STUDENTS' BELIEFS AND MATHEMATICAL LEARNING AND PROBLEM SOLVING: A REVIEW

To develop a more comprehensive understanding of the different kinds of students' beliefs considered to determine mathematical learning and problem solving, and the way they relate to each other, we conducted a search of the literature for the period 1984 to 2000 in the ERIC and PSYCINFO databases, entering the key words "mathematics" and "beliefs", and their derivatives. 119 articles were found. The content of the articles was analyzed and a further selection took place focusing on articles that dealt mainly with students' beliefs and/or presented a model of different beliefs (for a more general oriented review, see Leder & Forgasz, this volume). This yielded 50 articles. Through the so-called snowball-method the references of these articles were checked for other relevant studies and the search was broadened to include the most recent proceedings of the conferences of the International Group for the Psychology of Mathematics Education. This review of representative publications revealed that, when studying the various student beliefs that influence mathematical learning and problem solving, scholars discussed topics as different as:

(a) Beliefs about the nature of mathematics and mathematical learning and problem solving (e.g., Carter & Yackel, 1989; Frank, 1988; Mtetwa & Garofalo, 1989; Schoenfeld, 1985a, 1985b; Spangler, 1992);
(b) Beliefs about the self in the context of mathematics learning and problem solving, i.e., motivational beliefs (e.g., Carter & Norwood, 1997; Kloosterman, Raymond, & Emenaker, 1996; Pajares & Miller, 1994; Seegers & Boekaerts, 1993; Stipek & Gralinski, 1991; Vanayan, White, Yuen, & Teper, 1997)
(c) Beliefs about mathematics teaching and the social context of mathematics learning and problem solving (e.g., Cobb, Yackel, & Wood, 1989; de Abreu, Bishop, & Pompeu, 1997; Lampert, 1996)
(d) Epistemological beliefs, i.e., beliefs about the nature of knowledge and the processes of knowing (e.g., Hofer, 1999; Schommer, Crouse, & Rhodes, 1992)

Although the studies mentioned above are listed as dealing with only one of the four categories of beliefs, some of them in fact addressed different kinds of student beliefs. Carter and Yackel (1989), for example, studied beliefs about what it means to do mathematics, one's ability to do mathematics, and the origins of mathematical knowledge. However, in such investigations certain beliefs always received more attention than others. We based our tentative categorization on the beliefs most central to the research of the different authors. This limited overview certainly showed that there is a growing body of research on students' beliefs related to mathematical learning and problem solving over the last decade. What seems to be missing, until now, is a joint effort to develop a comprehensive categorization of all the relevant beliefs studied. Very few scholars presented an overall picture of students' mathematics-related beliefs that could function as a framework to situate the different beliefs and studies in relation to each other. In the next section, we discuss the research efforts most relevant to developing such an overall categorization (see Figure 1).

Underhill (1988) has summarized research on learners' beliefs in four areas: beliefs about mathematics as a discipline (U1); beliefs about learning mathematics (U2); beliefs about mathematics teaching (U3); and beliefs about self within a social context in which mathematics teaching and learning occur (U4).

Beliefs about mathematics as a discipline refer to the beliefs students have about the nature of mathematics; for example, mathematics is about addition, subtraction, multiplication and division, plus a collection of routine problems. *Beliefs about learning mathematics* include the beliefs students hold about what are productive and counterproductive learning strategies in mathematics; for example, learning mathematics is mainly memorizing. *Beliefs about teaching* consist of students' views on what effective teaching strategies are; for example, mathematics teaching is about adding, subtracting, multiplying and dividing fractions and decimals, not about silly things such as rolling dice. Underhill further points to the important role played by students' beliefs about motivation and about the self. He defines beliefs about motivation and the self as inherently linked with beliefs about learning and teaching, implying that he does not recognize beliefs about motivation and the self as a distinct category.

More generally, he stresses the close relationships between learners' beliefs about motivation and the first three categories of beliefs. Learners' conceptions of mathematics and their motivational beliefs tend to determine strongly their beliefs about learning and teaching.

The last category of beliefs he mentions, i.e., *beliefs about the social context*, explicitly takes into account the social nature of beliefs and student behavior. After all, a student's learning is influenced by the group behavior norms of the class and, more importantly, the learner's perception of what counts as appropriate behavior (in class). These beliefs about the social context refer, for example, to the fact that learners will construct representations of the beliefs and purposes of the community of others in which they reside, and be influenced by these.

McLeod (1992), studying beliefs from an affective perspective, distinguished between beliefs about mathematics (M1); beliefs about self (M2); beliefs about mathematics teaching (M3); and beliefs about the social context (M4).

Beliefs about mathematics include the first two categories of Underhill's model, i.e., students' beliefs about mathematics as a discipline and students' beliefs about mathematics learning. In addition, the learner's beliefs on the perceived usefulness of mathematics are also mentioned as a belief about mathematics. *Beliefs about self* refer, for example, to students' self-concept, and to their confidence and causal attributions in relation to mathematics. More broadly, "beliefs about self include much of the literature on motivational issues" (McLeod, 1992, p. 581). *Beliefs about mathematics teaching* point to the relevance of students' beliefs about mathematics instruction. McLeod argues for more research on these beliefs, since to date there is little information about them. *Students' beliefs about the social context* is another area that is of much relevance in understanding students' problem-solving behavior, more particularly some of its affective aspects (e.g., emotional reactions in class). This aspect refers mainly to students' perceptions of the social norms in the classroom (see also, U4), but McLeod also stresses the influence of the social context of the school and the home environment on students' beliefs.

Kloosterman (1996) presents a model of students' beliefs that aims to explain many of the motivational decisions students make in mathematics. He tries to integrate, critically, McLeod's four kinds of beliefs into two basic types of beliefs: beliefs about mathematics (K1) and beliefs about learning mathematics (K2). But he differentiates the latter category into three subcategories: beliefs about oneself as a learner of mathematics (K21); beliefs about the role of the teacher (K22), and other beliefs about learning mathematics (K23).

Categories	
Underhill	a) beliefs about mathematics as a discipline (U1) b) beliefs about learning mathematics (U2) c) beliefs about mathematics teaching (U3) d) beliefs about self within a social context (U4)
McLeod	a) beliefs about mathematics (M1) b) beliefs about self (M2) c) beliefs about mathematics teaching (M3) d) beliefs about the social context (M4)
Kloosterman	a) beliefs about mathematics (K1) b) beliefs about learning mathematics (K2) – beliefs about oneself as a learner of mathematics (K21) – beliefs about the role of the teacher (K22) – Other beliefs about learning mathematics (K23)
Pehkonen	a) beliefs about mathematics (P1) – beliefs on the nature of mathematics as such (P11) – beliefs on the subject of mathematics (P12) – beliefs on the origins of mathematical tasks (P13) – etc. b) beliefs about oneself within mathematics (P2) – – c) beliefs about mathematics teaching (P3) – – d) beliefs about mathematics learning (P4) – –

Figure 1: Different categorizations of students' beliefs

Although Kloosterman claims that his first category, i.e., *beliefs about mathematics*, is basically the same as McLeod's first, a closer analysis shows that this is not entirely the case. In Kloosterman's model this category refers only to beliefs about the nature of mathematics, such as - for example – "mathematics is computation" or "mathematics topics are integrated". In contrast with McLeod, this first category does not include students' beliefs about mathematics learning. These beliefs are situated in the second type of beliefs Kloosterman distinguishes, i.e., *beliefs about learning mathematics*, and more specifically in the third subcategory, i.e., *other beliefs about learning mathematics*.

Examples of *Other beliefs about learning mathematics*, are: "memorization is important in mathematics" and "making mistakes is part of the learning process". Kloosterman also argues that many of McLeod's beliefs about the social context fall into the subcategories that belong to this second general type of beliefs. After all, "when these factors [i.e., social norms] are seen through the eyes of the student, many fall into the categories of beliefs about how mathematics is learned" (p. 135, our addition between brackets). The subcategory *beliefs about oneself* captures the same beliefs as McLeod's beliefs of self, while *beliefs about the role of the teacher* capture some aspects of McLeod's students' beliefs about mathematics teaching, for example, beliefs that are related to central ideas such as the "teacher is the transmitter of knowledge" or "the teacher is a source of answers".

Finally, Pehkonen (1995) differentiates between four main categories of beliefs that constitute an individual's view of mathematics: beliefs about mathematics (P1); beliefs about oneself within mathematics (P2); beliefs about mathematics teaching (P3); beliefs about mathematics learning (P4). We will not get into a systematic discussion of each of these four categories or look for similarities and differences with the other models. His categorization tends to stay close to Kloosterman's, although he certainly develops a different hierarchy of categories and subcategories. We want to highlight Pehkonen's attempt to define for each category the different subcategories it contains. For example, *Beliefs about mathematics* comprises, in his view, beliefs concerning the nature of mathematics as such, the subject of mathematics, the nature of mathematical tasks, the origins of mathematical knowledge, and the relationships between mathematics and the empirical world. Similarly, he describes the other categories, referring to their different subcategories.

These pioneering contributions deserve credit for putting students' beliefs on the research agenda, and they certainly have extended our understanding of the role of beliefs in mathematical learning and problem solving. However, as shown above, the different models are not always easy to reconcile. None of them is structured identically and although at a first glance it might seem that they all cover mostly the same student beliefs, this is not always true. For instance, the same beliefs are sometimes categorized differently (e.g., U2, M1, K23), and some identical categories contain, to a certain degree, different beliefs (e.g., M1, K1, P1). But, more importantly, some beliefs that are dealt with in one model are not mentioned at all in another (e.g., beliefs about the social context). In some cases this is due to the fact that the authors have a different view on the functioning of these beliefs. Kloosterman, for example, does not include a category of beliefs about the social context, because he is convinced that students' perceptions of the specific social context and social norms immediately influence (and are present in) their beliefs about learning mathematics. According to this view, a student's perception that students are not sanctioned when they make mistakes during exercises in *his mathematics class* is immediately represented at a higher cognitive level as "making mistakes is part of the learning process". From this perspective there is indeed no need to make a distinction between beliefs about the social context and beliefs about mathematical learning. However, if one holds the opinion that students in some cases might differentiate between what happens in their class and what they in general believe to be true with regard to mathematics learning, then it makes sense to distinguish between those two kinds of beliefs. Notwithstanding this example, the differences in categorizations for the most part do not derive from distinct views on how beliefs function, and this is due to the fact that most models are not theory based. Rather, they have been developed on more pragmatic grounds with the purpose of enabling an organized presentation of the research in the area. The authors usually did not set out to present an overall model of students' mathematics-related beliefs, or as Kloosterman explains with regard to his model:

> the model should not be viewed as a model of "the" key beliefs that lead to motivation in mathematics. Rather, it is a model of a number of potentially important factors. I have tried to justify these factors, but justification for alternative factors is also possible. Note that there is considerable overlap among the factors in the model. (p. 143)

Nevertheless, such an overall model might be just what the research field needs. McLeod (1992, p. 581) stresses that "mathematics education needs to develop a more coherent framework for research on beliefs". Although the models presented certainly fostered communication and stimulated research on beliefs, the spawning of terms without a thorough analysis of the underlying assumptions and the mutual relations can be a serious obstacle for the advancement of research and practice. Indeed, the way in which researchers implicitly or explicitly define beliefs, the different categories of beliefs they acknowledge, and the way they perceive the relation with knowledge, are all elements that influence the measures they develop, the research methods they use, and the conclusions they draw. It is crucial for constructive communication between researchers as well as between researchers and teachers that the different beliefs are unambiguously defined and that their influence on learning and problem solving are clearly explained.

3. STUDENTS' BELIEFS: CONCEPTUAL ISSUES

From a conceptual point of view educational researchers have, until now, not dealt with the notion of beliefs in a substantial way. Thompson (1992) has pointed to the difficulty of distinguishing between beliefs and knowledge to explain this dearth of reasoned discourses on beliefs in the educational literature. She further indicated that some researchers are not at all convinced of the value for educational research of making such a distinction. Also in research on mathematics education, discussions on the nature of beliefs are limited (for a few exceptions see, Pehkonen & Törner, 1996; Thompson, 1992) and the difference between beliefs and knowledge has never been a major topic. One additional explanation for this might be that, historically, studies on students' mathematics beliefs have mainly dealt with students' misconceptions or naive and incorrect beliefs about mathematics (see e.g., Greeno, 1991; Lampert, 1990). Obviously, the distinction between incorrect beliefs and knowledge is much less an issue, if at all, than the conceptual distinctions and/or similarities between (correct) beliefs and knowledge. However, the emergence of a new perspective that stresses the important role for learning of positive, as well as negative, conative and affective factors in interaction with (meta)cognitive factors, forces the mathematics educational community to take a broader view of students' beliefs and their influence on mathematics learning and problem solving. From this perspective, a discussion on the nature and structure of beliefs and their relation to knowledge can no longer be avoided (see also Furinghetti & Pehkonen, this volume; Törner, this volume). Therefore, as Thompson (1992) has pointed out:

> Researchers interested in studying (teachers') beliefs should give careful consideration to the concept, both from a philosophical as well as a psychological perspective. Philosophical works can be helpful in clarifying the nature of beliefs. Psychological studies may prove useful in interpreting the nature of the relationship between beliefs and behavior as well as in understanding the function and structure of beliefs. (p. 129,our brackets)

In the following two sections we will analyze both the nature and the structure of beliefs.

3.1. The Nature of Beliefs

Richardson (1996, p.103) characterizes beliefs as "psychologically held understandings, premises, or propositions about the world that are felt to be true". A proposition is believed when the proposition's meaning is represented in a mental system and is treated as if it is true. Students believe that "mathematics is computation" the moment they perceive mathematics to be that way and/or they are told that mathematics is like that. Indeed, the first time students perceive and understand a notion they implicitly accept it as true. It seems to be one of the basic mechanisms characterizing the functioning of our mind that we start by uncritically accepting, believing, everything we see or hear (Gilbert, 1991). Young children's uncritical acceptance of everything they see and are told is a prototypical example. "We begin by believing everything; whatever is, is true" (Bain, 1859, p. 511). Only in a second phase, in conflict with other propositions or new situations, we might question it and maybe consciously change our mind (Spinoza, 1677/1982).

This shows how at a very basic level beliefs are grounded in the social contexts in which one functions. Beliefs are the product of social life (de Abreu, Bishop, & Pompeu, 1997). They are determined by the socio-cultural environment one lives and works in. Students' beliefs are a function of the classroom practices in which they participate. However, this does not imply that the individuality, the subjectivity of the student, totally merges into the intersubjectivity of the classroom. As a member of different social contexts (i.e., family, peers) students are subjected to a very complex and diverse network of influences, that determines the "unique" way in which they find themselves and look at the classroom context (Pehkonen & Törner, 1996; Underhill, 1988). The ways in which humans view the world and interact with it reflect their understanding of the basic beliefs and fundamental knowledge shared with members of their family, the intellectual discipline and other groups in which they function (Alexander, Schallert, & Hare, 1991). The conscious or unconscious re-construction of the "at first sight" accepted beliefs in a second phase (cf. supra), is always a function of the beliefs and knowledge acquired in other contexts and/or at earlier times. Consequently, the same context can lead to different beliefs in different students. For example, getting the message from his mathematics teacher that he failed a test on mathematical word problems can get one student to believe that he can do no good for this teacher, whereas another might be convinced that he is not good in mathematical word problems, and a third one probably comes to believe that he is no good in mathematics at all. The different beliefs students hold and the prior knowledge they have determine their interaction with the social context, their respective definitions of the situation and its consequences.

It becomes obvious that when we take a close look at students' thinking and learning, beliefs and knowledge operate together. Students' problem-solving behavior is always directed by what they believe to be true, referring to knowledge as well as to beliefs. As such, from a psychological perspective, beliefs and knowledge are closely related constructs. They both determine, in close interaction, students' understanding of specific mathematical problems and situations. This understanding implies always more than what is made explicit in the knowledge about it (Power & Dalgleish, 1997). A higher order construct that captures the integrated functioning of knowledge and beliefs more adequately fits with the complex functioning of the human mind. The use of constructs as schematic models (Teasdale & Barnard, 1993) or mental models (Johnson-Laird, 1983) better capture the nature of our understanding of phenomena and situations. Ernest (1989) takes the same perspective when he describes the three most important elements that influence the practice of mathematics teachers. One of these elements is the teacher's mental contents or schemas which incorporate his or her knowledge of mathematics, as well as beliefs about mathematics and its teaching and learning.

Although knowledge and beliefs are psychologically integrated in schemas or models, most authors still distinguish between both constructs. After all, although an individual's beliefs result from the different social contexts in which he participates, it is the subject who owns the belief and he is the only referent for its truthfulness. Beliefs refer to what "I" believe to be true, regardless of the fact that others agree with me or not, regardless of the fact that others "know" it to be true or not. From an epistemological perspective, beliefs are an individual construct, while knowledge is essentially a social construct. Knowledge requires a truth condition. Thompson (1992) refers to Scheffler (1965) explaining that "a claim to knowledge must satisfy a truth condition, whereas beliefs are independent of their validity" (p. 129).

This truth condition entails an agreement in a community that a certain proposition is true, and has met the criteria of truthfulness in that social context (Green, 1971). This consensus gives it its higher epistemic standing (Fenstermacher, 1994). Knowledge goes beyond the individual and is situated in communities of practice. It is distributed in the world among individuals, the tools and the books they use (Greeno et al., 1996; Salomon & Perkins, 1998). Of course, this is not to say that individuals cannot possess knowledge. However, the ultimate epistemological criteria for discriminating between belief and knowledge are not situated in the individual, but in the social context. Consensuality is not a transparent feature of beliefs, and thus, one has to transcend the subjective point of view to find out if a proposition one holds to be true is knowledge or belief (Abelson, 1979). This only makes sense for those propositions that have the structure of knowledge claims. For example, the belief "mathematics is computation" can be evaluated against socially shared “scientific” criteria of truthfulness, and as a result it can be described as an incorrect or correct belief, i.e., knowledge. However, the belief “I find mathematics important”, can in principle never be judged on its validity on the same grounds. We will never be able to compare the model that is socially constructed, based on behaviors of the subject, with what actually goes on in the subject’s head, with what he actually thinks is important. Such a belief does not have the structure of a knowledge claim in our current society and can never be anything else than a belief.

By way of conclusion, we will return to the working definition of students’ mathematics-related beliefs formulated in the introduction and will further clarify it. We proposed that

> *Students’ mathematics-related beliefs are the implicitly or explicitly held subjective conceptions students hold to be true, that influence their mathematical learning and problem solving.*

It seems correct to state that, to hold a belief like “mathematics is computation” presupposes at least some tentative notion of concepts as “mathematics” and “computation”. As Perkins (1995) points out, an implicit or explicit subjective understanding of relevant concepts is constitutive for the acquisition of beliefs. Some of these subjective conceptions, i.e., beliefs, will have the structure of knowledge claims and can be evaluated against socially shared criteria the moment these conceptions go public. If it is accepted that they meet the relevant criteria for truthfulness, they become knowledge. From an individual’s perspective then, knowledge can be experienced as correct beliefs. Besides these, students have subjective conceptions (e.g., motivational beliefs) on a conscious (explicit) or preconscious (implicit) level, that are as influential with regard to learning and problem solving, but that are strictly subjective. They can never be judged to be correct or incorrect but only to be positive (e.g., I like mathematics) or negative beliefs (e.g., I don't like mathematics) according to subjective criteria. In our opinion, this line of argument further clarifies how students’ mathematics-related beliefs are understood to be (subjective) conceptions of an explicit (conscious) or implicit (preconscious) nature. Moreover, it already provides some insight into the nature of the relation between beliefs and knowledge. In the next section, we elaborate further on the close relations between beliefs and knowledge, but even more on the relations between the different categories of beliefs.

3.2. The Structure of Beliefs: Belief Systems

The analysis of the nature of beliefs has already revealed some of their structural aspects. When introducing a schema structure to capture fully the integrated way in which beliefs and knowledge determine how we perceive and understand situations, we pointed to one of the structural relations between both concepts. In this section we will discuss in greater depth the characteristics of belief systems, further clarifying how they are related to knowledge systems, but also where they differ.

Since the end of the 1980's the importance of contextual factors for knowledge and learning has been highlighted by the situated cognition and learning paradigm that emerged in reaction to the mentalistic/computational view of learning and thinking (Brown, Collins, & Duguid, 1989). Situativity theory has quite rightly stressed that learning is enacted essentially in interaction with the social and cultural contexts, resulting in clusters of situated knowledge. Several researchers have pointed out that, like knowledge, beliefs are also organized in clusters around specific situations and contexts (Green, 1971). What we hear, perceive and comprehend in a certain situation, for example, a class situation, is accepted as true *in that context* (Bogdan, 1986).

As explained above, what we believe to be true refers both to the knowledge we possess and to our beliefs. The grounds for differentiating between them are situated elsewhere, i.e., in the social context. The criteria offered there allow a differentiation between beliefs and knowledge, between a knowledge system and a beliefs system, each with their own characteristics. Whereas they share the cluster structure, they differ on other structural dimensions (see e.g., Abelson, 1979; Törner, this volume; Törner & Pehkonen, 1996). Green (1971) discusses three dimensions of belief systems: *cluster structure, quasi-logicalness, and psychological centrality* (see also, Furinghetti & Pehkonen, this volume).

As constituting elements of a person's entire belief system, belief clusters are framed in a person's subjective rationality (i.e., *quasi-logicalness*). People always strive for a coherent belief system; only then are they able to function in an intelligible way. Beliefs that are perceived as incompatible will be changed, and consequently also the clusters of which they are a part. Depending on the *centrality* of these changing beliefs in a person's entire belief system, this will also affect other clusters.

Some (central) beliefs are held more strongly than others (peripheral). Consequently, modifying the first will have more far-reaching consequences than changing the latter. Carter and Yackel (1989) compare the change process in a person's central and fundamental beliefs with a paradigm shift. The equilibrium a belief system is trying to hold, or, for that matter, the driving force behind a change in beliefs is, however, not primarily logical in nature (as in knowledge systems), but rather psychological (or quasi-logical).

Snow, Corno, and Jackson (1996) rightfully acknowledge that "human beings in general show tendencies to form and hold beliefs that serve their own needs, desires and goals; these beliefs serve ego-enhancement, self-protective, and personal and social control purposes and cause biases in perception and judgment in social situations as a result" (p. 292). Consequently, beliefs do not only carry denotative, but also connotative, evaluative meaning (Snow et al., 1996). Students' networks of beliefs provide not only the context within which they perceive and understand the world, but also play an emotional and motivational role in their learning and problem solving (McLeod, 1992; Power & Dalgleish, 1997).

The subjective rationality underlying the organization of a belief system is rarely consciously known by the student. Rather, it reveals itself in students' different mathematics-related beliefs, the way they actually perceive and experience specific mathematical problems, and how they solve them. In recognizing and interpreting the different beliefs students hold and the relations between these respective beliefs, we always have to be aware of the above-mentioned structural characteristics of a belief system, i.e., the cluster structure, the quasi-logicalness (subjective rationality), and the psychological centrality.

This implies that what to an outsider might appear as "contradictory beliefs", probably will not be perceived as such by the individual herself. For example, Schoenfeld (1985b) reports data on student' beliefs from which he concludes that "on the one hand, there was a tendency to regard mathematics learning largely as a matter of memorization. On the other hand, the students expressed significant support for the idea that mathematics is interesting and challenging, allowing a great deal of room for discovery" (p.14). It might well be that, in the former case, they were expressing their beliefs about school mathematics, while in the latter they were giving their view on the use of mathematics in society. As long as students are convinced that what they learn in school has little or no resemblance with or relevance to what happens in the real world, they can happily live with these two beliefs about mathematics without perceiving any contradiction (Verschaffel, Greer, & De Corte, 2000).

4. A FRAMEWORK OF STUDENTS' MATHEMATICS-RELATED BELIEFS

An analysis of the nature and structure of beliefs indicates, first, that students' beliefs are grounded in their social life and are as such fundamentally social. They are a function of the broad social-historical context in which students find themselves. Finding themselves in a specific class context, students will interpret its rules and practices on the basis of their prior beliefs and knowledge and as such develop their own, to a large extent shared, conceptions about it.

Second, as the former point already illustrates, beliefs and knowledge operate in close interaction. Schemas or mental models are presented as higher-order constructs that characterize on a conceptual level the integrated functioning of knowledge and beliefs. They seem to stay much closer to the complex way our mind actually works.

Third, although closely related in their functioning, there are fundamental differences between the structure of belief and knowledge systems, one of the distinctive characteristics being that a belief system has a quasi-logical structure, where a knowledge system has a logical structure. Indeed, the equilibrium a belief system is trying to achieve is psychological in nature. The underlying rationale are the needs, desires and goals of the self. An analysis of the nature and the structure of beliefs and belief systems shows that the social context, the self, and the object in the world that the beliefs relate to, are constitutive for the development and the functioning of these systems. The constitutive dimensions of students' mathematics-related belief systems can then be represented as a triangle (see Figure 2).

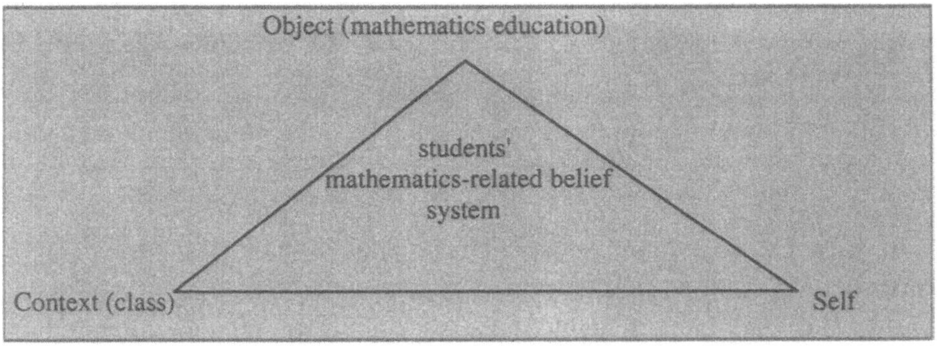

Figure 2: Constitutive dimensions of students' mathematics-related belief systems

Students' beliefs about mathematics education are situated in, and determined by, the context in which they participate as well as by their individual psychological needs, desires, goals etc. Framed in another way, students' mathematics-related beliefs are constituted by their beliefs about the class context, beliefs about the self, and, of course, beliefs about mathematics education. The developed framework of students' mathematics-related beliefs, grounded in what we know about the nature and the functioning of beliefs, is in line with Schoenfeld's more general view on the different kinds of beliefs that determine a person's cognitive actions in research settings (Schoenfeld, 1983). He points out that cognitive actions are

> often the result of consciously or unconsciously held beliefs about (a) the task at hand, (b) the social environment within which the task takes place, and (c) the individual problem-solver's perception of self and his or her relation to the task and the environment. (p. 330)

Grounded in these insights on the key dimensions and the functioning of belief systems, and the broad categories of beliefs that turned out to be constitutive, we can come to a more elaborated and concrete definition of students' mathematics-related beliefs.

> *Students' mathematics-related beliefs are the implicitly or explicitly held subjective conceptions students hold to be true about mathematics education, about themselves as mathematicians, and about the mathematics class context. These beliefs determine in close interaction with each other and with students' prior knowledge their mathematical learning and problem solving in class.*

In accordance with this definition, we developed a framework of students' mathematics-related beliefs in which we tried to integrate the major components of the models presented above, as well as the research already done on the role of students' beliefs. The different categories and subcategories of the framework are presented in Figure 3.

A framework of students' mathematics-related beliefs
1. Beliefs about mathematics education a) beliefs about mathematics as a subject b) beliefs about mathematical learning and problem solving c) beliefs about mathematics teaching in general
2. Beliefs about the self a) self-efficacy beliefs b) control beliefs c) task-value beliefs d) goal-orientation beliefs
3. Beliefs about the social context a) beliefs about social norms in their own class – the role and the functioning of the teacher – the role and the functioning of the students b) beliefs about socio-mathematical norms in their own class

Figure 3: A framework of students' mathematics-related beliefs

4.1. Beliefs about Mathematics Education

Students' general beliefs about mathematics education are located in this category. We differentiate between the following subcategories and indicate between brackets how they relate to (sub)categories in the models discussed earlier:

1. Beliefs about mathematics, such as "formal mathematics has little or nothing to do with real thinking or problem solving" *(see U1, K1, P1)*
2. Beliefs about mathematical learning and problem solving; for instance, "mathematics learning is memorizing" *(see U2, K23, P4)*.
3. Beliefs about mathematics teaching; for example, "a good teacher first explains the theory and gives an example of an exercise before he asks to solve mathematical problems" *(see U3, M3, K22, P3)*.

In line with Schoenfeld's (1985a) definition, students' beliefs about mathematics education reflect their view on what mathematics is like: the perspective with which they approach mathematics and mathematical problems and tasks. "What is mathematics" is a question often asked throughout history and the answers given differed over time (De Corte, Greer, & Verschaffel, 1996). Recently, the view of mathematics as a body of absolute facts and procedures, dealing with quantities and forms, with certain knowledge, is under attack. Several authors (e.g., Ernest, 1991; Schoenfeld, 1992; Tymoczko, 1986) advocate a conceptualization of mathematics as an activity grounded in human practices, a science of patterns with problem solving at the heart of it. Such fundamental issues are captured by students' beliefs about mathematics. Furthermore, what we fundamentally think about mathematics and mathematical knowledge is closely related to what we think mathematics learning, on the one hand, and mathematics teaching, on the other, are like (Hofer & Pintrich, 1997). Therefore, we consider these three kinds of beliefs to be closely related. Clustered, they seem to constitute three interrelated subsets of beliefs about mathematics.

Based on empirical research, several scholars have stressed the importance of beliefs about mathematics and about mathematical learning and problem solving (e.g., Garofalo, 1989; Greeno, 1991; Kloosterman, 1996; McLeod, 1992). Descriptive in nature, most of these studies reveal the kind of beliefs about mathematics and mathematics learning that students hold at different ages. They all deliver converging evidence concerning the prevailing beliefs students hold about the nature of mathematics and about mathematical learning and problem solving. An extended list of typical student beliefs derived from the literature, is given by Schoenfeld (1992; see also Mtetwa & Garofalo, 1989; Stodolsky, Salk, & Glaessner, 1991):

- Mathematics problems have one and only one right answer.
- There is only one correct way to solve any mathematics problem, usually the rule the teacher has most recently demonstrated to the class.
- Ordinary students cannot expect to understand mathematics; they expect simply to memorize it and apply what they have learned mechanically and without understanding.
- Mathematics is a solitary activity, done by individuals in isolation.
- Students who have understood the mathematics they have studied will be able to solve any assigned problem in five minutes or less.
- The mathematics learned in school has little or nothing to do with the real world.
- Formal proof is irrelevant to processes of discovery or invention. (p. 359).

Students' beliefs about how mathematics should be taught in class, and what the teacher should do, have been studied much less. Frank (1988) pointed out that junior high students traditionally believe that the role of the mathematics teacher is to transmit mathematical knowledge and to verify that the students have received this knowledge. This view is confirmed in an interview study by Kloosterman et al. (1996). Using data from 29 pupils, Kloosterman et al. (1996) investigated their beliefs about group work as a method for studying mathematics in class. Interestingly, they found a substantial variation among pupils' beliefs, but also significant fluctuations from positive to negative beliefs about group work and vice versa over the three years of the longitudinal study. The authors attribute this at least in part to the wide variety of teacher beliefs about the effectiveness of group work as well as to the kind of group activities students actually experienced in class (see also Carter & Norwood, 1997). Whereas we basically agree with this perspective, one also needs to be aware of the fact that teachers do not always teach as they preach (see e.g., Thompson, 1992). Teachers may regularly stress the importance of group work when asked for it or even in discussions with their students, but only occasionally allow their students actually to work in groups during the mathematics lessons. Some students will "buy" what their teachers say and believe them but only at a rhetorical level (Schoenfeld, 1985b). They will in general believe that group work is an effective way to work and learn in the mathematics class, but at a practical level, as far as their behavior in their specific class is concerned, they will believe the opposite and act accordingly. As already mentioned, students can happily hold such seemingly contradictory beliefs.

4.2. Beliefs about the Self

Students' beliefs about the self *(see M2, K21, P2)* in relation to mathematics refer to what in the motivational research literature is labeled as motivational beliefs (e.g., Pintrich & Schrauben, 1992). We differentiate between:

1. Goal orientation beliefs, such as "the most satisfying thing for me in this mathematics course is trying to understand the content as thoroughly as possible".
2. Task value beliefs, for instance, "it is important for me to learn the course material in this mathematics class".
3. Control beliefs, for example, "if I study in appropriate ways, then I will be able to learn the material in the course".
4. Self-efficacy beliefs, for instance, "I am confident I can understand the most difficult material presented in the readings of this mathematical course".

The differentiation in beliefs about the self is based on a general socio-cognitive model of motivation that proposes three basic motivational constructs (Pintrich, 1989): expectancy, value and affect. Expectancy components refer to students' beliefs that they can accomplish a task, i.e., self-efficacy beliefs and control beliefs. Value components focus on the reasons why students engage in learning and problem solving; these become apparent in their goal orientation beliefs and task value beliefs (Pintrich, Smith, Garcia, & McKeachie, 1993). The affective component includes students' emotional reactions to tasks and to their performance (Pintrich & Schrauben, 1992). Although important, these reactions are more a consequence of beliefs than beliefs as such. Emotions and emotional acts are perceived as expressions of beliefs (Carter & Yackel, 1989). They are not beliefs as such. After all, "the processing of social information important to the person is intrinsically affect laden so that such cognitions as beliefs about the self and one's personal future are themselves 'hot' and 'emotional'" (Mischel & Shoda, 1995, p. 252). Therefore, we do not perceive affect as a separate category of beliefs about the self. Several researchers (e.g., Ames & Archer, 1988; Fennema, 1989; Stipek & Gralinski, 1991; Vermeer, 1997) have addressed the influence of self-beliefs about mathematical problem solving, often reporting gender differences in relation to differences in performance. For instance, Fennema (1989) found that

> males have more confidence in their ability to do mathematics, report higher perceived usefulness, and attribute success and failure in a way that has been hypothesized to have a more positive influence on achievement. (p. 211)

In most studies, questionnaires or interviews were used to describe the self-beliefs students hold in relation to mathematical learning and problem solving (e.g., Kloosterman & Cougan, 1994; Vanayan et al., 1997). Yet, several researchers have made rather fine-grained analyses of the influence of students' beliefs about the self on mathematical learning and problem solving using path analysis and covariance analysis (e.g., Pajares & Miller, 1994; Seegers & Boekaerts, 1993). Seegers and Boekaerts (1993), for instance, investigated the influence of more general motivational beliefs (goal orientation, attributional style, self-efficacy) on task-specific appraisals (subjective competence, task attraction, personal relevance) in a group of eight graders, as well as the way in which these general and task-specific variables determine performance on mathematics tasks. They found that motivational beliefs had only an indirect influence on performance through the task-specific appraisals, and more specifically, through the perceived competence in the task. The advancement of our understanding of the relevance of beliefs about the self in relation to mathematics can greatly benefit from such continued in-depth analyses (for another interesting in-depth approach see Leder & Forgasz, this volume). In stressing the role of variables at the task level they clarify the exact ways in which these beliefs influence mathematical learning and problem solving. More studies investigating students' self-beliefs at the subject level together with their task-specific appraisals are needed to further unravel the mechanisms through which students' self-beliefs determine mathematics learning and problem solving.

4.3. Beliefs about the Social Context, i.e., the Class Context

This category of beliefs about the social context of mathematics education refers to students' views and perceptions of the classroom norms, including the social and the socio-mathematical norms, that direct teachers' and students' behavior *in their specific classroom* (Cobb & Yackel, 1998). It includes their perceptions about the role of the teacher as well as their own and their fellow students' role in the mathematics classroom, but also students' beliefs about aspects of the class culture that are specific to mathematical activity. The latter refers, for example, to students' views on what counts as a different solution or as an acceptable explanation *in their class*. To summarize, beliefs about the social context imply (see also U4, M4):

1. Beliefs about social norms in their own class
   - beliefs about the role and the functioning of the teacher
   - beliefs about the role and the functioning of the students
2. Beliefs about socio-mathematical norms in their own class

Researchers on mathematics learning and teaching have shown an increasing interest in studying the so-called culture of the mathematics classroom (Cobb & Yackel, 1998; Nickson, 1992), and the influence of the social environment in which learning tasks and problem solving take place (see also e.g., de Abreu et al., 1997; McLeod, 1992). As argued and shown by Cobb and his coworkers, teacher's and students' individual beliefs about their own role, and the role of the others in the class are the psychological correlates of the classroom's social norms relating to, for instance, the way in which a whole-class discussion takes place (Cobb et al., 1989). Furthermore, sociomathematical norms, i.e., norms that are specific to mathematics (e.g., what counts as a good solution?), have an impact on specifically mathematical beliefs (Yackel & Cobb, 1996; see also Cobb and Yackel, 1998). Both classroom social norms and sociomathematical norms, determine the interaction patterns that teacher and students mutually establish, in which implicit definitions are embedded about what mathematics is like, about how a problem should be solved, about the criteria for being a good student, etc. (see e.g., Verschaffel, Greer, & De Corte, 2000). Students develop their sense of what it means to do mathematics and what they and the others are expected to do in mathematics lessons from their actual experiences and interactions during the classroom activities in which they engage (Henningsen & Stein, 1997). Thus, as argued by de Abreu et al. (1997), beliefs and attitudes are the product of social life rather than being located in an autonomous individual. Beliefs about the nature of mathematics education, about oneself and about the class context are constructed in an attempt to make sense of classroom life during mathematics instruction.

Until now there has been little research that addresses these beliefs about the specific social context of the class, and how they relate to the more general beliefs. These general beliefs are abstracted from one's experiences and from the classroom culture in which one is embedded (Schoenfeld, 1992); but this culture is such a complex phenomenon of rules and interactions that there is clearly no linear relation between "a class context" and more general beliefs about mathematics and the self. Therefore, in order to fully understand the influence of mathematics-related beliefs on students' learning and problem solving, it is necessary to focus in future inquiry not only on their general beliefs about mathematics education and the self, but also on their beliefs about the mathematics class context in which they have to perform.

5. FINAL COMMENTS

In this chapter we have developed a framework of students' mathematics-related beliefs that enables us to categorize much of the research done on this topic. Although the framework, as such, is not theory-based, we have taken a first step in that direction, by grounding it in an analysis of the nature of beliefs and the functioning of belief systems.

The relevance of most of the categories and subcategories identified for mathematical learning and problem solving has already been illustrated in numerous studies. This is less the case for students' beliefs about the social context. Much more research is needed in this area to explain the influence of these beliefs on mathematical learning and problem solving and to unravel the relationships between students' beliefs about the specific social context and their more general beliefs about mathematics education and the self.

More importantly, a major gap in the existing body of research is the fact that, at the moment, there is little empirical evidence supporting the internal structure of the categorizations or frameworks that are presented. We need, for example, more questionnaire studies that show through factor analysis whether the categories and subcategories of students' beliefs and the way they are structured in a framework are empirically valid. Our frameworks need to be adequate representations of the relevant beliefs students hold in relation to mathematical learning and problem solving, on the one hand, and of the way they are related and structured in students' minds, on the other hand. After all, it is the internal dynamic of the belief systems that characterizes how students' beliefs influence learning and problem solving, much more than the individual beliefs. More research is needed to unravel the structure and the internal dynamics of students' mathematics-related belief systems.

Finally, we have tried to clarify in this chapter the differences, but also the close relationship, between knowledge and beliefs. Some might perceive this to be only a conceptual discussion with little relevance to teachers and students in classrooms. However, it is important to differentiate these two also in classrooms. Teachers need to be aware that there are elements influencing students' mathematical behavior in class other than their domain-specific knowledge, the heuristics and self-regulation strategies they master, and the meta-knowledge they possess. What students fundamentally believe mathematics is like, how they relate to it and how they perceive the class context, has been proven to be as influential, and is usually not captured by one of the other categories of a mathematical disposition mentioned above. Therefore, it is important to communicate to teachers the relevance of students' beliefs, as well as knowledge, and to stimulate further research that will allow us to develop a better understanding of the development and the influence of students' mathematics-related beliefs on learning and problem solving.

6. REFERENCES

Abelson, R. (1979). Differences between belief systems and knowledge systems. *Cognitive Science, 3*, 355-366.

Alexander, P. A., Schallert, D. L., & Hare, V. C. (1991). Coming to terms: How researchers in learning and literacy talk about knowledge? *Review of Educational Research, 61*(3), 315-343.

Ames, C., & Archer, J. (1988). Achievement goals in the classroom: Students' learning strategies and motivation processes. *Journal of Educational Psychology, 80*, 260-267.

Bain, A. (1859). *The emotions and the will*. London: Longmans, Green.

Bereiter, C., & Scardamalia, M. (1993). *Surpassing Ourselves: An inquiry into the nature and implications of expertise*. Chicago: Open Court.

Bogdan, R. J. (1986). The importance of belief. In R. J. Bogdan (Ed.), *Belief: Form, content, and function* (pp. 1-16). New York: Oxford University Press.

Brown, J. S., Collins, A., & Duguid, P. (1989). Situated cognition and the culture of learning. *Educational Researcher, 18*(1), 32-42.

Carter, C. S., & Yackel, E. (1989). *A constructivist perspective on the relationship between mathematical beliefs and emotional acts.* Paper presented at the annual meeting of the AERA, San Francisco.

Carter, G., & Norwood, K. S. (1997). The relationship between teacher and student beliefs about mathematics. *School Science and Mathematics, 97*(2), 62-67.

Cobb, P., & Yackel, E. (1998). A constructivist perspective on the culture of the mathematics classroom. In F. Seeger, J. Voigt, & U. Waschescio (Eds.), *The culture of the mathematics classroom* (pp. 158-190).

Cobb, P., Yackel, E., & Wood, T. (1989). Young children's emotional acts while engaged in mathematical problem solving. In D. B. McLeod, & V. M. Adams (Eds.), *Affect and Mathematical Problem Solving: A new perspective* . New York: Springer-Verlag.

Connell, J. P., & Wellborn, J. G. (1990). Competence, autonomy, and relatedness: A motivational analysis of self-system processes. In M. Gunnar, & L.A. Sroufe (Eds.), *Minnesota Symposium on Child Psychology*, Vol. 23. Hillsdale, N.J.: Erlbaum.

de Abreu, G., Bishop, A. J., & Pompeu, G. Jr. (1997). What children and teachers count as mathematics. In T. Nunes, & P. Bryant (Eds.), *Learning and teaching mathematics: An international perspective* (pp. 233-264). Hove, UK: Psychology Press Ltd.

De Corte, E., Greer, B., & Verschaffel, L. (1996). Mathematics teaching and learning. In D. C. Berliner, & R. C. Calfee (Eds.), *Handbook of Educational Psychology* (pp. 491-549). New York: Simon & Schuster Macmillan.

De Corte, E., Verschaffel, L., & Op 't Eynde, P. (2000). Self-regulation: A characteristic and a goal of mathematics learning. In M. Boekaerts, P. R. Pintrich, & M. Zeidner (Eds.), *Handbook of self-regulation* (pp. 687-726). San Diego: Academic Press.

DeBellis, V. A. (1996). *Interactions between affect and cognition during mathematical problem solving: A two year case study of four elementary school children.* Unpublished doctoral dissertation, Rutgers University: (University Microfilms No. 96-30716).

Ernest, P. (1989). The impact of beliefs on the teaching of mathematics. In P. Ernest (Ed.), *Mathematics teaching: The state of the art* (pp. 249-254). Basingstoke: Falmer Press.

Ernest, P. (1991). *The philosphy of mathematics education.* London: Falmer.

Fennema, E. (1989). The study of affect and mathematics: A proposed generic model for research. In D. B. McLeod, & V. M. Adams (Eds.), *Affect and mathematical problem solving: A new perspective* (pp. 205-219). New York: Springer-Verlag.

Fenstermacher, G. (1994). The knower and the known: The nature of knowledge in research on teaching. In L. Darling-Hammond (Ed.), *Review of research in education*, Vol. 20 (pp. 1-54). Itasca, IL: Peacock.

Frank, M. L. (1988). Problem solving and mathematical beliefs. *Arithmetic Teacher, 35*, 32-35.

Garofalo, J. (1989). Beliefs and their influence on mathematical performance. *Mathematics Teacher, 82*, 502-505.

Gilbert, D. T. (1991). How mental systems believe. *American Psychologist, 46*(2), 107-119.

Green, T. (1971). *The activities of teaching.* New York: McGraw-Hill.

Greeno, J. G. (1991). A view of mathematical problem solving in school. In M. U. Smith (Ed.), *Toward a unified theory of problem solving. View from the content domains* (pp. 69-98). Hillsdale, NJ: Lawrence Erlbaum Associates.

Greeno, J. G., Collins, A. M., & Resnick, L. B. (1996). Cognition and learning. In D. C. Berliner, & R. C. Calfee (Eds.), *Handbook of educational psychology* (pp. 15-46). New York: Simon & Schuster Macmillan.

Henningsen, M., & Stein, M. K. (1997). Mathematical tasks and student cognition: Classroom-based factors that support and inhibit high-level mathematical thinking and reasoning. *Journal for Research in Mathematics Education, 28*(5), 524-549.

Hofer, B. K. (1999). Instructional context in the college mathematics classroom: Epistemological beliefs and student motivation. *Program and Organization Development, 16*(2), 73-82.

Hofer, B. K., & Pintrich, P. R. (1997). The development of epistemological theories: Beliefs about knowledge and knowing and their relation to learning. *Review of Educational Research, 67*(1), 88-140.

Isoda, M., & Nakagoshi, A. (2000). A case study of student emotional change using changing heart rate in problem posing and solving Japanese classroom in mathematics. In T. Nakahara, & M. Koyama (Eds.), *Proceedings of the 24th conference of the International Group for the Psychology of Mathematics Education* (pp. 3-87 – 3-94). Hiroshima: Hiroshima University.

Johnson-Laird, P.N. (1983). *Mental models: Towards a cognitive science of language, inference, and consciousness.* Cambridge, UK: Cambridge University Press

Klinger, E. (1996). Emotional influences on cognitive processing, with implications for the theories of both. In P. M. Gollwitzer, & J. A. Bargh (Eds.), *The psychology of action: Linking cognition and motivation to behavior* . New York: The Guilford Press.

Kloosterman, P. (1996). Students' beliefs about knowing and learning mathematics: Implications for motivation. In M. Carr (Ed), *Motivation in mathematics* (pp. 131-156). Cresskill,NJ: Hampton Press.

Kloosterman, P., & Coughan, M. C. (1994). Students' beliefs about learning school mathematics. *The Elementary School Journal, 94*(4), 375-388.

Kloosterman, P., Raymond, A. M., & Emenaker, C. (1996). Students' beliefs about mathematics: A three-year study. *The Elementary School Journal, 97*(1), 39-56.

Kouba, V. L., & McDonald, J. L. (1986). Children's and teachers' perceptions and beliefs about the domain of elementary mathematics. In D. Lappan, & R. Even (Eds.), *Proceedings of the eight annual meeting of the North American chapter of the International Group for the Psychology of Mathematics Education* (pp. 250-255). East Lansing: Michigan State University.

Lampert, M. (1990). When the problem in not the question and the solution is not the answer: Mathematical knowing and teaching. *American Educational Research Journal, 27*, 29-63.

Lampert, M. (1996). Agreeing to disagree: Developing sociable mathematical discourse. In D. R. Olson, & N. Torrance (Eds.), *The handbook of education and human development: New models of learning, teaching and schooling* (pp. 731-764). Oxford, UK: Blackwell Publishers, Inc.

McLeod, D. B. (1988). Research on learning and instruction in mathematics: The role of affect. In E. Fennema, T. P. Carpenter, & S. J. Lamon (Eds.), *Integrating research on teaching and learning mathematics. Papers from the First Wisconsin Symposium for Research on Teaching and Learning Mathematics* (pp. 60-89). Wisconsin: Wisconsin Center for Education Research, University of Wisconsin.

McLeod, D. B. (1992). Research on affect in mathematics education: a reconceptualization. In D. A. Grouws (Ed.), *Handbook of research on mathematics teaching and learning: a project of the National Council of Teachers of Mathematics* (pp. 575-596). New York: Macmillan.

Mischel, W., & Shoda, Y. (1995). A cognitive-affective system theory of personality: Reconceptualizing situations, dispositions, dynamics, and invariance in personality structure. *Psychological Review, 102*(2), 246-268.

Mtetwa, D., & Garofalo, J. (1989). Beliefs about mathematics: An overlooked aspect of student difficulties. *Academic Therapy, 24*, 611-618.

National Council of Teachers of Mathematics. (1989). *Curriculum and evaluation standards for school mathematics*. Reston, VA: National Council of Teachers of Mathematics.

National Council of Teachers of Mathematics. (2000). *Principles and standards for school mathematics.* Reston, VA: NCTM.

Nickson, M. (1992). The culture of the mathematics classroom: an unknown quantity? In D.A. Grouws (Ed.), *Handbook of research on mathematics teaching and learning: a project of the National Council of Teachers of Mathematics* (pp. 101-114). New York: Macmillan.

Niemivirta, M. (1996). *Intentional and adaptive learning modes: The self at stake*. Paper presented at the second European Conference on Education, September 1996, Sevilla, Spain.

Ortony, A., Clore, G. L., & Collins, A. (1988). *The cognitive structure of emotions*. New York: Cambridge University Press.

Pajares, F., & Miller, M. D. (1994). Role of self-efficacy and self-concept beliefs in mathematical problem solving: A path analysis. *Journal of Educational Psychology, 86*(2), 193-203.

Pehkonen, E. (1995). *Pupils' view of mathematics: Initial report for an international comparison project.* University of Helsinki, Department of Teacher Education. Research report 152

Pehkonen, E., & Törner, G. (1996). Mathematical beliefs and different aspects of their meaning. *Zentralblatt Für Didaktik Der Mathematik, 4*, 101-108.

Pekrun, R. (1990). Emotion and motivation in educational psychology. In P. J. D. Drenth, J. A. Sergeant, & R. J. Takens (Eds.), *European perspectives in psychology: Theoretical, psychometrics, personality, development, educational, cognitive, gerontological* (Vol. 1pp. 265-295). Chisester, UK: John Whiley & Sons.

Perkins, D. (1995). *Outsmarting IQ. The emerging science of learnable intelligence*. New York: The Free Press.

Pintrich, P. (1989). The dynamic interplay of student motivation and cognition in the college classroom. In C. Ames, & M. Maehr (Eds.), *Advances in motivation and achievement: Motivation-enhancing environments* (Vol. 6pp. 117-160). Greenwich, CT: JAI Press.

Pintrich, P. R., & Schrauben, B. (1992). Students' motivational beliefs and their cognitive engagement in academic tasks. In D. Schunk, & J. Meece (Eds.), *Students' perceptions in the classroom: Causes and consequences* (pp. 149-183). Hillsdale, NJ: Lawrence Erlbaum Associates.

Pintrich, P. R., Smith, D. A. F., Garcia, T., & McKeachie, W. J. (1993). Reliability and predictive validity of the Motivated Strategies for Learning Questionnaire (MSLQ). *Educational and Psychological Measurement, 53,* 801-813.

Power, M., & Dalgleish, T. (1997). *Cognition and emotion: Form order to disorder*. Sussex, UK: Erlbaum Taylor & Francis LTD.

Richardson, V. (1996). The role of attitudes and beliefs in learning to teach. In J. Sikula (Ed.), *Handbook of research on teacher education* (second ed., pp. 102-119). New York: Simon & Schuster Macmillan.

Salomon, G., & Perkins, D. N. (1998). Individual and social aspects of learning. In P. D. Pearson, & A. Iran-Nejad (Eds.), *Review of Research on Education* Vol. 23 (pp. 1-25). Washington, DC: AERA.

Scheffler, I. (1965). *Conditions of knowledge: An introduction to epistemology and education*. Chicago: Scott Foresman.

Schiefele, U., & Csikszentmihalyi, M. (1995). Motivation and ability as factors in mathematics experience and achievement. *Journal for Research in Mathematics Education, 26*(2), 163-181.

Schoenfeld, A. H. (1983). Beyond the purely cognitive: Belief systems, social cognitions, and metacognitions as driving forces in intellectual performance. *Cognitive Science, 7,* 329-363.

Schoenfeld, A. H. (1985a). *Mathematical problem solving*. Orlando,Florida: Academic Press.

Schoenfeld, A. H. (1985b). Students' beliefs about mathematics and their effects on mathematical performance: A questionnaire analysis. Paper presented at the Annual Meeting of the American Educational Research Association, March 31 - April 4, 1985, Chicago, Illinois.

Schoenfeld, A. H. (1992). Learning to think mathematically: Problem solving, metacognition, and sense making in mathematics. In D. A. Grouws (Ed.), *Handbook of research on mathematics teaching and learning* (pp. 334-370). New York: Macmillan.

Schommer, M., Crouse, A., & Rhodes, N. (1992). Epistemological beliefs and mathematical text comprehension: Believing it is simple does not make it so. *Journal of Educational Psychology, 84*(4), 435-443.

Seegers, G., & Boekaerts, M. (1993). Task motivation and mathematics in actual task situations. *Learning and Instruction, 3* , 133-150.

Snow, R. E., Corno, L., & Jackson III, D. (1996). Individual differences in affective and conative functions. In D. C. Berliner, & R. C. Calfee (Eds.), *Handbook of Educational Psychology* (pp. 243-310). New York: Simon & Schuster Macmillan.

Spangler, D. A. (1992). Assessing students' beliefs about mathematics. *Arithmetic Teacher, 40*(3), 148-152.

Spinoza, B. (1982). *The etics and selected letters_.* (S. Feldman, Ed., and S. Shirley, Trans.) Indianapolis, IN: Hackett.

Stipek, D. J., & Gralinski, H. (1991). Gender differences in children's achievement-related beliefs and emotional responses to succes and failure in mathematics. *Journal of Educational Psychology, 83*(3), 361-371.

Stodolsky, S. S., Salk, S., & Glaessner, B. (1991). Student views about learning math and social studies. *American Educational Research Journal, 28,* 89-116.

Teasdale, J., & Barnard, P. (1993). *Affect, cognition and change*. Hove, UK: Lawrence Erlbaum Associates, LTD.

Thompson, A. G. (1992). Teachers' beliefs and conceptions: A synthesis of the research. In D. A. Grouws (Ed.), *Handbook of research on mathematics teaching and learning: a project of the National Council of Teachers of Mathematics* (pp. 127-146). New York: Macmillan.

Törner, G., & Pehkonen, E. (1996). On the structure of mathematical belief systems. *Zentralblatt Für Didaktik Der Mathematik, 4,* 109-112.

Tymoczko, T. (1986). (Ed.), *New directions in the philosophy of mathematics*. Boston: Birkhauser.

Underhill, R. (1988). Mathematics learners' beliefs: A review. *Focus on Learning Problems in Mathematics, 10*, 55-69.

Vanayan, M., White, N., Yuen, P., & Teper, M. (1997). Beliefs and attitudes toward mathematics among third- and fifth-grade students: A descriptive study. *School Science and Mathematics, 97*(7), 345-351.

Vermeer, H. J. (1997). *Sixth-grade students' mathematical problem-solving behavior: Motivational variables and gender differences*. Leiden: UFB, Leiden University.

Verschaffel, L., Greer, B., & De Corte, E. (2000). *Making sense of word problems*. Lisse, The Netherlands: Swets & Zeitlinger.

Yackel, E., & Cobb, P. (1996). Sociomathematical norms, argumentation, and autonomy in mathematics. *Journal for Research in Mathematics Education, 27*, 458-477.

CHAPTER 3

FULVIA FURINGHETTI AND ERKKI PEHKONEN

RETHINKING CHARACTERIZATIONS OF BELIEFS

Abstract. In this chapter we consider beliefs and the related concepts of conceptions and knowledge. From a review of the literature in different fields we observe that there is a diversity of views and approaches in research on these subjects. We report on a small research project of our own attempting to clarify the understanding of beliefs among specialists in mathematics education. A panel of 18 mathematics educators participated in a panel that we termed "virtual", since the participants communicated with us only by e-mail. We sent nine characterizations related to beliefs, selected from the literature, to the panelists, asked them to express their agreement or disagreement with the statements, and also asked each to give their own characterization of the term. The answers were analyzed, searching for the elements around which the concept of beliefs has developed along the years. We discuss issues on which there was agreement and disagreement and conjecture what lies behind the differences. As a final step we make some suggestions relating to characterization of the term belief and ways of dealing with it in future research.

1. INTRODUCTION

The purpose of this chapter is to draw attention to theoretical deficiencies in belief research. First, the concept of belief (and other related concepts) is often left undefined (e.g., Cooney, Shealy, & Arvold, 1998) or researchers give their own, possibly contradictory, definitions (e.g., Bassarear, 1989; Underhill, 1988). A second important problem is the inability to clarify the relations between belief and knowledge. At the end of the chapter, we describe the results of an empirical research study in which we tried to characterize the concept of belief based on views that emerged from written reports of mathematics education specialists in the field of beliefs.

2. BELIEFS AND CONCEPTIONS

Beliefs and belief systems began to be examined, to some extent, at the beginning of this century, mainly in social psychology (Thompson, 1992). But before long, behaviorism began to dominate research in the psychological domains. The focus turned to the observable aspects of human behavior, and beliefs were nearly forgotten. New interest in beliefs and belief systems emerged mainly in the 1970s, through developments in cognitive science (Abelson, 1979).

Individuals continuously receive signals from the world around them. According to their perceptions and experiences based on these messages, they draw conclusions

G. C. Leder, E. Pehkonen, & G. Törner (Eds.), Beliefs: A Hidden Variable in Mathematics Education? 39-57.

about different phenomena and their nature. Individuals' subjective knowledge, i.e., their beliefs (including affective factors), is a compound of these conclusions. Furthermore, they compare these beliefs with new experiences and with the beliefs of other individuals, and thus their beliefs are under continuous evaluation and may change. When a new belief is adopted, this will automatically form a part of the larger structure of their subjective knowledge, i.e., of their belief system, since beliefs never appear fully independently. Thus, an individual's belief system is a compound of her conscious or unconscious beliefs, hypotheses or expectations and their combinations (Green, 1971).

2.1. Different Understandings of Beliefs

As discussed in other chapters of this book (Leder & Forgasz; Op't Eynde, De Corte, & Verschaffel; Törner), there are many variations of the concepts belief and belief system used in studies in the field of mathematics education. As a consequence of the vague characterization of the concept, researchers have often formulated their own definition of belief which might even be in contradiction with others. For example, Schoenfeld (1985, p. 44) states that in order to give a first rough impression "belief systems are one's mathematical world view". He later adds explanations of his position, interpreting beliefs "as an individual's understandings and feelings that shape the ways that the individual conceptualizes and engages in mathematical behavior" (Schoenfeld, 1992, p. 358). Hart (1989, p. 44) – under the influence of Schoenfeld's (1985) and Silver's (1985) ideas – uses the word belief "to reflect certain types of judgments about a set of objects". Lester, Garofalo, and Kroll (1989, p. 77) explain that "beliefs constitute the individual's subjective knowledge about self, mathematics, problem solving, and the topics dealt with in problem statements". Törner and Grigutsch (1994) label their research object as the "mathematical world view", as is done in Schoenfeld (1985). In a recent paper by Grigutsch, Raatz, and Törner (1998) this concept is elaborated further, and anchored into the theory of attitudes, as explained, for example, in Olson and Zanna (1993). Other researchers, Underhill (1988) for one, think that beliefs are some kind of attitudes. Yet another different explanation is given by Bassarear (1989) who sees attitudes and beliefs on the opposite extremes of a bipolar dimension.

When looking at these different, and in some case even contradictory (Underhill, 1988; Bassarear, 1989) characterizations of beliefs, one observes that most of them (Underhill, 1988; Lester et al., 1989; Thompson, 1992; Furinghetti, 1996; Lloyd & Wilson, 1998) refer to the static part of beliefs saying: beliefs are, constitute, are contained etc. The definition given by Schoenfeld (1992) stresses the dynamic part of beliefs, i.e., how beliefs function. The definition proposed by Hart (1989) puts forward the aspect of judgments. The place of beliefs on the dimension affective – cognitive may be seen in different ways. If we were to stress the connections between beliefs and knowledge, we would see beliefs mainly as representatives of the cognitive structure of individuals. However, to see beliefs as a form of reactions toward a certain situation means that we consider beliefs to be linked to the affective part of individuals.

In research, there are representatives of both viewpoints. Some researchers consider beliefs as a real part of cognitive processing. Most researchers acknowledge that beliefs contain some affective elements, since the birth of beliefs happens in the social environment in which we live (McLeod, 1989, 1992). Among the six definitions of belief given above, those of Underhill (1988), and Lester et al. (1989) stress the affective component, whereas the definitions of Bassarear (1989) and Thompson (1992) are more on the cognitive side. In this book the different orientations are present in Goldin's chapter (affective orientation), in the chapters by Op't Eynde et al. and by Törner (cognitive orientation), while the chapter of Leder and Forgasz has a more marked mixed orientation (affective/cognitive).

In his study, Saari (1983) tried to structure the central concepts of the affective domain. He grouped them using three categories: feelings, belief systems, and optional behavior. Belief systems are seen as being developed from simple perceptual beliefs or authority beliefs – via new beliefs, expectations, conceptions, opinions and convictions – to a general conception of life. Such a viewpoint, that attitude has a component structure, seems to be commonly accepted in psychology today. One may find the following definition in the dictionary of psychology (e.g., Statt, 1990, p. 11): Attitude is "a stable, long-lasting, learned predisposition to respond to certain things in a certain way. The concept has a cognitive (belief) aspect, an affective (feeling) aspect, and a conative (action) aspect". The same threefold structure is found in many definitions on attitudes within research on mathematics education (e.g., Hart, 1989; Olson & Zanna, 1993; Ruffell, Mason, & Allen, 1998).

2.2. Different Characterizations of Conceptions

Conceptions belong to the same group of concepts as beliefs, which are also used in different ways in mathematics education (and the wider) literature. For example, Thompson (1992) understands beliefs as a sub-class of conceptions. But she claims that "the distinction [between beliefs and conceptions] may not be a terribly important one" (p. 130). Thompson's idea is taken up by Furinghetti (1996) who explains an individual's conception of mathematics as a set of certain beliefs. A different understanding is given by Pehkonen (1994) who, in accordance with Saari (1983), characterizes conceptions as conscious beliefs.

Some trials for a definition of conceptions are given, among others, by Freire and Sanches (1992), Ponte (1994), and Lloyd and Wilson (1998). For example, Lloyd and Wilson (1998, p. 249) connect beliefs with conceptions saying: "We use the word conceptions to refer to a person's general mental structures that encompass knowledge, beliefs, understandings, preferences, and views". But there are other researchers who clearly distinguish the meaning of these two terms. For example, this is the position emerging from the following passage of Ponte (1994) who has used Pajares (1992) as an authority:

> They [beliefs] state that something is either true or false, thus having a prepositional nature. Conceptions are cognitive constructs that may be viewed as the underlying organizing frames of concepts. They are essentially metaphorical. (p. 169)

Freire and Sanches (1992) describe the conception of a discipline and of its teaching as a set of ideas, beliefs, understandings and interpretations of pedagogical practices concerning the nature and content of the discipline, the students and the way they learn, the teachers and the role they play in the classroom, and the context in which pedagogical practices occur. This meaning of the word conception is adopted also in the study on the implementation of new curricula encompassing computer science reported in Bottino and Furinghetti (1996). This characterization of conceptions is close to the description of the knowledge for teaching given in Elbaz (1983).

Additionally, the term conception is used in a global sense. For example, Thompson (1992) uses the term conception not to refer to a single mathematical idea, but to the whole of mathematics. For her, the conception of the nature of mathematics

> may be viewed as that teacher's conscious and subconscious beliefs, concepts, meanings, rules, mental images, and preference concerning the discipline of mathematics. Those beliefs, concepts, views, and preferences constitute the rudiments of a philosophy of mathematics, although for some teachers they may not be developed and articulated into a coherent philosophy. (p. 132)

3. RELATIONS BETWEEN BELIEFS/CONCEPTIONS AND KNOWLEDGE

The method often used to study the relations between beliefs and knowledge is to consider the properties of the structures encompassed by each of them. However, it should be noted that not all researchers take the distinctions so seriously. Some researchers have argued that it is not important to distinguish between knowledge and beliefs. They consider it to be more or less an academic philosophical problem, see for example Audi (1988) and Thompson (1992). Knowledge, indeed, is a theme on which philosophers, from Plato onwards, have focused their research efforts. In philosophical studies a central theme has been the relation of knowledge to beliefs (e.g., Hintikka, 1962). Recent research has focused on the connections of Plato's theory to mathematics education (e.g., Rodd, 1997; Gardiner, 1998; Lindgren, 1999). Furthermore, there are researchers who confine themselves to studying how beliefs/knowledge systems influence teachers' and pupils' behaviors in mathematics classes (e.g., Thompson, 1992).

3.1. What is Knowledge?

The concept of knowledge will be discussed very briefly here, stressing its close connections to beliefs/conceptions. Those interested in a broader discussion are referred to the work of, for example, Fennema and Franke (1992), Goldin (1990), or Steffe (1990).

Rodd (1997) pointed out that a useful definition of knowledge is quite elusive. The classical definition of knowledge that originates from Plato states that "knowledge is justified true belief" (Mc Dowell, 1987, p.94, 201d). But Lindgren (1999) shows, with the aid of passages from *Theaetetus*, that Plato was aware of the difficulty of proving whether a judgment is true or not.

For the purposes of the present study we consider two different aspects of knowledge: objective (official) knowledge that is accepted by a community and subjective (personal) knowledge that is not necessarily subject to an outsider's evaluation. Beliefs belong to individuals' subjective knowledge, and when expressed as sentences they might be (or might not be) logically true. Knowledge always has this truth-property (Lester et al., 1989). We can describe this property with probabilities: knowledge is valid with a probability of 100%, whereas the corresponding probability for belief is usually less than 100 %. Therefore, this is one of the distinguishing properties between knowledge and beliefs.

Thompson (1992) summarizes three dimensions which distinguish beliefs from knowledge:

- the degree of intersubjective consensus,
- the type of argument needed for the acceptance of beliefs and knowledge respectively,
- knowledge is related to truth and certainty, while belief is more associated with doubts and disputes.

Sfard (1991) offers another explanation for the connection between beliefs and knowledge. She considers conceptions as the subjective/private side of the term concept defined as follows:

> The word 'concept' (sometimes replaced by 'notion') will be mentioned whenever a mathematical idea is concerned in its 'official' form as a theoretical construct within 'the formal universe of ideal knowledge'.

She further explains that

> the whole cluster of internal representations and associations evoked by the concept - the concept's counterpart in the internal, subjective 'universe of human knowing' - will be referred to as 'conception'. (p. 3)

The distinction between conception and knowledge is complicated by the fact that an individual's conception of a certain concept can be considered as a picture of that concept. As a picture and its object are not the same, and usually the picture shows only one view of the object, similarly a conception represents only partly its object (concept). In this kind of discussion, in general, Sfard does not use the word belief.

3.2. Properties of Belief Systems[1]

In order to deepen the discussion of the distinction between knowledge and beliefs, some structural differences between belief systems and knowledge systems have been noticed. For example, Rokeach (1968) organized beliefs along a dimension of centrality to the individual. The beliefs that are most central are those for which the individual has complete consensus; those beliefs about which there is some disagreement would be less central. Green (1971) introduces three dimensions, which are characteristic of belief systems: quasi-logicalness, psychological centrality (the degree of conviction) and cluster structure. These dimensions of Green will be discussed more closely below. Also, Thompson (1992) emphasized two of Green's dimensions as characteristics of beliefs: the degree of conviction, and the clustering aspect.

Quasi-logicalness. Knowledge systems are usually formed logically from premises and from conclusions deduced from them. The relationships between beliefs within a belief system, on the other hand, cannot be said to be logical, since beliefs are arranged according to how the believer sees their connections. In other words, each person has in her belief system her own logic. This may be described as a structure, called quasi-logical, which can also contain some primary beliefs and derivative beliefs. This quasi-logical order is unique for each person, and it reflects the thinking and valuing of the person in question.

Also, Abelson (1979) pointed to this lack of logic in belief systems: Within a belief system, beliefs are not necessarily held in consensus with other beliefs. Therefore, one could have beliefs which contradict other beliefs held by the same person at the same time. Furthermore, the believer is usually aware that others may have different beliefs, whereas one important feature of knowledge systems is that it cannot contain contradictions. An example, in the case of a teacher's mathematical beliefs, is given in Pehkonen (1994).

On the theme of the meaning of computers and calculators for mathematics teaching, one may easily generate an example of a set of teachers' beliefs where one belief is a primary one, and the others are derived beliefs. For example, a teacher may hold the following belief: "Technology helps remarkably in mathematics teaching". From this primary belief, a person may have conducted some derived beliefs for teaching practice, such as: "A teacher should allow pupils to use pocket calculators in classrooms whenever suitable", "Pupils should have time for computer exercises", "The school should invest money in a computer class", etc.

Psychological centrality. Some beliefs are more important for an individual than others. The more important could be said to be psychologically more central, and the others are peripheral in the individual's belief system; for examples see Pehkonen (1994). Thus, beliefs have their own psychological strength; i.e., the degree of conviction with which they are held (probability of validity ascribed by the individual). The degree of conviction may vary from belief to belief. The most central beliefs are held most strongly. They are usually considered to be 100% sure, whereas the peripheral ones may be changed more easily. These assumptions can be compared with Rokeach's concept of centrality (Rokeach, 1968).

An example of the coexistence of central and peripheral beliefs held by the same individual was illustrated in a teaching experiment, fully studied in Paola (1999), in which teaching mathematics was carried out through discussion. The teacher involved in the experiment was very enthusiastic about this method of working. It was her belief that this method is efficient and suitable for making students learn mathematics with understanding. On the other hand, she also held the belief that the input to be used in the mathematical activity should come only from the teacher. Since this latter was a central belief, while the former was peripheral, during the experiment the former belief was abandoned in favor of the latter and the discussion-style lesson shifted to the traditional teacher-driven lesson.

The dimension of psychological centrality is lacking in knowledge systems. When speaking of one's beliefs, it is possible to say "I am 60% sure that we will some day find an intellectual life form, other than human beings, in the stars". Thus one is open to accepting other statements. In the case of knowledge, one is always

100% sure. So if a person *knows* a certain situation, s/he is not prepared to accept a contrasting situation.

Teachers' beliefs about good mathematics teaching have been so deeply rooted that surface changes (changes in peripheral beliefs) – such as changing outer conditions, e.g., curriculum, teaching materials – cannot influence them. If teachers are compelled to undergo change, they will adapt to the new curriculum, possibly by interpreting their teaching in a new way[2], and absorb some of the ideas of the new teaching material into the old style of teaching. In fact, there seems to be a gap between teachers' expressed beliefs and their teaching practices (Jones, Henderson, & Cooney, 1986; Ernest, 1991). For example, a teacher may express a belief that exploring mathematical situations is more important than rote practice, yet often assign nearly 50 exercises for pupils to work during class (Shaw, 1989). Another teacher may believe that he is allowing pupils' ideas to guide classroom discourse, but in reality will only recognize those ideas that fit into the prepared plan (Paola, 1999). This discrepancy between accepted beliefs and beliefs in action generates what Furinghetti (1996) terms ghosts in classroom, e,g., hidden beliefs in action.

Cluster structure. Beliefs are held in clusters. Or as Green (1971, p. 41) puts it: "Nobody holds a belief in total independence of all other beliefs. Beliefs always occur in sets or groups". This cluster structure enables individuals even to hold conflicting beliefs within their own belief system (cf. quasi-logicalness). The clustering property may help to explain some inconsistencies found in an individual's belief system.

Let us take an example from research done by Hasemann (1987, p. 29–30). He described a girl (Yvonne) interviewed by him about her fraction skills. He noticed that she "added fractions by using the rule 'numerator plus numerator, denominator plus denominator', but she did the diagrammatic solutions to the problems correctly". Both algorithms seem to belong in her belief system to totally different clusters, since she accepted two different answers for the same task – and "she believed that they both were correct". This contradictory situation did not disturb her. The disturbance occurred when the interviewer took a realistic situation (a cake in the oven), and she grasped that, in reality, two different solutions could not exist.

In addition to the cluster structure, Abelson (1979) pointed out that belief systems rely heavily on evaluative and affective components. A belief system typically has extensive categories of judgments, which are grouped into "good", and "bad". As a typical example, those who support so-called "green values", also usually believe that nuclear power is bad, materialism and waste are bad, natural alternative energy sources are good, re-cycling is good. Knowledge systems lack such evaluations.

4. EMPIRICAL RESEARCH

In order to make as clear as possible the theories developed about beliefs we have identified the following elements of reflection:

- Terminology: What are the meanings researchers give the terms involved in research on beliefs?

- Methodology: How do we detect and analyze beliefs, which in their nature are entities hidden and elusive? (compare the metaphor used by Berger (1998) on reconstructing the skeleton of dinosaurs from some fossils which were found.)
- Effect: How may studies on beliefs/conceptions affect classroom practice and the strategies applied in teacher education?

The present study concerns the first element; i.e., we will try to investigate what researchers in the field intend when they work on beliefs and conceptions. Based on our earlier observations, this implies also that we take into consideration the concept of knowledge, which has strong historical and epistemological links with the terms in question. We are aware that our way of approaching the topic is closer to the way used in mathematics than the one used in psychology and mathematics education. Mathematicians develop their theories starting from well-stated axioms and definitions, while psychologists often keep their assumptions and meaning of the terms quite fuzzy. Our choice originated from a difficulty in dealing with studies in which it may happen that authors use different terms to express the same objects or the same terms to express different matters.

4.1. The Objectives of Research

As we have seen, the studies we have to consider when dealing with beliefs are in different disciplinary domains, e.g., philosophy, psychology, and education. In the present study we confined ourselves to the domain of mathematics education. Using an international specialist panel we looked for common background suitable to describe the characteristics of the concept of beliefs and the mutual relationships in the critical triad "beliefs - conceptions – knowledge".

The idea was to clarify some core elements, in studies relating to beliefs, which almost all specialists could accept or, if the different assumptions of researchers would make it impossible to reach complete agreement, to stress the existence of different positions. In the case of this second circumstance we felt that our study would contribute to convincing researchers about the necessity of making their assumptions explicit. To be clear, we point out that it was not our aim to introduce a democratic pattern according to which definitions are necessarily assessed as good if the majority of the researchers in the field of beliefs accept them. We simply want to stress the pitfalls generated by inadequately clarified or ambiguous assumptions in research.

5. METHOD

Influenced by the previous considerations we worked out a questionnaire in which we presented the nine characterizations shown in Table 1. Our list does not claim to be comprehensive because we have only considered which characterizations in the recent literature (1987–98) were suitable to focus on one or more terms of the triad beliefs-conceptions-knowledge which we are discussing in this chapter. The characterizations we proposed in our questionnaire were rather general. Only three statements contain the word mathematics, and only one statement refers to a precise category of persons (teachers) instead of referring generically to "persons" and "individuals". We were aware that the context in which the characterizations originally appeared might suggest a different interpretation than just the single statement presented in isolation. Notwithstanding, we considered the use of characterizations taken from the literature more efficient and fair than self-created characterizations.

Table 1. The nine characterizations of belief included in the questionnaire

Characterization #1 (Hart, 1989, p. 44)	"we use the word belief to reflect certain types of judgments about a set of objects"
Characterization #2 (Lester et al., 1989, p. 77)	"beliefs constitute the individual's subjective knowledge about self, mathematics, problem solving, and the topics dealt with in problem statements"
Characterization #3 (Lloyd & Wilson, 1998, p. 249)	"we use the word conceptions to refer to a person's general mental structures that encompass knowledge, beliefs, understandings, preferences, and views"
Characterization #4 (Nespor, 1987, p. 321)	"Belief systems often include affective feelings and evaluations, vivid memories of personal experiences, and assumptions about the existence of entities and alternative worlds, all of which are simply not open to outside evaluation or critical examination in the same sense that the components of knowledge systems are"
Characterization #5 (Ponte, 1994, p. 169)	"Beliefs and conceptions are regarded as part of knowledge. Beliefs are the incontrovertible personal 'truths' held by everyone, deriving from experience or from fantasy, with a strong affective and evaluative component."
Characterization #6 (Pehkonen, 1998, p. 44)	"we understand beliefs as one's stable subjective knowledge (which also includes his feelings) of a certain object or concern to which tenable grounds may not always be found in objective considerations"
Characterization #7 (Schoenfeld, 1992, p. 358)	"beliefs - to be interpreted as an individual's understandings and feelings that shape the ways that the individual conceptualizes and engages in mathematical behavior"
Characterization #8 (Thompson, 1992, p. 132)	"A teacher's conceptions of the nature of mathematics may be viewed as that teacher's conscious or subconscious beliefs, concepts, meanings, rules, mental images, and preferences concerning the discipline of mathematics."
Characterization #9 (Törner & Grigutsch, 1994, p. 213).	"Attitude is a stable, long-lasting, learned predisposition to respond to certain things in a certain way. The concept has a cognitive (belief) aspect, an affective (feeling) aspect, and a conative (action) aspect."

In the questionnaire the authors of the characterizations were not indicated. Each characterization was accompanied by the sentences

- "Do you consider the characterization to be a proper one?
- Please, give the reasons for your decision!"

Moreover there was the following final item:
"Your characterization: Please, write your own characterization for the concept of 'belief'."

In March 1999 we sent our questionnaire, via e-mail, to the 22 mathematics education specialists in the field of beliefs who had been invited to the international meeting "Mathematical Beliefs and their Impact on Teaching and Learning of Mathematics" held in November 1999 in Oberwolfach. The specialists were asked to respond within two weeks. Altogether 18 researchers sent us their responses within that time, commenting on all characterizations. However, only half of them gave us their own characterization. The specialists responding to our e-mail questionnaire were from seven different countries: Australia, Canada, Cyprus, Germany, Israel, UK, and USA. We expected to collect data on the following points: agreement or disagreement with the given characterizations, possible improvements, reasons for agreement or disagreement, and personal characterizations.

The panel was virtual in the sense that only e-mail was used for communication and that in the discussion there was no stage of communication between the specialists. This last stage was realized during the meeting, when the presentation of the results that we collected allowed the specialists attending the meeting to give comments and further information.

6. RESULTS

The following gives an overview of the results, showing the extent of difference among the specialists in their comments on the characterizations, and identifying points of agreement in relation to some of the characteristics which distinguish beliefs.

6.1. An Overview of Results

When reading the specialists' reactions to the nine characterizations of belief, we confronted a large variety of ideas, and had difficulties in finding patterns in it. Even though it was not our intention to work on statistical data we needed to have a starting point for our analysis. In order to have an overview, our first task was to group the responses according to a five-step scale: Y (= full agreement), P+ (= partial agreement with a positive orientation), P (= partial agreement), P- (= partial agreement with a negative orientation), N (= full disagreement).

The results are reported in Table 2. In the last column (#10: own characterization), the sign * means that an answer was given, and the sign - means that no answer was provided. In the first column, R.n stands for "respondent n". We both classified all the answers, independently, according to our five-step scale.

Afterwards we compared the classifications, prolonging our discussion until we reached consensus.

Table 2. Responses to the 10 items by individual respondents

Respondents	1	2	3	4	5	6	7	8	9	10
R.1	Y	Y	P	N	N	N	Y	Y	Y	-
R.2	P+	P	P	P-	N	N	Y	Y	Y	-
R.3	P	P	Y	P	N	N	P+	P+	P+	-
R.4	P+	P	P	Y	Y	Y	Y	P	P+	
R.5	Y	Y	N	N	N	Y	Y	Y	Y	-
R.6	P	N	N	P	N	N	P	Y	P	*
R.7	N	Y	Y	Y	Y	N	Y	Y	P	-
R.8	N	P+	P+	N	P+	P+	N	Y	Y	*
R.9	P+	Y	P+	P	N	Y	Y	Y	Y	*
R.10	Y	Y	Y	N	N	Y	Y	Y	P+	-
R.11	Y	Y	Y	N	N	Y	Y	N	N	*
R.12	N	P	Y	P-	N	P	P	Y	P-	-
R.13	P-	P	P+	N	N	Y	P	P	Y	*
R.14	Y	N	Y	Y	N	N	N	N	P-	*
R.15	P+	P	P	P	N	Y	Y	Y	P+	-
R.16	Y	N	Y	Y	N	N	N	N	N	*
R.17	Y	Y	Y	N	N	N	Y	Y	Y	*
R.18	N	P	Y	N	N	N	Y	N	N	*

In order to get a better overview of the situation, we give the number of responses to each item, grouped according to the five-step scale, in Table 3.

Table 3. Degree of agreement/disagreement of the respondents with the characterizations

	1	2	3	4	5	6	7	8	9
Y = Yes	7	7	9	4	2	7	11	11	7
P+ = Partly Yes	4	1	3	-	1	1	1	1	4
P = Partly	2	7	4	4	-	1	3	2	2
P- = Partly No	1	-	-	2	-	-	-	-	2
N = No	4	3	2	8	15	9	3	4	3

The first surprising result was that in the responses of the specialists, there was no clear pattern to be observed. But on some points there was some regularity. The answers were most unified in relation to characterization #5 (by Ponte, 1994) where 15 of the 18 specialists disagreed with the statement, and only three agreed.

In the following discussion of the results we must remember that the respondents did not know who was the author of each characterization - they had only the characterizations.

- The response of the panel to characterization #5 (Ponte, 1994) was a clear No. According to its author, this characterization is based on the views expressed by Pajares (1992), a researcher outside the mathematics education community.
- The next largest frequencies were in relation to characterizations #7 (Schoenfeld, 1992) and #8 (Thompson, 1992). In these cases, most of the panel members (11 of 18) were in agreement with the characterizations (i.e., the answer was Yes). This is not surprising, since the papers of Schoenfeld and Thompson are much used as reference literature in research on beliefs. But these were not accepted by consensus: there were 3–4 specialists who responded clearly No, and 2–3 others who agreed only partly.
- In three cases, we estimated the orientation of the panel to be positive, since the sum of Yes and Part Yes answers was larger than the negative answers. These were characterization #1 (Hart, 1989), characterization #3 (Lloyd & Wilson, 1998), characterization #9 (Törner & Grigutsch, 1994). For the higher level of agreement here, we can find reasons easily. One can say that Hart follows the ideas of Schoenfeld, and Lloyd and Wilson those of Thompson. And the last one (Törner & Grigutsch) is actually the commonly accepted three-component definition of attitudes where beliefs form the cognitive component (Olson & Zanna, 1993).
- In relation to characterization #6 (Pehkonen, 1998) agreements and disagreements were divided almost fifty-fifty (Yes–No). In this characterization, the word "stable" has caused confusion, since in the case of beliefs it can be understood in different ways.
- In relation to characterization #4 (Nespor, 1987), the majority of the responses were negative. Therefore, we can say that the answer was an almost No. This characterization also comes from outside the mathematics education community.
- Additionally, it is interesting that in one case, characterization #2 (Lester et al., 1989), there were many Partly answers.

In the following, we are not trying to identify any of the characterizations as right or good or proper. We concentrate on the points of agreement and/or disagreement in the responses in order to reveal some characteristic difficulties in defining the concepts of belief and of conception which might help other researchers to explain their position in belief studies.

6.2. Common Points of Disagreement with the Characterization

We focus here on the results of those items that most clearly give us an orientation. We have individuated the key concepts around which the disagreement with characterization #5 (Ponte, 1994) was centered. In the 15 negative answers to characterization #5, we have identified two central features determining the disagreement: the adjective *incontrovertible* and the *relation between beliefs and knowledge*. Another point of disagreement originated from the use of the term *conception*, present in items #3, #5 and #8.

Incontrovertible. The main difficulty seemed to be in the use of the term "incontrovertible" as an attribute to beliefs. Altogether ten specialists (R.1, R.5, R.8, R.10, R.11, R.12, R.14, R.15, R.17, R.18) out of the 15 disagreeing stressed this term. The response of R.1 gives the general feeling: "The 'incontrovertible [...] everyone' doesn't make sense to me." The dictionary gives "indisputable" and "undeniable" as synonyms for incontrovertible. The majority of the specialists did not consider beliefs as something that could be said to be indisputable or undeniable.

"Stable" is another word which was commented on in the responses. In relation to the statement in #6 this word was the main point of disagreement. However there was recognition that the concept of stability, in relation to beliefs, is not clear-cut. The comment of one respondent (R.18) on characterization #6 was:

> My reasons here are already stated in prior comments regarding use of the word 'subjective' and the notion of stability. ... However, it may be appropriate to discuss 'stability' in a local sense. That is, at any given time, it may make sense to say that there is some stability in an individual's beliefs.

Another respondent (R.6) begins his own characterization with the sentence "Beliefs, in a particular context/situation, are..." stressing the instant/local character of beliefs.

Relation between beliefs and knowledge. In characterization #5 beliefs are clearly included as a part of knowledge. This evokes the old dilemma, faced by Plato, between objective knowledge, which is accepted by the community, and subjective knowledge, which does not need an evaluation. Eight specialists (R.2, R.3, R.6, R.9, R.10, R.12, R.14, and R.17) pointed out problems in the relationship between beliefs and knowledge in connection to characterization #5. The different responses were grouped as follows:

- Four respondents (R.2, R.3, R.10, and R.14) stated that beliefs are not a part of knowledge. In the following, there is a response (R.2) which gives the flavor of the answers in this group: "We can speak about personal truths but I think knowledge is not an appropriate term in which to embed beliefs in the way it is presented here".
- Two specialists (R.6, R.17) considered beliefs and knowledge as synonyms, e.g., "I see beliefs and conceptions as synonymous with knowledge rather than subsets of it. I cannot think of knowledge that is not also a belief". (R.6)
- One specialist only wanted to separate beliefs and knowledge from each other: "it is useful to distinguish between beliefs and knowledge". (R.9)

An explanation for these reservations and doubts may be that characterization #5 had its origin outside the world of mathematics education, which may influence the way mathematics educators perceive it. But the crucial point in these different positions is expressed in the following answer: "Depends on the definition of knowledge". (R.7) The answer points out that the problem of defining beliefs relies on the problem of defining knowledge.

Conception. The term conception appears in #5, #3 and #8. None of the respondents focused on the term "conception" in characterization #5, but 11 respondents commented on this term in relation to characterizations #3 and #8. Those who considered the use of this term stressed the role of it as an umbrella term.

Only one respondent considered the term as almost synonymous with beliefs, and commented on characterization #8 with the following words: "To the extent that the word conception is used to mean that which is mentally conceived, beliefs are conceptions. They are an individual's understandings of ..." (R.18).

As in the case of beliefs, in relation to conceptions there were respondents (R.10, R.11) who stressed the distinction between subjective conceptions and socially accepted (or proven) conceptions. For example, respondent R.10, in relation to characterization #3, wrote as follows: "No, conceptions cannot be perceived to encompass knowledge. It's a question of defining knowledge. In my view knowledge has to be true, i.e., experimentally tested, proved, and/or socially accepted."

6.3. Common Points for Constructive Discussion

From the responses, the following issues of special interest for researchers emerged: the origin of beliefs, the affective component of beliefs, and the effect of beliefs on an individual's behavior, reaction, etc. These issues tend to overlap.

Beliefs might have many origins, as expressed in the following three responses: "Beliefs, in a particular context/situation, are part of a person's identity (or better, identities), which is (are) formed through learning, interacting, goals, needs and desires, and therefore also affective." (R.6), "Beliefs can also be adapted from others, especially from those in authority." (R.9), and "Finally sometimes people believe things because they have noticed them in personal experience, but very often they believe 'propaganda' instead (mathematics is hard, useful, dull, fun, etc.)." (R.13)

In two responses (R.9, R.17) the affective component of beliefs was mentioned. As an example, we give the following: "I think of beliefs as primarily cognitive with a significant affective component [...] and especially related to values [...] I also try to separate beliefs from more affective or attitudinal responses to mathematics (enjoyment of problem solving or preferences for certain mathematical topics)." (R.9). This position is close to the spirit of McLeod's chapter on affective factors in the *Handbook on Mathematics Teaching and Learning* (McLeod, 1992).

That beliefs have an effect on an individual's behavior, reaction, etc. is quite a common assumption for researchers of beliefs (e.g., Schoenfeld, 1992). This is expressed, for example, by the following quotation: "A belief is an attempt, often deeply felt, to make sense of and give meaning to some phenomenon. It involves cognition and affect, and guides action." (R.11). Some researchers (e.g., R.4 and R.17) stressed the importance of context or a particular situation in shaping beliefs or behavior.

7. SUGGESTIONS FOR A CHARACTERIZATION OF BELIEFS

From the criticisms and the constructive parts of the responses given there are points on which future research may be based. In the following we comment on some of these points.

Firstly we would drop the idea of a multipurpose characterization suitable for all the possible fields of application (mathematics education, philosophy, general education, psychology, and sociology) and refer our considerations to a given context, a specific situation and population. Also it is useful to link a given characterization to the goals that we have in mind when using the concept we are characterizing. *Contextualization* and *goal-orientation* make the characterization an *efficient* one.

There is also the need to specify concepts used in research. It seems to us that part of the previous discussion could be avoided, if we distinguish in mathematics between objective knowledge (which is accepted by the mathematical community), and subjective knowledge (which is constructed by an individual). Individuals have access to objective knowledge and construct (in the language of Sfard, 1991) their own conceptions of mathematical concepts and procedures, i.e., they construct some pieces of their subjective knowledge. In an ideal case, these conceptions and mathematical concepts in question correspond isomorphically to each other. In such a sense the two domains may be overlapping, but not be coincident. In the domain of objective knowledge, there are parts which may not be accessible to the individual, or in which the individual has no interest. Also what individuals take from objective knowledge becomes part of their subjective knowledge. This happens after an operation of processing information, in which the existing knowledge and their earlier beliefs intervene. In the domain of subjective knowledge, there are elements that are strictly linked to the individual: they are beliefs in a broad sense that includes affective factors. The distinction of two types of knowledge (objective and subjective) allows us to escape the trap of linking beliefs to judgments of truth. Moreover our statements match a common way of speaking in classroom practice. We often hear statements such as "This student has a good [bad] knowledge of mathematics", which refer to the student's subjective knowledge. Other statements, such as "to generate knowledge", refer to the knowledge shared in the community of mathematicians.

Beliefs concern different fields: mathematics, classroom norms, an individual's personality, and so on. All may have consequences both in the individual's cognitive and affective domains. For example, those who think that mathematics is a boring list of rules to follow will not find enjoyment in doing mathematics. Also they will have difficulty in reaching understanding in mathematics. If one thinks oneself to be not mathematically minded, confidence in doing mathematics will be lacking. In the presentation of the rationale of our research we have observed that for researchers the boundary between affective factors and beliefs is often very fuzzy (see McLeod, 1992). Contextualizing and specifying the discourse on beliefs and knowledge allows us to clarify these boundaries.

Individuals are not always conscious of their beliefs. Thus we have to consider conscious and unconscious beliefs. Also, an individual may hide her beliefs from external scrutiny, because in her opinion they are not satisfying someone's expectations. Among the hidden beliefs which generate the ghosts in classroom considered by Furinghetti (1996) there are unconscious beliefs. In Furinghetti's paper (1996) the phenomenon of the "ghosts" in the classroom is discussed. These are the hidden or unconscious beliefs in action in the classroom.

The answers and comments suggest that it is worth retaining the distinction between beliefs that are deeply rooted and surface beliefs (Green, 1971; Kaplan, 1991). For this distinction we have different types of nuances of interpretations: *to be central* and *to be peripheral*, to be *stable* or *subject to changes*. This type of dual structure provides a frame within which the important subject of the inconsistency of beliefs may be discussed. In reference to this, respondent R.2 notes:

> Is it ever the case that beliefs and practice could be inconsistent? The answer to this must surely be 'Yes' in some sense. Certainly, the case of irrational behavior could result in inconsistency. More likely, there are two kinds of beliefs, central and peripheral, which are held with different levels of intensity. The verbal part could be seen as a peripheral belief in that the beliefs are not held with the same level of intensity as a central belief is held.

Actually we seem to need more levels of beliefs than these two (deep beliefs, surface beliefs), in order to keep our statements powerful. A solution might be that we accept that there exists a wide spectrum of beliefs: deep beliefs are at one extreme, and surface beliefs are at the other. Let us consider as an example an imaginary researcher's beliefs. If she has learned to use only statistical methods in her research, the positivist paradigm is for her self-evident. It is not a question for further discussion. Therefore, beliefs connected with this situation are very deep-rooted. Whereas the belief "(i)n the list of references, one should always use the APA guidelines" may be a surface belief. There are many levels of beliefs between these two extremes.

Again we stress the strong connection of centrality of beliefs with their degree of certainty. Deep-rooted beliefs (such as a research paradigm) are self-evident and, therefore, their certainty is 100%. They form the "axioms" in the individual's worldview. On the other hand, the individual might possess beliefs of which she is 100% sure, but she is ready to change when there is enough evidence for that. For example, the belief "Seville is situated in the middle of Spain" could be changed easily, since a look at a reliable map shows that Seville is almost on the seaside. The difference between these two types of beliefs held with a 100% certainty lies in the amount of affect in them and on the possibility of checking the truth by confrontation with information accepted by a community.

In the results, the terms "incontrovertible" and "stable" were disputed as attributes for beliefs. We suppose this depends on the fact that those working in education need to trust in the possibility of challenging and changing beliefs because, otherwise, didactic activities would not make sense. The intermediate solution of considering central and peripheral beliefs seems more flexible for describing how beliefs are modified.

We summarize the previous considerations by pointing out that, when dealing with beliefs and related terms, it is advisable

- to consider two types of knowledge (the objective and the subjective)
- to consider beliefs as belonging to subjective knowledge
- to include affective factors in belief systems, and distinguish affective and cognitive beliefs, if needed
- to consider degrees of stability, and to acknowledge that beliefs are open to change

- to take note of the context (e.g., population, subject, etc) and the research goal within which beliefs are considered.

We are aware of the fact that in characterizing beliefs and related concepts it is unlikely that complete agreement will be reached on the use of all the terms at issue. Nevertheless it can be asked that the authors of studies on beliefs reduce the terms and the concepts involved in their work to the minimum needed. Additionally, they have to make clear their assumptions, the meaning they give to basic words, and the relationship between the concepts involved.

8. NOTES

[1] Prof. Thomas Cooney (University of Georgia, USA) kindly pointed us to Green's (1971) analysis of belief systems.

[2] e.g., in Victoria (Australia), when the teaching of problem-solving was made obligatory in the curriculum, teachers interpreted most of their earlier routine tasks to be problems, see (Stacey, 1991). In Italy, when computer science was introduced into the programs of secondary school, teachers transferred their styles of teaching to the new subject (see Bottino & Furinghetti, 1996).

9. REFERENCES

Abelson, R. (1979). Differences between belief systems and knowledge systems. *Cognitive Science, 3,* 355–366.

Audi, R. (1988). *Belief, justification, and knowledge.* Belmot (CA): Wadsworth.

Bassarear, T. J. (1989). The interactive nature of cognition and affect in the learning of mathematics: two case studies. In C.A. Maher, G.A. Goldin, & R.B. Davis (Eds.), *Proceedings of the PME-NA-8* Vol. 1 (pp. 3–10). Piscataway, NJ.

Berger, P. (1998). Exploring mathematical beliefs – the naturalistic approach. In M. Ahtee & E. Pehkonen (Eds.), *Research methods in mathematics and science education* (pp. 25–40). University of Helsinki. Department of Teacher Education. Research Report 185.

Bottino, R. M., & Furinghetti, F. (1996). The emerging of teachers' conceptions of new subjects inserted in mathematics programs: The case of informatics. *Educational Studies in Mathematics, 30,* 109-134.

Cooney, T. J., Shealy, B. E., & Arvold, B. (1998). Conceptualizing belief structures of preservice secondary mathematics teachers. *Journal for Research in Mathematics Education, 29*(3), 306–333.

Elbaz, F. (1983). *Teacher thinking: A study of practical knowledge.* London: Croom Helm.

Ernest, P. (1991). *The Philosophy of mathematics education.* Hampshire (U.K.): The Falmer Press.

Fennema, E., & Franke, L. M. (1992). Teachers' knowledge and its impact. In D. A. Grouws (Ed.), *Handbook of research on mathematics learning and teaching* (pp. 47-164). New York: Macmillan.

Freire, A. M., & Sanches, M. C. (1992). Elements for a typology of teachers' conceptions of physics teaching. *Teaching and Teacher Education, 8,* 497-507.

Furinghetti, F. (1996). A theoretical framework for teachers' conceptions. In E. Pehkonen (Ed.), *Current State of Research on Mathematical Beliefs III.* Proceedings of the MAVI-3 Workshop (pp.19–25). University of Helsinki. Department of Teacher Education. Research Report 170.

Gardiner, T. (1998). The art of knowing. *The mathematical gazette, 82*(495), 2-20.

Goldin, G. A. (1990). Epistemology, constructivism, and discovery learning in mathematics. In R. B. Davis & al. (Eds.), Constructivist views on the teaching and learning of mathematics. *JRME Monograph Number 4* (pp. 36-47). Reston (VA): NCTM.

Green, T. F. (1971). *The activities of teaching.* Tokyo: McGraw-Hill Kogakusha.

Grigutsch, S., Raatz, U., & Törner, G. (1998). Einstellungen gegenüber Mathematik bei Mathematiklehrern. *Journal für Mathematik-Didaktik, 19*(1), 3-45.

Hart, L. E. (1989). Describing the affective domain: Saying what we mean. In D. B. McLeod & V. Adams (Eds.), *Affect and mathematical problem solving* (pp. 37-45). New York: Springer-Verlag.

Hasemann, K. (1987). Pupils' individual concepts on fractions and the role of conceptual conflict in conceptual change. In E. Pehkonen (Ed.), *Articles on mathematics education* (pp. 25–39). Department of Teacher Education. University of Helsinki. Research Report 55.

Hintikka, J. (1962). *Knowledge and belief.* Ithaca: Cornell U. P.

Jones, D., Henderson, E., & Cooney, T. (1986). Mathematics teachers' beliefs about mathematics and about teaching mathematics. In G. Lappan & R. Even (Eds.), *Proceedings of PME-NA-8* (pp. 274–279). East Lansing (MI): Michigan State University.

Kaplan, R. G. (1991). Teacher beliefs and practices: a square peg in a square hole. In R. G. Underhill (Ed.), *Proceedings of PME-NA-13 (2)* (pp. 119-125). Blacksburg, VA: Virginia Tech.

Lester, F. K., Garofalo, J., & Kroll, D. L. (1989). Self-confidence, interest, beliefs, and metacognition: Key influences on problem-solving behavior. In D. B. McLeod & V. Adams (Eds.), *Affect and mathematical problem solving: A new perspective* (pp. 75-88). New York: Springer-Verlag.

Lindgren, S. (1999). Plato, knowledge and mathematics education. In G. Philippou (Ed.), *MAVI-8 Proceedings* (pp. 73-78). Nicosia: University of Cyprus.

Lloyd, G. M., & Wilson, M. (1998). Supporting innovation: The impact of a teacher's conceptions on functions on his implementation of a reform curriculum. *Journal for Research in Mathematics Education, 29,* 248–274.

Mc Dowell, J. (1987). *Plato, Theaetetus.* (Translation). Oxford: Calrendon.

McLeod, D. B. (1989). Beliefs, attitudes, and emotions: New views of affect in mathematics education. In D. McLeod & V. Adams (Eds.), *Affect and mathematical problem solving* (pp. 245–258). New York: Springer-Verlag.

McLeod. D. B. (1992). Research on affect in mathematics education: a reconceptualization. In D. A. Grouws (Ed.), *Handbook of research on mathematics learning and teaching* (pp. 127-146). New York: Macmillan.

Nespor, J. (1987). The role of beliefs in the practice of teaching. *Journal of Curriculum Studies, 19,* 317–328.

Olson, J. M., & Zanna, M.P. (1993). Attitudes and attitude change. *Annual of Reviews in Psychology, 44,* 117-154.

Pajares, M .F. (1992). Teachers' beliefs and educational research: Cleaning up a messy construct. *Review of Educational Research, 62,* 307-332.

Paola, D. (1999). Communication et collaboration entre practiciens et chercheurs: étude d'un cas. F. Jacquet (Ed.), *Proceedings of CIEAEM 50* (pp. 217-221). Neuchâtel.

Pehkonen, E. (1994). On teachers' beliefs and changing mathematics teaching. *Journal für Mathematik-Didaktik, 15*(3/4), 177–209.

Pehkonen, E. (1998). On the concept 'mathematical belief'. In E. Pehkonen & G. Törner (Eds.), *The state-of-art in mathematics-related belief research. Results of the MAVI activities* (pp. 37–72). University of Helsinki. Department of Teacher Education. Research Report 195.

Ponte, J. P. (1994). Knowledge, beliefs, and conceptions in mathematics teaching and learning. In L. Bazzini (Ed.), *Proceedings of the fifth International Conference on Systematic Cooperation between Theory and Practice in Mathematics Education* (pp. 169-177). Pavia: ISDAF.

Rodd, M. M. (1997). Beliefs and their warrants in mathematics learning. In E. Pehkonen (Ed.), *Proceedings of PME 21 4* (pp. 64-71). Helsinki: University of Helsinki.

Rokeach, M. (1968). *Beliefs, attitudes, and values.* San Francisco (Ca): Jossey-Bass.

Ruffell, M., Mason, J., & Allen, B. (1998). Studying attitude to mathematics. *Educational Studies in Mathematics, 35,* 1-18.

Saari, H. 1983. Koulusaavutusten affektiiviset oheissaavutukset. [Affective Consequences of School Achievement] *Institute for Educational Research. Publications 348.* University of Jyväskylä. [in Finnish].

Schoenfeld, A. H. (1985). *Mathematical Problem Solving.* Orlando (FL.): Academic Press.

Schoenfeld, A. H. (1987). What's all the fuss about metacognition?. In A.H. Schoenfeld (Ed.), *Cognitive Science and Mathematics Education* (pp. 189–215). Hillsdale (NJ): Lawrence Erlbaum Associates.

Schoenfeld, A. H. (1992). Learning to think mathematically: problem solving, metacognition, and sense making in mathematics. In D.A. Grouws (Ed.), *Handbook of research on mathematics learning and teaching* (pp. 334–370). New York: Macmillan.

Sfard, A. (1991). On the dual nature of the mathematical objects. *Educational Studies in Mathematics, 22,* 1-36.

Shaw, K. L. (1989). *Contrasts of teacher ideal and actual beliefs about mathematics understanding: Three case studies.* Doctoral dissertation. Athens (GA): University of Georgia (unpublished).

Silver, E. A. (1985). Research on teaching mathematical problem solving: Some underrepresented themes and directions. In E.A. Silver (Ed.), *Teaching and learning mathematical problem solving: multiple research perspective* (pp.247–266). Hillsdale, NJ: Lawrence Erlbaum Associates.

Stacey, K. (1991). Linking application and acquisition of mathematical ideas through problem solving. *International Reviews in Mathematical Education (ZDM), 23*, 8-14.

Statt, D. A. (1990). *The concise dictionary of psychology*. London: Routledge.

Steffe, L. P. (1990). On the knowledge of mathematics teachers. In R.B. Davis et al. (Eds.), Constructivist views on the teaching and learning of mathematics. *JRME Monograph Number 4* (pp. 167–184). Reston, VA: NCTM.

Thompson, A. (1992). Teachers' beliefs and conceptions: A synthesis of the research. In A. D. Grouws (Ed.), *Handbook of research on mathematics learning and teaching* (pp. 127-146). New York: Macmillan.

Törner, G., & Grigutsch, S. (1994). 'Mathematische Weltbilder' bei Studienanfängern – eine Erhebung. *Journal für Mathematik-Didaktik, 15*(3/4), 211–251.

Underhill, R. G. (1988). Mathematics learners' beliefs: A review. *Focus on Learning Problems in Mathematics, 10 (1)*, 55-69.

Yackel, E., & Cobb, P. (1996). Sociomathematical norms, argumentation, and autonomy in mathematics. *Journal for Research in Mathematics Education, 27*(4), 458-477.

CHAPTER 4

GERALD A. GOLDIN

AFFECT, META-AFFECT, AND MATHEMATICAL BELIEF STRUCTURES [1]

Abstract. Beliefs are defined here to be multiply-encoded, internal cognitive/affective configurations, to which the holder attributes truth value of some kind (e.g., empirical truth, validity, or applicability). This chapter offers some theoretical perspectives on mathematical beliefs drawn from analysis of the affective domain, especially the interplay between meta-affect and belief structures in sustaining each other in the individual.

1. INTRODUCTION

Research in mathematics education has tended to focus principally on cognition, and far less on affect. This may be due, in part, to the popular myth that mathematics is a purely intellectual endeavor in which emotion plays no essential role. Valerie DeBellis and I, in developing a language for careful discussion of the affective domain in mathematics, seem to have introduced several rather uncommon ideas. This chapter offers some theoretical perspectives on mathematical beliefs, drawn from analysis of the affective domain, based on some of those ideas. My main assertion is that the stability of beliefs in individuals has much to do with the interaction of belief structures not only with affect (feelings), but with meta-affect (feelings about feelings) - that through their psychological interplay, meta-affect and belief structures sustain each other.

The chapter is organized as follows. First I mention several important perspectives on affect as a system of representation encoding information, intertwined with cognitive representational systems, and as a language for communication having an important cultural dimension. Next I consider the key construct of meta-affect, including affect about affect, affect about and within cognition that may again be about affect, monitoring of affect, and affect as monitoring. We shall see that powerful affective representation inheres not so much in the affect as in the meta-affect. A basic idea here is that affect stabilizes beliefs, and beliefs establish meta-affective contexts.

I then turn to a discussion of beliefs, belief structures, and belief systems. Beliefs are defined to be multiply-encoded cognitive/affective configurations, to which the holder attributes some kind of truth value (e.g., empirical truth, validity, or applicability). I distinguish among: working assumptions or conjectures; weakly- or

G. C. Leder, E. Pehkonen, & G. Törner (Eds.), Beliefs: A Hidden Variable in Mathematics Education? 59-72.

strongly-held beliefs; individual and shared beliefs; belief structures and systems of belief; warrants for beliefs; psychological functions of beliefs and belief systems; knowledge (beliefs that in some sense *apart* from the fact of belief or the fact of warranted belief, are true or valid); and individual and shared values and value systems. Then a preliminary typology of mathematically-related beliefs is offered, organized not by who holds them but by their content domain. Belief structures can intersect several of these categories.

Finally I consider how belief structures, warrants for belief, and meta-affect can establish and sustain each other - at the social level, as well as in the individual - and explore some warrants for mathematical beliefs in relation to affective structures. Changes in mathematical belief structures require, and entail, changes in affect as well as cognition.

2. SOME PERSPECTIVES ON AFFECT

Let me begin by referring briefly to a few of the perspectives taken here regarding affect. For related and sometimes contrasting perspectives, see Leder (1982,1993), McLeod (1988, 1989, 1992), McLeod and Adams (1989), Drodge and Reid (2000), and Gomez-Chacon (2000). A useful overview of research on affect in mathematics education is given in McLeod (1994).

First, affect is seen as one of several internal systems of representation in individuals (cf. Zajonc, 1980; Rogers, 1983; Goldin, 1987, 1988, 1998, 2000; Picard, 1997). That is, the affective system does not merely accompany cognition, or occur as an inessential response to cognitive representation, but affect itself has a *representational* function. Affect meaningfully encodes information. This includes information about the external physical and social environment (e.g., feelings of fear encoding danger), information about the cognitive and affective configurations of the individual herself or himself (e.g., feelings of bewilderment encoding insufficiency of understanding, feelings of boredom encoding absence of engagement, or feelings of loneliness encoding absence of intimacy), and information about the cognitive and affective configurations of others, including social and cultural expectations, as represented in and projected by the individual (e.g., feelings of pride encoding satisfaction taken by one's parents or teachers in one's achievements).

When individuals are doing mathematics, the affective system is not merely auxiliary to cognition - it is central. However affect as a representational system is *intertwined with cognitive representation.* Affective configurations can stand for, evoke, enhance or subdue, and otherwise interact with cognitive configurations in highly context-dependent ways. The very metaphors used in thinking may carry positive or negative affect.

Systems of cognitive representation, specifically the verbal/syntactic, imagistic, formal notational, and strategic/heuristic systems that I have discussed extensively elsewhere in formulating my developing model of mathematical problem-solving competence, function in part by evoking affect and the representational information that it encodes. In earlier work DeBellis and I focused particularly on the interplay between affective states and the heuristic or strategic decisions taken by students during problem solving (DeBellis & Goldin, 1991, 1993). For example, feelings of frustration while doing mathematics may encode (i.e., represent) the fact that a certain strategy has led down a succession of "blind alleys", and (ideally), these feelings might evoke a change of approach.

Emotions are biologically based, and there is evidence for emotional systems having evolved in other species. Affect does much more than inform and motivate individuals. It serves also as an extraordinarily powerful *evolutionary language for communication,* that is essentially human. Each individual person's affect interacts with that of other people (often quite tacitly or unconsciously, often very powerfully), so that crucial information is exchanged. The specifics of this communicative system, which functions through "body language", eye contact, facial expressions, tone of voice, and scent, as well as spoken language, cries, laughter, and other noises and interjections, seem to have evolved alongside the emergence of the human species. The sharing of affect among pairs or groups of people is generally *essential to human survival.* In discussing the influence of affect on beliefs, then, it is important to take note of, and distinguish between, individual affect and shared affect. Progress is being made in understanding the neuroscientific basis of affect, which allows informed discussions of the role it plays in our conscious awareness (cf. Damasio, 1999).

In the individual, we can distinguish certain subdomains of affective representation (McLeod, 1988, 1989; DeBellis, 1996; DeBellis & Goldin, 1997): (1) *emotions* (rapidly changing states of feeling, mild to very intense, that are usually local or embedded in context), (2) *attitudes* (moderately stable predispositions toward ways of feeling in classes of situations, involving a balance of affect and cognition), (3) *beliefs* (internal representations to which the holder attributes truth, validity, or applicability, usually stable and highly cognitive, may be highly structured), and (4) *values, ethics, and morals* (deeply-held preferences, possibly characterized as "personal truths", stable, highly affective as well as cognitive, may also be highly structured).

Likewise, shared affect refers not only to transient, shared emotions (e.g., intimate excitement between lovers, pleasant laughter among a group of friends, tension in a mathematics class before an examination, or swelling group enthusiasm during a well-led problem solving discussion), but also to complex, shared, and possibly very powerful structures of feeling that are culturally embedded (e.g., religious reverence, or nationalist fervency), involving attitudes, beliefs, and values.

It is important to be able to discuss affective competencies and affective structures, in a way somewhat analogous to discussions of cognitive competencies and structures. Among the important constructs here are the distinction between local (transient, special-context) and global (long-term, multi-context) affect; the notion of affective pathways and networks (recurring sequences and links among emotional states), with accompanying meanings to the affective configurations; defense mechanisms (affective structures that serve to protect the individual from experiences of emotional hurt or pain); and processes of change in global affect (e.g., passage by an individual from long-held anger to forgiveness).

Constructs especially important to the doing of mathematics include the notions of *mathematical intimacy,* involving vulnerability, personal caring, private experience, and possibly creative expression; and *mathematical integrity,* including, e.g, self-acknowledgment of inadequate understanding (DeBellis & Goldin, 1997, 1999; DeBellis, 1998; Vinner, 1997). A further, fundamental construct that I think has received insufficient attention is that of *mathematical self-identity.* By this I mean the spectrum of related affect and cognition, growing as the child learns and grows, that eventually may take the form of answers to the question, "Who am I?" in relation to mathematics.

3. AFFECT AND META-AFFECT

A central notion that DeBellis and I are developing is that of *meta-affect.* We introduced this term (DeBellis, 1996; DeBellis & Goldin, 1997) to refer to affect about affect, affect about and within cognition that may again be about affect, the monitoring of affect, and affect itself as monitoring. In many situations, the meta-affect is actually the most important aspect of the affect.

Consider, for example, the emotion of fear. One thinks first of fear as a negative state of feeling, signaling danger. In the absence of actual danger in the environment, fear might be seen as a counterproductive state, an incorrect encoding, a feeling to be avoided or soothed. A young child may be terrified of the dark, or of being alone. An adolescent may experience fear of rejection or of failure. Some people are terribly and involuntarily afraid of crowds, of heights, of flying in airplanes, or of public speaking. Some, of course, fear mathematics. In these situations, our first impulse is to try to assuage the feeling.

But a moment's reflection reminds us that in the right circumstances, individuals can find fear highly pleasurable. People flock to horror movies. They enjoy amusement park rides, where the more terrifying the roller coaster experience, the more exhilarating and "fun" it is. Why is this? The cognition that the person is "really safe" on the roller coaster *permits the fear to occur in a meta-affective context of excitement and joy.* The more afraid the rider feels, the more wonderful she feels *about* her fear. She may experience a satisfying sense of her own bravery, of having conquered fear, and the anticipatory joy (Vorfreude) of stepping onto the solid earth again to be regarded with admiration by friends.

Imagine, however, that a cable breaks during such a ride, and the roller coaster swerves uncontrollably. The experience changes entirely! Now the rider is "truly" afraid, as the danger is (believed to be) actual. This fear feels entirely different, because the *meta-affect* has changed. Even if the person is really in no danger, the removal of the *belief* that she is safe changes the nature of the affective state - no longer does it feel wonderful to feel so afraid. The terrifying ride, not fun any more, has become horrible.

What makes the different meta-affect possible? It might seem that the "cognitive" belief, that the ride is in fact safe, is the main essential to the joyful meta-affect. In this sense, the belief stabilizes the meta-affect. But a straightforward influence of cognitively-based belief on meta-affect cannot be the whole story.

Other beliefs and values (tacit or overt) play key roles, influencing the *ecological function* of the fear in the individual's personality - values of life and safety, of approval by peers or authority, of personality traits such as bravery. Yet even conscious cognitions, beliefs, and values do not suffice to account for all the meta-affect: an adult having a "panic attack" in a crowd or in an airplane may well "know intellectually" he is "really safe", not *want* to feel fear, but experience it involuntarily. *Unconscious defense mechanisms* prevent the person from "really" believing in the fact of his safety. Here, the meta-affect stabilizes some level of belief in the actuality of the danger.

Fear of mathematics - or even fear of a particular topic in mathematics, such as fractions, algebra, or word problems - is a common phenomenon. One student may experience fear immediately on being given a mathematical problem to solve; another upon realizing that he does not know how to proceed with the problem. Some may be afraid of the test, of the teacher, of the computer, of failure, or of disapproval at home. This fear may feel quite involuntary. The knowledge that one has studied hard and is well-prepared may or may not remove the feeling of fear, or embed it in positive meta-affect. Even - or perhaps I should say, *especially* - advanced graduate students of mathematics may fear exposure of (self-perceived) mathematical inadequacies.

The meta-affect of fear in doing mathematics is not usually joyful, though occasionally it can be. For instance, a bright high school student might be fearfully nervous before an interscholastic mathematics contest, with positive meta-affect - the "contained fear" can enhance the experience, as she already anticipates being able to say, "I was really nervous, really afraid ... I'm always that way at these contests ... and I did great!"

When we consider less extreme feelings, such as frustration during mathematical problem solving, there is a wide range of commonly occurring meta-affect. For some, the frustration signals anticipation of failure, with attendant negative emotions, so that the meta-affective context is one of anxiety or fear. But for another student solving the same problem, the experience of frustration may involve meta-affect that is positive. The student anticipates success, or at least a satisfying learning experience. The local affect of frustration signals (i.e., represents) that the problem is nontrivial, deep, or interesting, and heightens the anticipation of joy in success. The student's "cognitive" belief in her high likelihood of success, her confidence that mathematics yields to insightful processes, along with the high personal value she places on meeting challenges, may contribute to her feeling quite positive *about* the frustration - a very different meta-affective context.

Powerful affective representation that fosters mathematical success inheres not so much in the surface-level affect, as it does in the meta-affect.

4. BELIEFS, BELIEF SYSTEMS, KNOWLEDGE, AND VALUES

Furinghetti and Pehkonen (chapter 3, this volume) discuss alternative definitions and interpretations of *beliefs* and *knowledge.* Törner (chapter 5, this volume) stresses the absence of consensus, and proceeds to elaborate considerably on the notion of beliefs. Here I propose to define *beliefs* as multiply-encoded cognitive/affective configurations, usually including (but not limited to) propositional encoding, to which the holder attributes some kind of *truth value.* The latter term is not taken in the technical sense of symbolic logic, but as a term that may variously refer to logical truth, empirical truth, validity, applicability to some degree of approximation, metaphysical truth, religious truth, practical truth, or conventional truth. Within mathematics, the "truth value" might be logical truth in the sense of deducibility from specific formal assumptions, or it might sometimes be conventional truth in the sense of satisfying some agreed-upon, arbitrary rules of definition or notation.

A *belief structure* is a set of mutually consistent, mutually reinforcing, or mutually supportive beliefs and warrants (see below) in the individual, mainly cognitive but often incorporating supportive affect. A *belief system* is an elaborate or extensive belief structure that is socially or culturally shared. Since I may be employing these terms rather differently from their casual uses, let me stress that my intent is to distinguish *individuals'* belief structures from *socially or culturally shared* belief systems.

A belief is, to begin with, individually held: for example, I believe that guiding children to discover logical patterns for themselves generally fosters their enjoyment and learning of mathematics. Such a belief may or may not be shared - some may agree with me, others may not. Belief structures, like cognitive structures, refer here to the *individual's* complex, personal, internal representational configurations: my belief about guiding children to discover patterns for themselves does not stand in isolation in my head; it is part of a structure of mutually reinforcing beliefs that I hold, together with a variety of reasons - or warrants - I have for holding them. Beliefs and belief structures are important in understanding individuals' mathematical problem solving heuristics and strategies (Schoenfeld, 1985; Lester, Garofalo, & Lambdin Kroll, 1989).

Belief systems, on the other hand, refer to socially or culturally shared belief structures, that are sufficiently broad to warrant the term. Shared beliefs, or belief systems are, in turn, not exactly the same as *normative* beliefs. The latter are idealized, approximate descriptions, at the societal or cultural level, of beliefs that one "should" hold (but may or may not *actually* be held by very many people). Relations among social norms and shared beliefs, the beliefs of individuals, and emotions are discussed further by Cobb, Yackel, & Wood (1989) and Cobb & Yackel (1996).

Note that it is the attribution of *truth* (of some kind) that turns mere propositions, conjectures, stories, or hypotheses, into beliefs. This attribution is by the holder, and not necessarily by others. It is not to be assumed, even when a belief is shared and normative, that the believer or believers are correct in their attributions of truth or validity to it. That is, some beliefs may - in fact - be false ones.

There is an unfortunate tendency among cultural relativists to use the word "knowledge" as if it were synonymous with "belief", "shared belief", "normative belief", "warranted belief", or some combination of these. For example, according to Confrey (2000), constructivism - which, in its radical formulation, has significantly influenced quite a few mathematics education researchers - holds as one of its "key concepts" the tenet that "knowledge is justified belief":

> In posing this claim I am not requiring that all knowledge is justified, but rather that if and when challenged, it can be justified. Insisting on only potential challenge is necessary so that in a stable body of knowledge, I can claim such statements as 3+7=10 as knowledge without requiring that I have actually justified it at the time. However, should someone ask me, how do I know that, I am obliged to produce a trajectory of acceptable reasoning and argument. If this does not result in convincing my audience, then the statement's status as knowledge is in question. Its validity will remain in doubt until an appeal to a larger and/or more qualified group of experts can be made successfully or until the previous challenge is resolved.
>
> ... That is, constructivism entails a rejection of assured transcendent truth in our knowledge. (Confrey, 2000, p. 12)

Using the term "knowledge" this way leaves no convenient word to distinguish beliefs that are *in fact* true, correct, good approximations, valid, insightful, rational, or veridical, from those that are *in fact* false, incorrect, poor approximations, invalid, mistaken, irrational, or illusionary. I do not propose here to address the profound philosophical issues glossed over by the many different forms that relativism has taken. However, rejecting on first principles the very possibility of any such "in fact" distinction among competing, well-justified beliefs is in my view ultimately destructive of reasoned discourse - in mathematics and the empirical sciences, precisely because this distinction is essential to the integrity of the subject. The philosophical problems associated with "truth" do not disappear by rejecting it from the start as a requisite characteristic of knowledge. I have intentionally suggested a variety of possible interpretations, of which most do not require the "truth in our knowledge" to be either assured or transcendent.

It is important to characterize some related notions, that differ from or elaborate on the notion of belief. A *working assumption,* or a *conjecture,* is an internal configuration, often propositionally encoded, taken as a basis for exploration or discussion but (at least temporarily or provisionally) *without* attribution of truth, validity, or applicability. A *hypothesis* is such a proposition put forth with the explicit intent of investigating its possible truth or falsity (e.g., by gathering empirical data that tend to confirm or disconfirm it). A *viable* conjecture, hypothesis, or belief is one that has possibly proven useful or empowering, and has not to this point been disconfirmed in the judgment of the person or people making or holding it.

Viability is *not* the same thing as validity. A belief in the uncanny accuracy of astrological forecasting is personally viable for many people, and socially viable today in a large subculture (very likely, larger than the professional scientific subculture). But it is not valid.

A belief may be *weakly- or strongly-held* according to two different characteristics: the magnitude of the *importance* that is attributed to its being true (i.e., the believer may have a lot at stake), and the degree of *certainty* with which its truth value is attributed (i.e., the believer may have very convincing evidence).

Warrants for a belief consist in the believer's reasons or justifications for attributing truth, validity, or applicability to it. (Thus, in my language, Confrey's discussion pertains entirely to warrants for belief.) Warrants, like the beliefs that they warrant, may be personal or shared. In ordinary life, personal warrants (whether or not shared) include direct observations, indirect reports, plausible inferences from observations and reports, rational deduction from or compatibility with other beliefs, assertions of authority, etc. Some personal warrants may never be intended to convince anyone else (e.g., "My parents raised me to believe it.") In mathematics, the usual shared warrants include verification and proof making use of agreed-upon reasoning processes. In the natural sciences, they include goodness of fit with observation and measurement, compatibility with the outcomes of controlled and repeatable experiments, theoretical coherence, and parsimony. Depending on their nature, some warrants may be stronger than others - one who attributes validity to a scientific hypothesis after it has been repeatedly verified through controlled experimentation normally has a stronger case than another who attributes validity based on informal observations and anecodotal reports.

Many beliefs have *psychological functions* in the believer, and these are most often of an affective nature. A saying attributed to the U.S. journalist Henry Louis Mencken (1880-1956), "People will believe what they want to believe", has considerable descriptive validity. A belief or belief system may contribute essentially to the holder's self-identity, to the coherence of the believer's world view, or to the sense of certainty in his or her values. Thus it is important to be able to speak of the affective consequences of beliefs, and the affective contexts in which beliefs are held. This applies to all of us. If my belief structure that includes the invalid nature of astrology were to crumble, some of my self-identity as a scientist might be called into question. Of course, this provides not the slightest *scientific* warrant for my holding the belief, but it may help account psychologically for the importance I attach to it.

Knowledge (in the technical sense of this article) refers to beliefs that, in a sense *apart* from the fact of belief or the acceptance of warrants for belief by an individual or group, are true, correct, valid, veridical, good approximations, or applicable. Sometimes the term knowledge may be restricted further, to warranted or even well-warranted beliefs that have one or more of these characteristics.

Beliefs should also be distinguished from *values,* with which they are often closely entwined. The distinction can be subtle, because the latter are sometimes called beliefs - in ordinary speech, I might say that "I *value* learning", or (more or less equivalently) that "I *believe* learning is valuable". The desired distinction is psychological, not philosophical - values have to do with what is held to be good, worthy, or desirable (rather than with what is held to be logically or empirically true), and are thus fundamentally matters of personal choice. Of course, when an individual further accords "truth" to a statement of value, seeing it as validated by religion, authority, or social consensus, it becomes a belief as well as a value; but this does not always happen.

A complex, variegated system of shared values, morals, or ethics and expectations is a component of cultural representation. Developed in the individual from childhood (Kohlberg, Levine, & Hewer, 1983), values/ethics/morals comprise a component of the affective system of internal representation in the individual, and one of the most powerful motivators of human beings - driving us to define our life purposes, to feel right or wrong, justified or guilty, to judge others as right or wrong, and to engage in creative, altruistic, or extraordinarily destructive behavior.

The system of values/ethics/morals usually influences beliefs, and provides a partial (or, sometimes, total) basis for them. In doing mathematics, for instance:

> Following the rules, or following directions (including mathematical rules), may be regarded by the child as 'good', failing to do so as 'bad'. To us this is much more than a belief about what mathematics is, or what works to obtain solutions. Some students who do not follow established instructional procedures, as in addressing a non-routine problem, may actually be tacitly contravening their own moral values or self-expectations, while others (who value originality, rebellion, or self-assertiveness) may be acting consonant with them. Cheating in school may be considered evil or shameful, and doing mathematics with help may for the child be a form of cheating. The tacit commitments made by students to learn and to understand, their sense of goodness about themselves when they do as they 'should' do, and wrongness when they fail to do as they 'should', all fall within this component. (Goldin & DeBellis, 1997, p. 212)

Now, very different notions of truth or validity pertain in various contexts to beliefs. I have not been able to discover who first offered the observation, "The belief that something is so does not make it so", an assertion generally regarded as essential to scientific thinking. Nevertheless, this holds in some but not all contexts. With regard to the physical world, or with regard to mathematical questions *after* conventions have been established and axioms accepted, the mere fact that something is believed is *not* a valid warrant for it, and does not *per se* influence its truth. In some psychological contexts, a belief may have a *partial* influence on its own truth - for instance, with regard to an individual's estimation of his or her own mathematical ability. In still other contexts, where beliefs are based wholly on personal or shared values, beliefs can *create their own truth* in a self-referential manner – e.g., belief in one's own courage, in the beauty of the beloved, in the value of honesty, or in the elegance of the proof of a theorem.

5. SOME TYPES OF MATHEMATICAL BELIEFS

This preliminary typology of the kinds of beliefs of interest to mathematics educators is organized not by which individuals, groups, populations, or cultures might hold them, but by the nature of their content domain. It is included in order to lend specificity to the general points I have made. Op 't Eynde, De Corte, and Verschaffel (chapter 2, this volume) discuss similar categorizations by Underhill, McLeod, and others, offering a comprehensive overview. Belief structures in the individual, and belief systems occurring in social groups, often intersect several of the categories.

- Beliefs about the physical world, and about the correspondence of mathematics to the physical world (e.g., number, measurement);
- Specific beliefs, including misconceptions, about mathematical facts, rules, equations, theorems, etc. (e.g., the law of exponents, the quadratic formula, the idea that "multiplication always makes larger");
- Beliefs about mathematical validity, or how mathematical truths are established;

- Beliefs about effective mathematical reasoning methods and strategies or heuristics;
- Beliefs about the nature of mathematics, including the foundations, metaphysics, or philosophy of mathematics;
- Beliefs about mathematics as a social phenomenon;
- Beliefs about aesthetics, beauty, meaningfulness, or power in mathematics;
- Beliefs about individual people who do mathematics, or famous mathematicians, their traits and characteristics;
- Beliefs about mathematical ability, how it manifests itself or can be assessed;
- Beliefs about the learning of mathematics, the teaching of mathematics, and the psychology of doing mathematics;
- Beliefs about oneself in relation to mathematics, including one's ability, emotions, history, integrity, motivations, self-concept, stature in the eyes of others, etc.

For some of these kinds of belief, there exist fairly well-defined, culturally normative systems within the mathematical community. For others, the norms vary or scarcely exist. In all cases, there is the question of the interface between the individual and the social: How does the individual's affective representational system interact with those around her - as the belief or belief structure first forms, and as it evolves in the doing of mathematics?

6. INTERFACES BETWEEN THE INDIVIDUAL AND THE SOCIAL

We have *within* the individual (1) personal emotions, (2) personal attitudes, (3) personal belief structures, and (4) personal values/ethics/morals. Likewise, in the individual, there is the capacity to represent each of these in *other* individuals, especially when one person may be the object of another's intense feeling. Furthermore individuals have the capacity to represent the notion of *normative* or *appropriate* emotions, attitudes, beliefs, and values/ethics/morals, to evaluate their own feelings in relation to such norms, and to experience accompanying feelings about their feelings - guilt at having an inappropriate emotion, self-approval of an appropriate one.

Then, distinct from any one individual's affective representations, we have *external* to the individual a sociocultural environment that provides complex and often remarkably consistent feedback: (1) shared emotions, (2) prevailing or acceptable attitudes, (3) belief systems across the culture or subgroups in the culture, and (4) the values, ethics and morals communicated through schooling, peer groups, the examples of adult family members or authority figures, etc.

This perspective allows us to focus on affective *interactions* of the individual with the surrounding culture, without taking either the individual or the culture to be the sole level of analysis.

7. BELIEF STRUCTURES AND META-AFFECT

Affect stabilizes beliefs - human beings who feel good about their beliefs, proud of them or happy with them, are likely to continue to hold them. Belief structures are especially stable, partly because the repeated experience of one belief confirming another within the structure offers something to feel good about. But the psychology of belief is much more complicated than this. Beliefs, in turn, establish *meta-affective contexts* for the experience of emotion connected to the beliefs. Stable belief structures are comfortable, but this is not the same as saying they are pleasant. They may simply reinforce defenses against pain and hurt, unconsciously helping believers to feel a bit better about themselves. Such meta-affect can be strong enough to ensure that even careful, intelligent, rational believers will find warrants to retain their beliefs in the face of apparently contravening evidence or experience.

Let us consider, as an example, the value placed by the school culture on speed and accuracy of computation in school mathematics - specifically, arithmetic and algebra. Let us consider the related beliefs that these characteristics are good indicators of mathematical ability and potential, and that they are appropriate and sufficient measures of how well mathematics has been learned. (Just to be clear, I do not myself hold these beliefs.)

The context surrounding these values and beliefs may evoke personal emotions that range from pride and pleasure (in some students) to frustration and anger (in others). The resulting changing states of feeling, initially local, form meaningful affective pathways that encode cognitive information (e.g., regarding the individual's likelihood of mathematical problem-solving success). As such pathways become better established and interwoven with cognition, a meta-affective context for doing mathematics is created in the individual - and individual beliefs are constructed (for instance, about the person's own mathematical ability, or about the nature of mathematics as a system of applied rules) that serve to support and sustain the meta-affect.

A computationally successful child in elementary school, experiencing pride or pleasure in such activity, might come to believe in himself as mathematically able, and believe that speed and accuracy are indeed good measures of mathematical ability. If he is also able, spontaneously, to construct reasonably insightful internal mathematical representations, he may come to believe that training for speed and accuracy not only suffices to separate the mathematically able like himself from his less able classmates, but also lays an appropriate foundation for the more abstract, non-computational mathematics that requires intellectual effort later on in high school. In a culture where competitiveness among young boys is encouraged, this boy feels personally validated. A belief structure is formed, establishing a comfortable meta-affective context in which the student himself is hardworking as well as talented, a *deserving* member of an achieving elite. Even if he initially dislikes computational activity, he may come to see it as a kind of necessary pain, a rite of passage, or enjoy the competitive thrill of being better than the others - so that his pride of later achievement is enhanced.

The belief structure just described, and the satisfying emotions associated with it, may be experienced by the student as *socially appropriate,* at least in academic environments. Perhaps he later becomes a teacher of mathematics for whom speed and accuracy are central student objectives. Thus we have some of the makings of one part of a belief system widely (but far from universally) held in the mathematics community.

In contrast, a computationally less successful child, experiencing frustration and pain, might come to have a lot at stake in believing herself to be "not very good at mathematics". Although she may spontaneously generate insightful visual and spatial representations, find interesting patterns, and notice logical connections in nonroutine problems, these do not translate immediately into computational speed and accuracy. The rules of computation, to be followed without thinking about them, may be seen as derived from authority, not from reasoning. In a culture where competitiveness among young girls is not encouraged, she needs a way to assert her personal worth in the context of negative affective feedback. The belief in her own lack of ability *excuses* the self-perceived failure, to herself and to others (as she represents their evaluations of her). It is not her fault she is "slow"; it is simply her lack of ability. She may not feel exactly good about the frustration and disappointment she experiences with mathematics, but she need not feel so bad about it either. There is no need for shame. She has a belief that accounts for her performance, creating a meta-affective context that is reasonably comfortable. The belief contributes to her developing self-identity as a "non-math" person. She and others may take this as enhancing her attractiveness.

For this student, too, it has become important to see mathematics as consisting of computational activity, and speed and accuracy as valid measures of ability. Her belief structure has important elements in common with the belief structure of the first student, though their self-concepts in relation to mathematics are radically different. This student's growing belief structure may also influence the kinds of warrants she later accepts for particular beliefs in mathematics – e.g., appeal to authority or social acceptability, in place of mathematically illustrative examples, diagrams, and rational arguments. The reason is *not* that she is fundamentally unable to understand mathematical reasoning. Rather, the comfortable meta-affect created by her belief in her own relative mathematical ineptness does not permit her any pleasure in the *experience* of mathematical reasoning. She prefers the safety and security of the computational rules. Perhaps she later becomes an elementary school teacher for whom speed and accuracy in mathematics are central student objectives.

Of course these stories involve highly stereotyped characters, but the ingredients of their beliefs and feelings are real-life.

8. CONCLUSION

In considering how individuals develop, we must note that *prevailing belief structures in relation to mathematics are powerfully stabilized by meta-affect.* Such beliefs are unlikely to change simply because factual warrants for alternate beliefs are offered.

Mathematics educators who set out to modify existing, strongly-held belief structures of their students (e.g., future teachers) are not likely to be successful addressing only the content of their students' beliefs, or only the warrants for their beliefs. It will be important to provide experiences that are sufficiently rich, varied, and powerful in their emotional content to foster the students' construction of new meta-affect.

This is a difficult challenge indeed.

9. NOTES

[1] This chapter, partially based on joint work with Valerie A. DeBellis, is adapted from the author's presentations at the November 1999 meeting on "Mathematical Beliefs and their Impact on the Teaching and Learning of Mathematics" in Oberwolfach, Germany and at the March 2000 meeting on "Social Constructivism, Social Practice Theory and Sociocultural Theory: Relevance and Rationalisations in Mathematics Education" in Gausdal, Norway.

10. REFERENCES

Cobb, P., & Yackel, E. (1996). Constructivist, emergent, and sociocultural perspectives in the context of developmental research. *Educational Psychologist, 31*, 175-190.

Cobb, P., Yackel, E., & Wood, T. (1989). Young children's emotional acts while engaged in mathematical problem solving. In D. B. McLeod & V. M. Adams (Eds.), *Affect and mathematical problem solving: A new perspective* (pp. 117-148) New York: Springer-Verlag.

Confrey, J. (2000). Leveraging constructivism to apply to systemic reform. *Nordisk Matematik Didaktik* (Nordic Studies in Mathematics Education), *8*(3), 7-30.

Damasio, A. (1999). *The feeling of what happens: Body and emotion in the making of consciousness.* New York: Harcourt Brace & Co.

DeBellis, Valerie A. (1996). *Interactions between Affect and Cognition during Mathematical Problem Solving: A Two Year Case Study of Four Elementary School Children.* Rutgers University doctoral dissertation. Ann Arbor, MI: Univ. Microfilms 96-30716.

DeBellis, V. A. (1998). Mathematical intimacy: Local affect in powerful problem solvers. In S. Berenson et al. (Eds.), *Proceedings of the 20th annual meeting of PME-NA* Vol. 2 (pp. 435-440). Columbus, OH: ERIC.

DeBellis, V. A., & Goldin, G. A. (1991). Interactions between cognition and affect in eight high school students' individual problem solving. In R. G. Underhill (Ed.), *Proceedings of the 13th annual meeting of PME-NA* Vol. 1 (pp. 29-35). Blacksburg, VA: Virginia Tech.

DeBellis, V. A., & Goldin, G. A. (1993), Analysis of interactions between affect and cognition in elementary school children during problem solving. In J. R. Becker & B. Pense (Eds.), *Proceedings of the 15th annual meeting of PME-NA* Vol. 2 (pp. 56-62). Pacific Grove, CA: San Jose State Univ. Ctr. for Math. and Computer Sci. Educ.

DeBellis, V. A., & Goldin, G. A. (1997). The affective domain in mathematical problem solving. In E. Pehkonen (Ed.), *Proceedings of the 21st annual conference of PME* Vol. 2 (pp. 209-216). Helsinki, Finland: University of Helsinki Dept. of Teacher Education.

DeBellis, V. A., & Goldin, G. A. (1999). Aspects of affect: Mathematical intimacy, mathematical integrity. In O. Zaslavsky (Ed.), *Proceedings of the 23rd annual conference of PME* Vol. 2 (pp. 249-256). Haifa, Israel: Technion, Dept. of Education in Technology and Science.

Drodge, E. N., & Reid, D. A. (2000). Embodied cognition and the mathematical emotional orientation. *Mathematical Thinking and Learning, 2*(4), 249-267.

Goldin, G. A. (1987). Cognitive representational systems for mathematical problem solving. In C. Janvier (Ed.), *Problems of representation in the teaching and learning of mathematics* (pp. 125-145). Hillsdale, NJ: Erlbaum.

Goldin, G. A. (1988), Affective representation and mathematical problem solving. In M. J. Behr, C. B. Lacampagne, & M. M. Wheeler (Eds.), *Proceedings of the 10th annual meeting of PME-NA* (pp. 1-7). DeKalb, IL: Northern Illinois Univ. Department of Mathematics.

Goldin, G. A. (1998). Representational systems, learning, and problem solving in mathematics. *Journal of Mathematical Behavior, 17*(2), 137-165.

Goldin, G. A. (2000). Affective pathways and representation in mathematical problem solving. *Mathematical Thinking and Learning, 2*(3), 209-219.

Gomez-Chacon, I. M. (2000). *Matematica emocional.* Madrid: Narcea, S. A. de Ediciones.

Kohlberg, L., Levine, C., & Hewer, A. (1983). *Moral stages: A current formulation and a response to critics.* Basel: Karger.

Leder, G. (1982). Mathematics achievement and fear of success. *Journal for Research in Mathematics Education, 13,* 124-135.

Leder, G. (1993). Reconciling affective and cognitive aspects of mathematics learning: Reality or a pious hope? In I. Hirabayashi et al. (Eds.), *Proceedings of the 17th annual meeting of PME* Vol. 1 (pp. 46-65). Tsukuba, Japan: Univ. of Tsukuba.

Lester, F. K., Garofalo, J., & Lambdin Kroll, D. (1989). Self-confidence, interest, beliefs, and metacognition: Key influences on problem-solving behavior. In D. B. McLeod & V. M. Adams (Eds), *Affect and mathematical problem solving: A new perspective* (pp. 75-88). New York: Springer-Verlag.

McLeod, D. B. (1988). Affective issues in mathematical problem solving: Some theoretical considerations. *Journal for Research in Mathematics Education, 19,* 134-141.

McLeod, D. B. (1989). Beliefs, attitudes, and emotions: New views of affect in mathematics education. In D. B. McLeod & V. M. Adams (Eds.), *Affect and mathematical problem solving: A new perspective* (pp. 245-258).. New York: Springer-Verlag.

McLeod, D. B. (1992). Research on affect in mathematics education: A reconceptualization. In D. Grouws (Ed.), *Handbook of Research on Mathematics Teaching and Learning* (pp. 575-596). New York: Macmillan.

McLeod, D. B. (1994). Research on affect and mathematics learning in the JRME: 1970 to the present. *Journal for Research in Mathematics Education, 25*(6), 637-647.

McLeod, D. B. & Adams, V. M., Eds. (1989). *Affect and mathematical problem solving: A new perspective.* New York: Springer-Verlag.

Picard, R. W. (1997). *Affective computing.* Cambridge, MA: The MIT Press.

Rogers, T. B. (1983). Emotion, imagery, and verbal codes: A closer look at an increasingly complex interaction. In J. Yuille (Ed.), *Imagery, memory, and cognition: Essays in honor of Allan Paivio* (pp. 285-305). Hillsdale, NJ: Erlbaum.

Schoenfeld, A. (1985). *Mathematical problem solving.* Orlando, FL: Academic Press.

Vinner, S. (1997). From intuition to inhibition—mathematics, education, and other endangered species. In E. Pehkonen (Ed.), *Proceedings of the 21st annual conference of PME* Vol. 1 (pp. 63-78). Lahti, Finland: University of Helsinki Dept. of Teacher Education.

Zajonc, R. B. (1980). Feeling and thinking: Preferences need no inferences. *American Psychologist, 35,* 151-175.

CHAPTER 5

GÜNTER TÖRNER

MATHEMATICAL BELIEFS – A SEARCH FOR A COMMON GROUND: SOME THEORETICAL CONSIDERATIONS ON STRUCTURING BELIEFS, SOME RESEARCH QUESTIONS, AND SOME PHENOMENOLOGICAL OBSERVATIONS

Abstract. A range of research and theory from different sources is reviewed in this chapter, in an attempt to understand better the construct of mathematical beliefs. Definitions of mathematical beliefs in the literature are not consistent and thus working out the core elements of a definition is one aspect of the chapter. Specifically, a four-component definition of beliefs is presented. The model focuses on belief object, range and content of mental associations, activation level or strength of each association, and some associated evaluation maps. This framework is not empirically derived but is based on common characteristics of the literature on didactics, particularly mathematics didactics. This effort towards achieving a precise definition can provide new understandings of fundamental issues in research on mathematical beliefs and give rise to new research questions. In particular, it allows description of the term "belief systems" allowing clustering of individual beliefs into a system across each of the four components. Furthermore, it makes sense to distinguish between global beliefs, domain-specific beliefs and subject-matter beliefs. The question immediately arises as to what interdependencies exist between the individual beliefs. Some observations from a survey of mathematical beliefs of students studying calculus are also included.

So, my hypothesis is: whatever the notion of belief is, it may solve our problem.
(Bogdan, 1986, p. 2)

1. THE STARTING POINT: THE LACK OF CONSENSUS ABOUT A DEFINITION OF MATHEMATICAL BELIEFS

The lack of consistency in definitions of the term mathematical beliefs has often been noted. Standard references on this issue are contributions by Pajares (1992) and Thompson (1992), which themselves feature a host of other references. In this chapter the focus is not so much a philosophical analysis of the term "beliefs" (see Berger, 2001; Bogdan, 1986), as an "inventory check" of this term in the context of research questions in mathematics didactics. This effort is made more difficult because previous work has not adequately distinguished between knowledge and beliefs (Abelson 1979; Pajares, 1992). However, attention here is focused on the question of what we choose to term a belief and, only to a lesser extent, on the

G. C. Leder, E. Pehkonen, & G. Törner (Eds.), Beliefs: A Hidden Variable in Mathematics Education? 73-94.

distinction between belief and knowledge. The latter discussion has suggested that beliefs can be viewed as at the periphery of knowledge (Ryan, 1984).

1.1. The Empirical Relevance of Beliefs

Going beyond the understanding of cognitive processes (see Schoenfeld, 1985), it is my intention to make the individual accomplishment of mathematical tasks understandable. It is now widely recognized that research into the knowledge component alone does not lead to satisfactory results. It therefore comes as no surprise that the "knowledge" side has been only one portal into research on beliefs. Ryan (1984) conceptualizes, in close alliance to Perry (1970), differing subjective theories on the nature of knowledge. The dualistic- and fact-orientated concept of knowledge (in which a clear distinction between right and wrong is made) is juxtaposed in a dichotomy with a relativistic or context-oriented concept. Ryan demonstrates that different knowledge concepts possessing differing subjective-individual causes correlate with differing information-processing and learning strategies.

The results of the *Third International Mathematics and Science Study* [TIMSS III], recently published in Germany, include 40 pages on the effects of beliefs ("epistemological beliefs") on understanding in mathematics lessons (see Köller, Baumert & Neubrand, 2000). It is noteworthy that scales of mathematical world views (i.e., beliefs in the widest sense) include directly or indirectly mediated effects on mathematical achievement. Similar links between epistemological beliefs and mathematical achievement can be proven for the academic discipline of physics.

1.2. Issues Related to Definitions of Belief

> It will not be possible for researchers to come to grips with ... beliefs, however, without first deciding what they wish belief to mean and how this meaning will differ from that of similar constructs. (Pajares, 1992, p. 308)

On numerous occasions, beliefs have been, and still are, related to notions of misconceptions. It is time to reconsider this dominating, biased view of the role of beliefs and to view their functions soberly and in a productive fashion. Fischbein (1987) convincingly points out in his book that

> ... for a long time, reasoning has been studied mainly in terms of prepositional networks governed by logical rules. Consequently, the instructional process, especially in science and in mathematics, has tended to provide the learner with a certain amount of information (principles, laws, theorems, formula) and to develop methods of formal reasoning adapted to the respective domain. (p 206)

And Fischbein continues:

> What has been shown [...] is that, beyond the dynamics of the conceptual network, there is a world of stabilized expectations and beliefs, which deeply influence the reception, and the use of mathematical and scientific knowledge.

Further:

> The dynamics of mathematical reasoning – and generally, of every kind of scientific reasoning – include various psychological components like beliefs .... These are not mere residuals of more primitive forms of reasoning. They are genuinely productive, active ingredients of every type of reasoning. (p. 212)

There is, however, another way of looking at beliefs, the mental side of beliefs. Beliefs serve as the basic modules for the perception of virtual entities. New perception research has shown that our visual perception has objective capacity limits and that relatively small amounts of information suffice to recognize pictures. As perception psychologists have been able to prove, our perception system must inevitably reduce the volume of information by a factor of 10^8 to 10^9 (for example, when watching a television film) (see Klaus & Liebscher, 1979). Thus, when developing virtual entities - the "world of mathematics" or the "world of geometry" - we cannot manage without filter processes that restrict our perceptions. A debatable issue is the extent to which the information reduction remains acceptable and still represents the original reality and the extent to which a reconstruction is possible. While on the one hand the filter processes are inescapable in virtual perception, it must be presumed that these processes are subjective. Contributions from belief research can be expected here, particularly in relation to the research on selective memory as an aspect of cognitive information processing (Olson & Zanna, 1993).

From this viewpoint we must define beliefs as precisely as possible without suggesting that a definition must be absolute and unchanging. As Hunter Lewis, quoted in Pajares (1992), suggests:

> ... the most fruitful concepts are those to which it is impossible to attach a well-defined meaning. (p. 308)

Given the variety of productive uses of the term "beliefs", clarifying the term is difficult. Borrowing from the title of a paper by Cooney (1994) *"in search of common ground"*, what, then, is the "common ground"? In the literature a great number of papers can be found concerning beliefs about mathematics as well as about the learning and teaching of mathematics (e.g., Calderhead, 1996; Thompson, 1992). However, there is still no consensus on a unique definition of the term belief, as convincingly demonstrated in a review of the literature by Furinghetti and Pehkonen (1999). Eisenhart, Shrum, Harding, and Cuthbert (1988, p.52) even speak of "definitional confusion among researchers". Many authors seem to be aware of this deficiency and thus establish their own terms: *conceptions* (e.g., Erlwanger, 1975; Pehkonen, 1988; Thompson, 1984), *philosophy* (e.g., Ernest, 1991; Lerman, 1983), *ideology, perception* (e.g., concept image as in Tall & Vinner, 1981), *world view* (e. g., Schoenfeld, 1985), *image* (e.g., Lim, 2000; Rogers, 1994), *disposition* (Kuhs & Ball, 1986) and so forth. This continued use of related but not necessarily well-defined terms has contributed to the lack of consistency in definitions of beliefs.

Further, there are many papers focusing on processes of learning and teaching that do not address beliefs explicitly, although the notion of belief is implicit. One reason might be that related theoretical frameworks and neighboring theories are widely accepted and strongly established. For example, the theory of attitudes and the theory of attributions or motivations integrate belief aspects a priori, and then apply these aspects directly without the necessity of mentioning beliefs.

1.3. The Fruitful Aspects of Definitional Confusion

> It seems that in some languages you cannot easily attribute belief to simple information processors. There may be a correlation between what a language allows you to say about belief and what you say philosophically about belief in that language. (Bogdan, 1986, p. 3)

Linguistic dependency, in terms of the cultural relativity of lexical items in each individual language, appears to be quite considerable with respect to beliefs (e.g., Alexander & Dochy, 1995). For example the word belief cannot be translated into the German language without being open to interpretation and thus Köller, Baumert, and Neubrand (2000) speak of "epistemological convictions" (epistemologische Überzeugungen) to avoid even greater terminological confusion. Moreover, only cursory attention is given to beliefs in the *Handbook of Educational Psychology* (Berliner & Calfee, 1996).

> Further contributing to the definitional confusion is the fact that researchers have different conceptions of the source of [...] beliefs.... First, it is clear that the concept of belief has been used to refer to different levels and aspects of ideology. No single definition of belief is widely accepted in the educational research community. Little cumulative development of the concept of belief is possible while those studying it hold such a variety of definitions. (Eisenhart, Shrum, Harding, & Cuthbert, 1988, pp. 52 - 53)

This sobering conclusion, however, is tempered by the observation that more than a few scientific papers in the field of mathematics education have provided significant results with respect to mathematical beliefs without first explicitly defining the term or specifically referring to an existing definition. Concepts which incorporate the definitions of beliefs of other researchers are being used tacitly (see, for example, Furinghetti & Pehkonen, 1999). Even the various authors who contributed to the book edited by McLeod and Adams (1989) failed to use a uniform terminology for beliefs, yet the individual articles are significant.

1.4. What can one Learn from Mathematical Definitions?

What is the purpose or the intended effect of clarifying the term mathematical beliefs? In scientific contexts, terms play a *functional* role. One measure of their appropriateness is the extent to which they facilitate the formation of pertinent research questions. This interplay between the creation of terminology on the one hand, and the resulting implications on the other, should be clarified for the field of *mathematical beliefs*. The various descriptions and studies of mathematical beliefs have a phenomenological character.

Finally, it should be asked whether it is realistic to search for an authoritative definition of beliefs. After all, there is no clear definition in arithmetic to indicate what is to be understood by a number, and yet man has successfully worked with numbers for many centuries in spite of this. In his important work: *What are numbers and what should they be?* Dedekind (1995/1988) discussed this question at some length. His analysis led the way to an axiomatic definition of numbers. In other words, the naïve number term is anchored in the perception of the (number) fields, which can be defined precisely in mathematical terms. The question "what is a vector" has similar overtones and cannot be answered in a simple sentence. However, the analysis of how one should operate with vectors led to the birth of linear algebra. It is also important to note that the definition of a vector space is by no means monomorphic, i.e., there are numerous models of vector space which are not isomorphic to each other. And finally there is a theory for vector spaces, which entails propositions for all or most vector space cases, independent of the individual models.

The situation may be similar when searching for a definition for beliefs. Differing concepts for the term beliefs exist quite legitimately. These concepts are in a state of coexistence with each other. But which of these can be classified as core definitions and which can be classed as more marginal and dependent on the individual research context? Authors have, over time, modified many of the definitions they propose (e.g., Schoenfeld, 1985, 1998). These observations speak for an open-ended process in the defining of what should be understood as beliefs.

2. THEORETICAL FRAMEWORK AND SIGNIFICANCE

In the sections that follow, some of the characteristics of common definitions of beliefs are discussed. More specifically, a *"four-component-definition"* of beliefs is presented.

Initially, however, some definitions from the literature are introduced to clarify the term beliefs, particularly its non-cognitive characteristics. Schoenfeld (1998) says:

> Beliefs are mental constructs representing the codification of people's experiences and understandings as beliefs. (p. 19)

A later section of this chapter attempts to describe what is meant by "mental constructs". It is also useful to integrate the known term "concept image" into the terminology of the discussion, as mental associations include pictures or perceptions. It has far too rarely been noticed that Tall and Vinner's (1981) discussion of "concept images" contains important elements of a definition of belief:

> We shall use the term concept image to describe the total cognitive structure that is associated with the concept, which includes all the mental pictures and associated properties and processes. It is built up over years through experiences of all kinds, changing as the individual meets new stimuli and matures. (p. 152)

Or even more explicitly,

> the visual representations, mental pictures, the impressions, and the experiences associated with the concept name. (Vinner, 1991, p. 61)

a description which recalls Schoenfeld's (1998) statement about beliefs, above.

2.1. Belief Objects O

When one discusses beliefs, it is necessary to consider the subject context, which provides a focus. A belief is generally a "belief about something". In the language of social psychology, entities that are evaluated are known as attitude objects (Eagly & Chaiken, 1992) and thus we speak of **belief objects**. In contrast to the approach in the theory of attitudes, we do not necessarily suggest that stimuli must stem directly from the belief objects.

Basically, anything that shares a direct or indirect connection to mathematics can function as a belief object. Some belief objects are abstract, for example the nature of mathematics (Lerman, 1990) or of science (Ledermann & Ziedler, 1987; Meichtry, 1993). Others are more concrete (e.g., the theorem of Pythagoras). Several examples are provided.

(a) subject-specific mathematical facts (mathematical objects) such as: division (Ball, 1990) and multiplication (Tirosh & Graeber, 1989), the binomial theorem, the definition of a square, the number Pi, mathematical procedures, the concept of area (Tierney, Boyd, & Davis, 1990); the isosceles triangle, angles, right angles, straight angles, altitude in a triangle (Vinner & Herskowicz, 1980), limit and continuity (Tall & Vinner, 1981), instantaneous speed (Azcarate, 1991), tangent (Tall, 1987), Taylor's series (Uriza, 1989), function (Vinner & Dreyfus, 1989), derivative (Zandieh, 1998); domains within mathematics such as geometry (Patronis, 1994), algebra (Pence, 1994), or calculus (Amit & Vinner, 1990); mathematics as whole, symbolism within mathematics (Stacey, 1994), mathematics as a discipline (school mathematics, mathematics at university, industrial mathematics, mathematics within society etc.);

(b) relations where mathematics or a subunit of mathematics (see (a)) is a substantial part: mathematics and application, mathematics and history, usefulness of mathematics; the role of definition (see Edwards, 1999) or the role of proof (Raman, 2001);

(c) relations where mathematics as well as the individual is a substantial part: self-concept as a learner of mathematics (Pajares & Miller, 1994), for example; self-concept as a teacher of mathematics, or personal anxiety and mathematics;

(d) the learning of mathematics itself (Thompson, 1989), the learning within a specific domain, the learning of special content or topic.

It is apparent that the belief objects have various "sizes", so that we refer to the **breadth** of a belief object.

2.2. The Content Set C_O Associated with a Belief

Associated with the belief object O is what we traditionally call "beliefs". Using the terminology of Schoenfeld (1998) and Abelson (1979), beliefs are the mental constructs of *some individual.* Accepting Schoenfeld's working definition, one needs to accept that "mental constructs" may include individual statements, suppositions, commitment and ideologies, and also attitudes, stances, comprehensive episodical knowledge, rumors, perceptions and finally even mental pictures. Pajares (1992) claims that the terms beliefs, values, attitudes, judgments, opinions, ideologies, perceptions, conceptions, conceptual systems, preconceptions, dispositions, implicit theories, and perspectives have frequently been used almost interchangeably. With respect to the breadth of the belief object, it is noteworthy that this set of beliefs can be small yet lead to a large network. We presuppose that the associated beliefs allow sufficient stability. I will refer to the range of these mental associations as the **content set C_O of a belief related to the object O**. Obviously, the content set describing the range of beliefs is usually highly "open" (see also Abelson, 1979).

Acknowledging the extension of the content set C_O of a belief leads in a similar direction as when Schommer (1990) proposed an inquiry into the dimensionality of belief systems. The belief objects in Schommer's question are the nature of knowledge and comprehension. This delimitation of dimensions is also important to Cooney, Shealy, and Arvold (1998):

> In particular, it makes sense to study the structures of teachers' beliefs for it is that structure that provides a certain dimensionality to what people believe. That dimensionality is paramount to understanding the process of conceptualizing the professional growth of teachers. It enables us to see teachers' beliefs as systems of beliefs and not as entities based on singular claims. (pp. 331 – 332)

Various "conflicting" elements of the content set C_O can be held simultaneously. For example, in a study of teachers' beliefs, Thompson (1984) described the responses of Jeanne, Kay, and Lyn when the belief object O consisted of *mathematics as a discipline at school*. The elements of C_O for Jeanne included

(c) Mathematics is mysterious...

(d) Mathematics is accurate, precise, and logical (p. 110).

Further examples are provided by teachers Greg, Sally, Henry and Nancy who were involved in a different study (see Cooney et al., 1998) and who held quite different views about the teaching of mathematics C_O. In this case, the belief object O addressed the *teaching of mathematics*.

It is clear that simply listing partly coherent but also partly inconsistent views and beliefs constitutes only a part of the relevant information. The set C_O, which is more than just a list of items in practice, can be complemented by further structures reflecting parts of reality that are certainly a part of a belief definition.

2.3. *The Content Set C_O as a Fuzzy Set – Different Membership Degrees μ_i as Belief Attributes*

As we have just pointed out, not all "elements" – that is diverse mental constructs – stand at the same level. Their meaning varies. Whereas in a traditional set all elements have the same value, let us say the value 1, a fuzzy set allows different membership degrees (Zimmermann, 1990). We adopt this basic thought here and thus we assign some **membership degree function(s)** μ_i, such that $\mu_i(x) \in [0,1]$ to each element x within the content set of beliefs C_O, with *i* numbering different interpretations of the membership degree function.

Fuzzy sets are of great significance in engineering sciences today and play a major role in the design of control technology and the modeling of expert knowledge. The membership degree scales in question, that is the value of the membership degree function, vary from 0 to 1; they are mostly interpreted linguistically and are reduced a number of times from the full breadth of an interval down to a few discrete values. Strictly speaking, however, a number of semantically differing membership degree functions are imaginable, and when viewed exactly in theory, they lead to differing fuzzy sets.

2.3.1. μ_i - *Measuring Levels of Certitude*

When Thompson (1992) and Abelson (1979) point out the importance of considering strength of beliefs, they are referring to a special membership degree function μ_i, namely the one which represents certitude.

> ... One feature of beliefs is that they can be held with varying degrees of conviction. (Thompson, 1992, p. 129)

> ... The believer can be passionately committed to a point of view, or at the other extreme could regard a state of affairs as more probable than not ... This dimension of variation is absent from knowledge systems. One would not say that one knew a fact strongly. (Abelson, 1979, p. 360)

Using the notation of Cooney et al. (1998), we may understand this membership degree also in the sense of:

> ... attention to the intensity with which beliefs are held, and the nature of the evidence that supports beliefs. (p. 331)

Also the "incontrovertibility" pointed out by Parajes (1992) and others cited in that paper could serve as an analog scale μ_i.

I would like to point out here that it is conceivable to measure "certainty" on a scale from 0 to 1, with "truth" described in the ideal case by the value 1. A similar approach can be used to quantify "knowledge" (see also Pajares, 1992, p. 309). For example, Rocket (1968) already referred to belief as a type of knowledge. If one therefore accepts that beliefs possess a fuzzy character, then knowledge can be understood as a special case with a certainty degree of $\mu_i = 1$. Hence, the foundation of "knowledge" can in a certain sense be integrated into a theory of beliefs or "beliefs" can be treated like knowledge, however, with a degree of certainty μ_I far less than 1.

2.3.2. μ_i – *Measuring Levels of Consciousness*

Ernest (1989) points out that beliefs held by an individual are characterized by different levels of consciousness. Thus we are able to use the membership degree functions as a modeling tool for *the levels of consciousness* of an individual's beliefs (Ernest, 1989). Higher consciousness is assumed to lead to a greater integration of beliefs and practice.

2.3.3. μ_i - Measuring Levels of Activation

Activation levels of a belief can also be modeled using the membership degree. To ensure completeness, it is often remarked that beliefs in differing contexts have differing strengths (see Schoenfeld, 1998). With this approach, pairs of beliefs (e.g., x_1, x_2) can have membership degrees where $\mu_i(x_1)$ equals nearly zero and $\mu_i(x_2)$ equals nearly one. In such cases, apparent contradictions between x_1 and x_2 may not become evident.

2.3.4. Green's Dualistic Categories

By considering the content set C_O as a fuzzy set described by means of a membership degree function, we are also in a position to integrate Green's dualistic categories. In his book *Activities of Teaching,* Green (1971) considered the role beliefs play in the learning process. Alongside the obvious postulate that beliefs distinguish clusters, Green distinguishes beliefs according to two features. He refers to quasi-logical and quasi-psychological dimensions of beliefs and allocates them to two polar states; in our terminology this means:

μ_i - *measuring the quasi-logical character of beliefs:* beliefs can be *primary* or *derivative*. This model can also be represented by a fuzzy set.

μ_i - *measuring the quasi-psychological character of beliefs*: Here again the membership degree function consists of two states: psychological primary or alternatively, peripheral.

Green (1971) argued that beliefs can be called primary and yet at the same moment be peripheral and vice versa (see also Cooney et al., 1998). At a first glance this 2 x 2 typification appears quite convincing. However, it proves to be problematic and finally open-ended for the identification of beliefs. The role of evidence is the critical part of Green's (1971) analysis according to Cooney et al. (1998). Only a few papers in the literature have previously offered convincing interpretations and contributions regarding which criteria should be correlated to each respective "value" (Cooney et al., 1998; Jones, 1990). An open question is the possible interaction patterns of the accordingly categorized beliefs.

2.4. Evaluation Maps

It is well accepted, and implicit in many definitions, that beliefs rely heavily on evaluative and affective components (e.g., Nespor, 1987).

> Finally, in addition to knowledge of and about mathematics, people's understanding of mathematics is colored by their emotional responses to the subject and their inclinations and sense of self in relation to it. Interviews with prospective and experienced teachers illustrate how mathematical understanding is a product of an interweaving of substantive mathematical knowledge with ideas and feelings about the subject. (p. 7-8)

For this reason we require as a further module one or more **evaluation map(s)** ε_j, defined for the range of a belief C_O and with a linguistic value scale. Similar to attitude theory (Eagly & Chaiken, 1992), evaluative responses are those that express approval or disapproval, favor or disfavor, liking or disliking, approach or avoidance, attraction or aversion, or similar reactions. Such reactions and feelings give rise to various (linguistic) scales. In fuzzy theory we describe these maps as "linguistic variables" (see Zimmermann, 1990, p. 132). However, the values do not necessarily have to constitute a continuous linear scale; bipolar scales are also conceivable. Therefore, social scientists often represent the hypothetical state that they assume underlies evaluative responding as a location on a bipolar continuum or dimension that ranges from extremely positive to extremely negative and that includes a reference point of neutrality.

> ... Laura, a prospective elementary teacher, responded: Zero is such a stupid number. ... In this tiny snapshot of Laura's understanding of mathematics, we see that what she does not know in this case is framed by her beliefs about mathematical knowledge and her feelings about its senselessness. (Ball, 1991, p. 7)

2.5. *Bounding the Modules Together*

With the terminology introduced in 2.1 to 2.4, the essential components necessary for a concluding definition are in place. It is now obvious that any belief definition must always take two basic variables into account, namely the ***person P*** who has *professed* the belief or to whom the belief is *attributed.* Second, beliefs are dependent on the ***time t of constitution***

In short, a belief B constitutes itself by a quadruple $B = (O, C_O, \mu_i, \varepsilon_j)$, where O is the debatable belief object, C_O is the content set of mental associations (what traditionally is called a belief), μ_i is the membership degree function(s) of the belief, and ε_j is the evaluation map(s).

Proceeding in the sense of Abelson (1979), we furthermore claim that B should fulfill the following characteristics in a probabilistic sense. Thus the properties describe a framework whereby in individual cases one or two claims can be judged irrelevant or simply incorrect.

(1) For different persons $P' \neq P$. That is, the content sets C_O of beliefs about the same belief object O is not necessarily consensual (non-consensuality).

(2) Beliefs are likely to include a substantial amount of episodic material from either personal experience, from folklore or from propaganda, which influences the evaluation map e_λ (episodic material and its evaluative impact).

(3) The content set C_O of a belief is a priori not necessarily bounded (unboundedness).

(4) Beliefs may not be anchored in authorities (external anchoring).

(5) Beliefs are directly or indirectly linked to the self-concept of the believer P at some level (self-linkage).

The five categories named here are in part oriented to the discussion about the border issue of the distinction between knowledge and beliefs in Abelson (1979); they are explicit as well as implicit for the construct belief.

I want to emphasize that the question of the distinction between beliefs and knowledge is an interesting academic one. However, for many individual persons no sharp borderline is drawn between knowledge and beliefs. There is enough evidence of this in qualitative studies. In particular as potential differentiations between beliefs and knowledge are irrelevant for any given individual, it is all the more necessary to pay attention to this (uncritical) equivalence of the two terms. The claims (1) – (5) are to be understood as criteria for critical self-reflection which an individual can conduct him/herself. Within the framework of such a discussion the terms "professed beliefs" and "attributed beliefs" acquire new relevance.

I conclude with some further remarks:

(a) It is self evident that even in the best defined contexts with given beliefs, it is nearly impossible to determine the exact value of the membership degree $\mu_i(x)$ or the value of the individual evaluation maps ε_j. Here we must be content to determine the "fuzzy" specifications. Moreover, fine precision would not be suitable given the linguistic character of the variable. In any case, these variables can also be understood as central information parameters used to discuss the individual beliefs. Determining the relevant underlying influencing and initiating variables (when, why, how much etc.), however, is always a research question and a potential dimension of scientific analysis.

(b) Beliefs of different persons about the same belief objects are not necessarily consensual (non-consensuality).

(c) It is known that knowledge systems are not necessarily dependent on episodical material and that the knowledge possibly carries a stamped date. However, this fact does not contradict the first point.

(d) "Openness" and "unboundedness" applies to the amount C_O. This can be accounted for by the situation in which the process of the integration of episodic material can never be perceived as fully completed.

(e) The issue of how authority influences beliefs can also be found in part in Abelson's work when he postulates that belief systems are in part concerned with the existence or nonexistence of certain conceptual entities. Cooney (1994, p. 628) points out this fact in a different context.

> Green (1971) and Rokeach (1960), in their analyses of belief systems, both point to the fact that the extent to which beliefs are isolated and the individual fails to see the world as a connected place is the extent to which the person relies on an external authority for verification of truth. Green differentiates beliefs that are evidentially held from those that are nonevidentially held. The former beliefs are based on rationality; the latter are based on the acceptance of what an authority dictates. Rokeach's notion of dogmatism is similarly focused on one's relationship to authority.

Here, these authorities might be also virtual authorities in a platonic sense. These might be teachers, colleagues, friends, parents, and so forth. Nevertheless beliefs may also anchor in empirical evidence, regardless of how relevant the situation in question may have been. It should be mentioned that for transforming a belief into knowledge, the warrants of the beliefs are crucial (see Rodd, 1995). This is valid in particular for situations perceived with one's own eyes.

(f) The property (5) is in some sense dual to (4): Abelson pointed out that knowledge systems usually exclude the Self, while beliefs do not.

Finally it should be noted that the definition frameworks mentioned here consciously avoid drawing connections to willingness to take action. This willingness to act is postulated in many definitional approaches to beliefs. One of Schoenfeld's (1998) accomplishments has been to dissolve the "a priori" dependence between belief and action and to make this a topic of research in the sense of behavioral response.

3. POSSIBLE CATEGORIZATIONS OF BELIEFS AND BELIEF SYSTEMS

With reference to the above definition of beliefs and its constituents, I would like to present possibilities for structuring beliefs. By holding individual parameters constant, or varying them, familiar domains of belief research are produced. The first differentiation is the one most often found in the literature, and thus the one that has been studied the most.

3.1. The Personal Parameter P as a Variable – Group-Specific Differentiations

Beliefs are often specified and then researched according to the various groups of subjects. Accordingly, beliefs about mathematics have been studied in various groups (e.g., students, primary teachers, secondary teachers, preservice teachers, professors). Because relationships in part exist between these groups (students – teachers, preservice teachers – mathematics educators, etc.), the question is raised of the extent to which even partial relations between the beliefs of these groups of persons are verifiable for one and the same belief object.

3.2. Belief Objects O as a Subject-Specific Variable

When mathematical beliefs are discussed in the literature, a first categorization is often found in the specification of the belief object O in our terminology. Examples include conception of the nature of mathematics, mathematics as whole, teaching of mathematics, learning of mathematics, one's self-concept, and so forth. These objects can be differentiated much further, for example when the field is divided into mathematics as a science subject, as a university subject, as a school subject, or as an engineering discipline. The learning or teaching of mathematics as a belief object requires an exact specification of beliefs to be investigated, including learning at a primary level, learning at a secondary level and learning within a university course.

Obviously, these specifications take the possible diversity of potential belief objects O into consideration. There are numerous indications that beliefs about single objects (e.g., mathematics) can not be discussed successfully when one ignores the relation to other objects (e.g., mathematics teaching). Thom's (1973) quotation demonstrates that cross-links between the above-mentioned fields cannot be ignored. In other words, in many contexts it is not sufficient to study beliefs; the analysis of belief systems must take priority.

3.3. Personal Clustering by Means of Similar Distributions of Membership Degrees μ_i as well as the Evaluation Maps ε_j

Pajares (1992), following Calderhead (1996), suggests that beliefs serve another important function in the ways in which schools operate. He argues that they help individuals identify with one another and form mutually supportive social groups. Belief systems reduce dissonance and confusion and teachers, he suggests, are able to gain confidence and clearer conceptions of themselves from belonging to groups that support their particular beliefs. In the language of our definition, what such groups have in common is the similar distribution of membership degrees μ_i as well as of the evaluation maps ε_j for the belief objects in question.

4. BELIEF SYSTEMS

We will forego a detailed definition of a **belief system**. Concerning the clustering of attitudes, there are numerous approaches and modellings (viz. Eagly & Chaiken, 1992), which we do not wish to discuss here. Because of the conceptional similarity to attitudes, we assume that beliefs have an internal structuring.

Early on, clusters were postulated by Green (1971) (see the discussion below). Nespor (1987) also suggested that beliefs tend to be organized in terms of larger belief systems, which are loosely bounded networks with highly variable and uncertain linkages to events, situations, and knowledge systems (Calderhead, 1996, p. 719). The conceptual detail of our definition of beliefs makes it possible to pinpoint variables that can constitute belief systems. The previous section has already demonstrated possibilities for cluster formation.

4.1. Belief Object-Induced Clustering (see also 3.2)

When assuming a connection between several belief objects, such as school mathematics as content, the learning of school mathematics, and combining several beliefs objects, the induced structure can be viewed as a (larger) belief system. The rational network of the objects O_1, O_2, O_3, ... can be mapped onto belief structures B_1, B_2, B_3, ... with respect to their corresponding content sets C_{O1}, C_{O2}, C_{O3}, ...

To present an example, think of the geometrical objects such as reflections, rotations, translations, and assume various beliefs associated with these objects. It is not surprising that these beliefs may be bound together, inducing some belief system around geometrical congruence transformations and symmetries.

4.2. Beliefs' Clustering through similar Membership Degree Function

With the integration of beliefs B_i, $i = 1,..., n$, whose membership degrees μ_i, $i = 1, ..., n$ possess similar contextual or social conditions, macrostructures can be constructed that can be understood as belief systems. That is, the aspect of proof or proving in various mathematical contexts may serve as an example.

To be precise, let us think of the role of mathematical proofs. Maybe, there are various situations in certain mathematics lessons where proofs are estimated to be peripheral by the teacher, for whatever reason. It seems to be evident that the belief system with the role of proofs as a common belief object is estimated to be peripheral.

To quote Pajares (1992):

> Beliefs are prioritized according to their connections or relationship to other beliefs or other cognitive and affective structures. (p. 397)

4.3. Social Clustering through Beliefs with Related Evaluation Maps

Finally, it is also conceivable that the beliefs referring to different objects O_1, O_2, O_3, ..., but with similar ranges R_1, R_2, R_3, ... and analogous evaluation maps ε_1, ε_2, ε_3, ... cluster to macrostructures. Possibly the occurrence of mathematical symbols, the occurrence of formulas and the perception of numbers evoke similar fear reactions: Is it mathematical stuff? If it is, the evaluation maps cluster and point to the existence of negative beliefs of mathematics as a whole. Baroody (1987) stated that some children are so overwhelmed by fear of mathematics that they become intellectually and emotionally paralyzed.

5. SUBJECT-SPECIFIC STRUCTURING OF BELIEFS AND BELIEF HIERARCHIES

The extent to which related sets of beliefs are structured appears to us (Törner & Pehkonen, 1996) to be of great importance. It can be assumed that cognitive memory patterns and their links are related via the internal network structures of beliefs, and thus understanding structures of belief networks is of central importance.

The terminological framework described above allows for a more detailed specification of beliefs. That is, belief objects can have various "sizes" and thus we can talk of the **breadth** of a belief object. It is therefore not surprising that beliefs of similar but differently sized objects are also attributed with different names.

5.1. Subject-Specific Structuring of Beliefs

This links up to the thoughts expressed in 3.2. and 4.3 with respect to the merging of beliefs on the factual level.

5.1.1. Global Beliefs

I will use the term *"global beliefs"* to describe very general beliefs including beliefs on the teaching or learning of mathematics, on the nature of mathematics, and on the origin and development of mathematical knowledge. Global beliefs may in some sense be synonymous with the terms philosophy or ideology, particularly beliefs about mathematics as a discipline (McLeod, 1989). In this context it may be impossible to make a distinction between beliefs systems and global beliefs.

5.1.2. Subject-Matter Beliefs

Analogous to the term subject-matter-knowledge used by Even (1993), we often speak of *"subject-matter beliefs"* which refer to the amount and organization of knowledge and beliefs in the mind of the subject (see also Lloyd & Wilson, 1998). I repeat myself here when I emphasize that each mathematical term and every mathematical object or mathematical procedure can be the object of a belief (cf. section 2.1).

5.1.3. Domain-Specific Beliefs

Any investigation of beliefs will indicate that the poles of global beliefs as compared to subject-matter-beliefs do not adequately explain all mathematics beliefs. In mathematics journals we classify mathematical subjects into fields and we also do this in the teaching of mathematics – one teaches calculus, one fears algebra, one works in geometry. Because the different fields of mathematics possess differing characteristics, and because there is reason to distinguish between subject-matter beliefs and global beliefs, we propose the term *"domain-specific beliefs"*.

My research (Törner, 2000) shows that mathematical domains such as geometry, stochastics or calculus are always associated with specific beliefs. For example, in the case of calculus, beliefs represent views on the role of logic, application, exactness, calculation, and so forth. Domain-specific beliefs should be classed hierarchically higher than, for example, notions of the term derivative or the term function, although on the whole they still touch on basic views on mathematics.

5.2. Is there an Implicit Structure within Subject-Matter-Specific Beliefs?

Do global beliefs overlay both domain-specific beliefs and subject-matter beliefs? Or are single subject-matter beliefs stronger in some situations than global beliefs? Is it possible that domain-specific beliefs or subject-matter beliefs come before global beliefs? To be precise one would have to at least determine whether an implicit structure is developed through an evolution of beliefs or whether professed beliefs as such develop a structure in another way. Obviously these questions have not yet been discussed comprehensively in the literature, even though this implicit structure is an issue when one wants to change beliefs. Thus the **research question** arises:

> What mental structure links global beliefs, domain-specific beliefs and subject-matter-beliefs? Do the sum of the beliefs from the individual fields of mathematics constitute beliefs on mathematics as a whole, or do general attitudes tend to imprint subjective perceptions more in the individual domains?

Viewed from one perspective, one can place the *"top-down-influence-structure"* in opposition to the *"bottom-up influence-structure"* (see Figure 1).

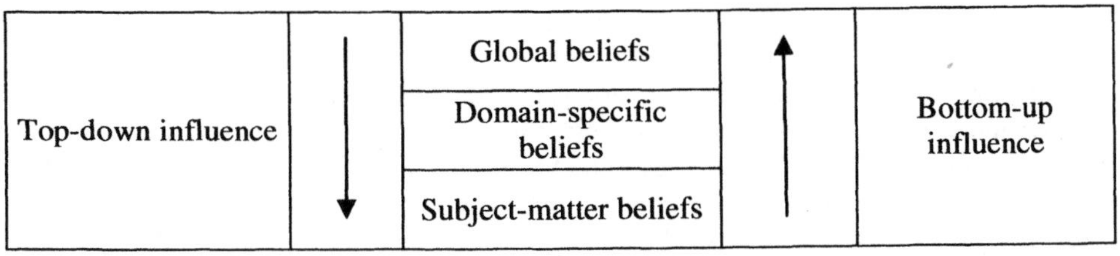

Figure 1: Different belief structures

It is to Bauersfeld's (1983) credit that he referred to the field specificity of thought structures when learning mathematics, which led to the theory of "fundamental realism" (subjective realms of experience, "*subjektive Erfahrungsbereiche*"). Analogous theories in the field of psychology in which "field" terminology was used were no rarity. In Lawler (1981) (cf. Bauersfeld, 1983) so-called microworlds emerge. Specializations for mathematical contexts gained momentum at the same time: the world of calculus, the world of geometry, and so forth were treated as distinctly different classes of mathematics. Thus, the question arises of whether a similar structuring concept also makes sense for beliefs.

6. SOME QUALITATIVE OBSERVATIONS

The present literature includes a number of references to the notion that global mental constructs about mathematics dominate the domain-specific or subject-specific beliefs. In another context Ball (1991) reports the following:

> In addition to the explicitness and connectedness of teachers' knowledge of concepts and procedures, another critical area of inquiry and analysis is the way in which their ideas about mathematics influence their representations of mathematics. What do they emphasize? What stands out to them about the mathematical issues they confront? ... Obviously the prospective teachers' ideas about mathematics **do not exist separately** from their substantive understandings of particular concepts or procedures.... Although people have many ideas about the nature of mathematics, these ideas are generally implicit, built up out of years of experiences in math classrooms and from living in a culture in which mathematics is both revered and reviled. (p. 20)

Further research indicates that short-term interventions do not result in any considerable attitude changes towards mathematics:

> There appears to be no evidence of associations between students' attitudes to mathematics and exposure to alternative teaching approaches or between students' attitudes to mathematics and new technology. (cp. Dungan Thurlow, 1989, p. 11)

We would like to include here some further observations to illuminate the interplay between beliefs about mathematics and beliefs about teaching this subject area.

6.1. Sources of Information and Mode of Inquiry

In a recently conducted study, I asked six preservice upper-secondary-school teachers (in their post-graduate phase) to describe their experiences with calculus lessons in the form of freely written essays. At the time of composing these essays, the students were still participants in a university-level didactics of mathematics course. Therefore, we had to rely on voluntary, anonymous participation for completing the questionnaire. The essay themes were "Calculus and me - how I experienced Calculus at school and university", "How I would have liked to have learned Calculus", and "How I would like to teach Calculus". This yielded a total of 3 x 6 = 18 usable, but partially anonymous, statements of two to four pages. These served as the basis of a study of the beliefs of these teachers (Törner, 2000).

Analysis of the data revealed the following main belief statements:

(1) Calculus is (reduced in school down to) calculating (not necessarily meaningfully) with functions;
(2) differential calculus is a craft – integral calculus is an art;
(3) logic is a central guideline for mathematics and in particular for calculus;
(4) exactness as a property of mathematics can be demonstrated in calculus in particular; and
(5) calculus has the special task of preparing pupils for subsequent university courses. Two additional statements dealt with aspects of learning mathematics:
(6) mathematical elegance and abstractness - liked by mathematicians - mean a loss of descriptiveness and understandability; and
(7) the recognition of application links facilitates learning.

In the following, statements (4) and (5) in the students' essays were assessed to explore if there was a possible interrelation with general views on mathematics. However, the data were limited to the essays, as subsequent research was restricted by the partly anonymous nature of the essays.

6.2. Some Results

Although a full discussion of the data to support the seven types of beliefs is beyond the scope of this chapter, several examples are appropriate. Lars, a prospective teacher student, made the most prominent statement on the aspects of logic in its relation to calculus. It is remarkable that his global view on mathematics is structurally dominated. In the words of Lars, *"logical material" can easily be worked with ... when you have acquired the rules, e.g., the transformation of fractions into decimal numbers.* According to Lars, calculus has a similar pattern, *as ... mathematical-logical thought was developed and deepened here ...* The university seminar he visited on this strengthened his belief: ... *This began in calculus with the foundations of logic which I found to be very helpful.* The consequence for him is a rigorous orientation to the aspects of logic: ... *if it were possible to do something on logic in school as early as sixth grade* (with the eleven to twelve-year-olds). Without going into details, the beliefs about logic expressed by Lars can be psychologically evaluated – referring to Green's dualistic categories - as central as well as primary.

Sascha, another student, also spoke of the central role of logic in calculus lessons. His view of mathematics was indirectly influenced by his assessment of lessons at secondary school in Germany in the mathematics courses in the Oberstufe, and it is his opinion that the schools should pay greater attention to the demands of the mathematics students to make studying the topics later at university possible, even attractive (in the calculus course) with the aid of formal logic.

Nicolas and Lars offered their perspectives concerning assessment of exactness as an important feature of mathematics, particularly calculus. Whereas Nicolas views exactness as an unavoidable difficulty, which can be didactically mastered, Lars views the aspect of exactness more fundamentally. Mathematics demands in his words ... *utmost precision and a lot of effort..., therefore one should start operating with exact terms as soon as possible. Calculus is suitable for this pursuit. ... For* *example the ε-δ-definition for continuity can be considered one of the greatest achievements in the cultural history of mathematics....*

6.3. Interpretation of Results

The students' quotations show that domain-specific beliefs must be considered in terms of global views on mathematics. A number of obvious conclusions can be drawn to this effect. Mathematics lessons, and also many university courses, do not necessarily induce a pluralistic world view of mathematics. There are a number of reasons for this. Mathematics is often taught in modules and for this reason is often perceived as such. Also, from a learning psychology viewpoint, the perception of unity is more dominant than perception of broad variation. Thus, global beliefs are oriented towards a more structural-axiomatic organization of mathematics, which in turn leads to aspects of logic being allocated a central role. In this sense, a perceptive student can experience a reinforcement of his or her assessment due to the content and the methodology of the university calculus course. Under the "axiom" that school mathematics classes are a preparation for the university, school lessons are also viewed one-sidedly.

A recently published paper by Kaldrimidou, Sakonidis, and Tzekaki (2000) shows similar features. The authors investigated the question of whether algebra and geometry have different epistemological features in the mathematics classroom. Although it is quite evident that these two domains differ epistemologically because of different patterns of thinking and learning, the authors observed that these differences disappear in the classroom.

> This suggests that the management of mathematical knowledge in the two contexts does not only prevent the differentiation of their epistemological elements (homogeneity of mathematical elements of different epistemological meaning), but moreover it unifies them. (Kaldrimidou et al., 2000, pp. 3-117)

There is the impression that Perry's stages theory (1970), presented by Ernest (1991) in another context, offers a possible explanation for understanding the strict dependency in Lars' beliefs: they can be understood as a dualism. From my viewpoint, there is evidence here of a multifaceted, pluralistic working with and understanding of mathematics. Central mathematization patterns have to balance scales with the multifaceted nature of mathematical phenomena and have to enrich each other in their interdependent nature. This ideal state could then be described, in the wording of Perry, as "relativism".

7. CONCLUSION

The purpose of this chapter was to search for "common ground" in definitions of beliefs. The proposed *four-components-model* proposed motivates one to specify the belief object, to reflect the breadth of the content set of beliefs, to trace possible interacting membership degree functions as attributes of beliefs, and to identify evaluation maps in question.

It is clear that only in rare cases can a final precise definition of all components of a belief definition be achieved in a specific context. However, striving towards a precise definition is in itself worth the effort, as it reveals or evokes deeper-lying research questions. As an example, Section 5 of this chapter, with its discussion about a hierarchical structure of subject-specific beliefs, bears witness to this. These experiences demonstrate once again the fruitfulness of this approach, which in turn legitimates the approach to this term used in this chapter.

This author is convinced that the four-component model developed in this chapter will lead to a better understanding of the belief discussion without claiming that final definitions and answers are given. Furthermore, it can be expected that the rather precise elaboration of various aspects and components of the definition model presented here will lead to a more vigorous and fruitful debate and will pose new research questions in the future.

8. NOTES

[1] I will not distinguish at this point between professed and attributed beliefs.
[2] In fact, whether one wishes it or not, all mathematical pedagogy, even if scarcely coherent, rests on a philosophy of mathematics.

9. REFERENCES

Abelson, R. (1979). Differences between belief systems and knowledge systems. *Cognitive Science, 3*, 355 - 366.

Alexander, P.A., & Dochy, F.J.R.C. (1995). Conceptions of knowledge and beliefs: A comparison across varying cultural and educational communities. *American Educational Research Journal, 32*(2), 413 - 442.

Amit, M., & Vinner, S. (1990). Some misconceptions in calculus. Anecdotes or the tip of an iceberg? In G. Booker, P. Cobb & T.N. de Mendicuti (Eds.), *Proceedings of the 14th Conference of the International Group For the Psychology of Mathematics Education (PME) with the North American Chapter 12th PME-NA Conference* Vol. 1 (pp. 3-10). México.

Azcarate, C. (1991). Instantaneous speed: Concept images at college students level and its evolution in a learning experience. In F. Furinghetti (Ed.), *Proceedings of the 15th International Conference of the International Group for the Psychology of Mathematics Education (PME).* Vol. 1 (pp. 96-103). Genoa (Italy): Dipt. di Matematica.

Ball, D. L. (1990). Prospective elementary and secondary teachers' understanding of division. *Journal for Research in Mathematics Education, 21*(3), 132-144.

Ball, D. L. (1991). Research on teaching mathematics: Making subject-matter knowledge part of the equation. In J. Brophy (Ed.), *Advances in research on teaching* Vol. 2. *Teacher's knowledge of subject matter as it relates to their teaching practice. A research annual* (pp. 1-48). Greenwich, CT: Jai Press.

Bauersfeld, H. (1983). Subjektive Erfahrungsbereiche als Grundlage einer Interaktionstheorie des Mathematiklernens und -lehrens. In H. Bauersfeld, H. Bussmann, G. Krummheuer, J. H. Lorenz, & J. Voigt (Eds.), *Lernen und Lehren von Mathematik. Untersuchungen zum Mathematikunterricht* (pp. 1-56). Köln: Aulis.

Baroody, A. J. (1987). *Children's mathematical thinking. A developmental framework for preschool, primary, and special education teachers*. New York: Columbia University, Teachers College Press.

Berger, P. (2001). *Computer und Weltbild. Habitualisierte Konzeptionen von Lehrern im Kontext von Informatik, Mathematik und Computerkultur*. Dissertation. Wiesbaden: Westdeutscher Verlag.

Berliner, D. C., & Calfee, R. (Eds.). (1996). *Handbook of Educational Psychology.* New York: Macmillan Library Reference USA.

Bogdan, R. J. (1986). The importance of belief. In R. J. Bogdan (Ed.), *Belief: Form, content, and function* (pp. 1-16). New York: Oxford University Press.

Calderhead, J. (1996). Teachers: Beliefs and knowledge. In: D. C. Berliner. & R. Calfee, R. (Eds.), *Handbook of Educational Psychology* (pp. 709-725). New York: Simon & Schuster Macmillan.

Cooney, T. (1994). Research and teacher education: in search of common ground. *Journal for Research in Mathematics Education, 25*, 608-636.

Cooney, T. J., Shealy, B. E., & Arvold, B. (1998). Conceptualizing belief structures of preservice secondary mathematics teachers. *Journal for the Research in Mathematics Education, 29*, 306-333.

Dedekind, R. (1995). *What are numbers and what should they be?* Orono, ME: Research Institute for Mathematics (first published in 1888).

Dungan, J. F., & Thurlow, G. R. (1989). Students' attitudes to mathematics: A review of the literature. *The Australian Mathematics Teacher, 45*(3), 8-11.

Eagly, A. H., & Chaiken, S. (1992). *The psychology of attitudes.* San Diego, CA: Harcourt Brace Janovich.

Edwards, B. (1999). Revisiting the notion of concept image / concept definition. In F. Hitt & M. Santos, (Ed.), *Proceedings of the 21st Annual Meeting of the North American Chapter of the International Group for the Psychology of Mathematics Education (PME)* Vol. 1 (pp. 205-210). Columbus, OH: ERIC Clearinghouse for Science, Mathematics and Environmental.

Eisenhart, M. A., Shrum, J. L., Harding, J. R., & Cuthbert, A. M. (1988). Teacher beliefs: Definitions, findings, and directions. *Educational Policy, 2*(1), 51-70.

Ernest, P. (1989). The knowledge, beliefs, and attitudes of the mathematics teacher: A model. *Journal of Education for Teaching, 15*(10), 13-33.

Ernest, P. (1991). *The philosophy of mathematics education.* Hampshire (UK): The Falmer Press.

Erlwanger, S. (1975). Case studies of children's conceptions of mathematics, Part 1. *Journal of Children's Mathematical Behavior, 1*, 157 – 283.

Even, R. (1993). Subject-matter knowledge and pedagogical content knowledge: Prospective secondary teachers and the function concept. *Journal for the Research of Mathematics Education, 24*, 94-116.

Furinghetti, F., & Pehkonen, E. (1999). A virtual panel evaluating characterizations of beliefs. In E. Pehkonen & G. Törner (Eds.), *Mathematical Beliefs and their Impact on Teaching and Learning of Mathematics. Proceedings of the Workshop in Oberwolfach*, November 21-27, 1999 (pp. 24-30). Schriftenreihe des Fachbereichs Mathematik, No. 457. Duisburg: Universität Duisburg.

Green, T. F. (1971). *The Activities of Teaching.* Tokyo: McGraw-Hill Kogakusha.

Jones, D. L. (1990). *A study of the beliefs systems of two beginning middle school mathematics teachers.* Dissertation. University of Athens, Georgia.

Kaldrimidou, M., Sakonidis, H., & Tzekaki, M. (2000). Epistemological features in the mathematics classroom: Algebra and geometry. In T. Nakahara, & M. Koyama (Eds.), *Proceedings of the 24th International Conference of the International Group for the Psychology of Mathematics Education (PME)* Vol. 3 (pp. 111-118). Hiroshima: Hiroshima University.

Klaus, G., & Liebscher, H. (1979). *Wörterbuch der kybernetik.* Frankfurt: Fischer Taschenbuch Verlag.

Köller, O., Baumert, J., & Neubrand, J.(2000). Epistemologische Überzeugungen und Fachverständnis im Mathematik- und Physikunterricht. In J. Baumert, W. Bos, & R. Lehmann, (Hrsg.), TIMSS / III - *Dritte Internationale Mathematik - und Naturwissenschaftstudie – Mathematische und naturwissenschaftliche Bildung am Ende der Schullaufbahn* (pp. 229–269). Band 2: *Mathematische und physikalische Kompetenzen am Ende der gymnasialen Oberstufe.* Opladen: Leske + Budrich.

Kuhs, T.M., & Ball, D.L. (1986). Approaches to teaching mathematics, unpublished paper, National Center for Research on Teacher Education, Michigan State University.

Lawler, W. (1981). The progressive construction of mind. *Cognitive Science, 5*, 1-30.

Lerman, S. (1983). Problem-solving or knowledge centered: The influence of philosophy on mathematics teaching. *International Journal of Mathematical Education in Science and Technology, 14*(1), 59-66.

Lerman, S. (1997). The psychology of mathematics teachers' learning in search of theory. In E. Pehkonen (Ed.), *Proceedings of the 21st International Conference of the International Group for the Psychology of Mathematics Education (PME)* Vol. 3 (pp. 200-207). Helsinki: University of Helsinki; Lahti Research and Training Center.

Lim Chap Sam. (2000). A comparison between Malaysian and United Kingdom teachers' and students' images of mathematics. In T. Nakahara & M. Koyama (Eds.), *Proceedings of the 24th International Conference of the International Group for the Psychology of Mathematics Education (PME)* Vol. 3 (pp. 323-330). Hiroshima: Hiroshima University

Lloyd, G. M., & Wilson, M. R. (1998). Supporting innovation: The impact of a teacher's conceptions of functions on his implementation of a reform curriculum. *Journal for Research in Mathematics Education, 29*(3), 248-274.

McLeod, D. B., & Adams, V. M. (Eds.). (1989). *Affect and mathematical problem solving - A new perspective.* New York: Springer.

McLeod, D. B. (1989). Beliefs, attitudes, and emotions: new views of affect in mathematics education. In D. B. McLeod & V. M. Adams (Eds.), *Affect and mathematical problem solving. A new perspective* (pp. 245-258). New York: Springer-Verlag.

Nespor, J. (1987). The role of beliefs in the practice of teaching. *Journal of Curriculum Studies, 19*(4), 317 – 328.

Olson, J. M., & Zanna, M. P. (1993). Attitudes and attitude change. *Annual Review of Psychology, 44*, 117-154.

Pajares, M. F. (1992). Teachers' beliefs and educational research: Cleaning up a messy construct. *Review of Educational Research, 62*(3), 307 – 332.

Pajares, M. F., & Miller, M. D. (1994). Role of self-efficacy and self-concept beliefs in mathematical problem solving: A path analysis. *Journal of Educational Psychology, 86*, 193-203.

Patronis. T. (1994). On students' conceptions of axioms in school geometry. In J. P. Ponte & J. F. Matos (Eds.), *Proceedings of the 18th International Conference of the International Group for the Psychology of Mathematics Education (PME)* Vol. 4 (pp. 9-16). Lisboa (Portugal): University of Lisboa.

Pehkonen, E. (1988). Conceptions and images of mathematics professors on teaching mathematics in school. *International Journal of Mathematical Education in Science and Technology, 30*, 389-397.

Pence, B. (1994). Teachers perceptions of algebra. In J. P. Ponte & J. F. Matos (Eds.), *Proceedings of the 18th International Conference of the International Group for the Psychology of Mathematics Education (PME)* Vol. 4 (pp. 17-24). Lisboa (Portugal): University of Lisboa.

Perry, W. (1970). *Forms of intellectual and ethical development in the college years: A scheme.* New York: Holt, Rinehoart, & Wilson.

Rodd, M. M. (1997). Beliefs and their warrants in mathematics learning. In E. Pehkonen (Ed.), *Proceedings of the 21st International Conference of the International Group for the Psychology of Mathematics Education (PME)* Vol. 4 (pp. 64-65). Helsinki: University of Helsinki; Lahti Research and Training Center.

Rogers, L. (1992). Images, metaphors and intuitive ways of knowing: The contexts of learners, teachers and of mathematics. In F. Seeger & H. Steinbring (Eds.), *The dialogue between theory and practice in mathematics education: Overcoming the broadcast metaphor. Proceedings of the Fourth Conference on Systematica Co-Operation between Theory and Practice in Mathematics Education* (SCTP). Brakel.

Rokeach, M. (1960). The organisation of belief-disbelief systems. In M. Rokeach (Ed.), *The open and closed mind.* New York: Basic Books.

Rokeach, M. (1968). *Beliefs, attitudes, and values: A theory of organization and change.* San Francisco: Jossey-Bass.

Ruffell, M., Mason, J., & Allen, B. (1998). Studying attitude to mathematics. *Educational Studies in Mathematics, 35*, 1-18.

Ryan, M. P. (1984). Monitoring text comprehension: Individual differences in epistemological standards. *Journal of Epistemological Psychology, 76*, 248 – 258.

Schoenfeld, A.H. (1985). *Mathematical problem solving.* Orlando (FL): Academic Press.

Schoenfeld, A.H. (1998). Toward a theory of teaching-in-context. *Issues in Education, 4*(1), 1-94.

Shulman, L.S. (1986). Paradigms and research programs in the study of teaching: a contemporary perspective. In M. C. Wittrock (Ed.), *Third Handbook of Research on Teaching* (pp. 3–36). New York: Macmillan.

Snow, R., Corno, L., & Jackson III, D. (1996). Individual Differences in affective and conative function. In: D. C. Berliner & R. Calfee (Eds.), *Handbook of Educational Psychology* (pp. 243-310). New York: Simon & Schuster Macmillan.

Stacey, K. (1994). Algebraic sums and products: Students' concepts and symbolism. In J.P. Ponte & J.F. Matos (Eds.), *Proceedings of the 18th International Conference of the International Group for the Psychology of Mathematics Education (PME)* Vol. 4 (pp. 289-296). Lisboa (Portugal): University of Lisboa.

Tall, D., & Vinner, S. (1981). Concept images and concept definition in mathematics with particular reference to limits and continuity. *Educational Studies in Mathematics, 12*, 151-169.

Tall, D. (1987). Constructing the concept image of a tangent. In J.C. Bergeron, N. Herscovics, & C. Kieran (Eds.), *Proceedings of the 11th International Conference of International Group for the Psychology of Mathematics Education (PME)* Vol. 3 (pp. 69-75). Montreal.

Thompson, A. G. (1984). The relationship of teachers' conceptions of mathematics and mathematics teaching to instructional practice. *Educational Studies in Mathematics, 15*(2), 105 – 127.

Thompson, A.G. (1989). Learning to teach mathematical problem solving: Changes in teachers' conceptions and beliefs. In R. I. Charles & E. A. Silver (Eds.), *The Teaching and Assessing of Mathematical Problem Solving* Vol. 3 (pp. 232-243). *Research Agenda for Mathematics Education.* Reston, VA: Lawrence Erlbaum & National Council of Teachers of Mathematics.

Thompson, A. G. (1992). Teachers' beliefs and conceptions: A synthesis of the research. In D. A. Grouws (Ed.), *Handbook of research on mathematics learning and teaching* (pp. 127-146). New York: Macmillan Publishing.

Tierney, C., Boyd, C., & Davis, G. (1990). Prospective primary teachers' conceptions of area. In G. Booker, P. Cobb, & T. N. de Mendicuti (Eds.), *Proceedings of the 14th Conference of the International Group For the Psychology of Mathematics Education (PME) with the North American Chapter 12th PME-NA Conference* Vol. 2 (pp. 307-318). México.

Tirosh, D., & Graeber, A.O. (1989). Preservice elementary teachers' explicit beliefs about multiplication and division. *Educational Studies in Mathematics, 20*(1), 79-96.

Törner, G., & Pehkonen. (1996). On the structure of mathematical belief systems. *International Reviews on Mathematical Education (ZDM), 28*(4), 109-112.

Törner, G. (2000). Domain specific beliefs and calculus. Some theoretical remarks and phemonological observations. In E. Pehkonen & G. Törner (Eds.). *Mathematical Beliefs and their Impact on Teaching and Learning of Mathematics. Proceedings of the Workshop in Oberwolfach*, Nov. 21 – 27, 1999 (pp. 127-137). Schriftenreihe des Fachbereichs Mathematik, No. 457. Duisburg: Universität Duisburg.

Uriza, R. C. (1989). Concept image in its origins with particular reference to Taylor's series. In C. A. Maher, G. A. Goldin & R. B. Davis (Eds.), *Proceedings of the 11th Annual Meeting of the North American Chapter of the International Group for the Psychology of Mathematics Education (PME)* Vol. 1 (pp. 55-60). New Brunswick (NJ): Rutgers-The State University of New Jersey.

Vinner, S. (1991). The role of definitions in the teaching and learning of mathematics. In D. Tall (Ed.), *Advanced mathematical thinking* (pp. 65-81). Dordrecht: Kluwer.

Vinner, S., & Dreyfus, T. (1989). Images and definition for the concept of function. *Journal for Research in Mathematics Education*, 20, 356-66.

Vinner, S., & Hershkowitz, R. (1980). Concept images and common cognitive paths in the development of some simple geometrical concepts. In R. Karplus (Ed.), *Proceedings of the 4th International Conference of the International Group for the Psychology of Mathematics Education (PME)* Vol. 1 (pp. 177-184). Berkeley (CA): University of California, Lawrence Hall of science.

Zandieh, M. J. (1998). The role of a formal definition in nine students' concept image of derivative. In Berenson, S.B. et al. Proceedings of the 20th Annual Meeting of the North American Chapter of the International Group for the Psychology of Mathematics Education Vol. 1 (pp. 136-141). Columbus, OH: ERIC Clearinghouse for Science, Mathematics and Environmental Education.

Zimmermann, H. J. (1990). *Fuzzy set theory and its applications.* Boston: Kluwer.

CHAPTER 6

GILAH C. LEDER AND HELEN J. FORGASZ

MEASURING MATHEMATICAL BELIEFS AND THEIR IMPACT ON THE LEARNING OF MATHEMATICS: A NEW APPROACH

Abstract. In this chapter we provide a brief overview of commonly used definitions of beliefs, ways in which beliefs are measured in general, and in mathematics education research in particular. Next we describe how the technique known as the Experience Sampling Method was used to infer students' attitudes to, and beliefs about a range of daily activities, including those related to their (mathematical) studies. Briefly, on receipt of a signal sent six times per day for six consecutive days, our sample of mature age students[1] was requested, through completion of a specially designed form, to record the activity in which they were currently engaged and their reactions to that activity. We argue that strengths of the approach adopted include the extended period of time used for data collection, the opportunity to gauge participants' attitudes, beliefs, and emotions about the wide range of activities tapped, and to compare these with their beliefs about mathematics and the learning of mathematics.

1. INTRODUCTION

It is now widely accepted that cognitive as well as affective factors - such as attitudes, beliefs, feelings, and moods - must be explored if our understanding of the nature of mathematics learning is to be enhanced. How students' beliefs and attitudes about mathematics influence their learning of this subject has attracted considerable research attention. Yet, finding ways to infer beliefs and attitudes from behaviors has continued to be a challenge to researchers. In an influential article, Schoenfeld (1992) argued: "The older measurement tools and concepts found in the affective literature are simply inadequate; they are not at a level of mechanism and most often tell us that *something* happens without offering good suggestions as to how or why" (p. 364). In this chapter we describe an instrument devised to gauge students' beliefs, feelings, and attitudes as they were engaged in a range of activities and thus we go some way towards providing the "meaningful integration of cognition and affect" (p. 364) for which Schoenfeld (1992) pleads.

G. C. Leder, E. Pehkonen, & G. Törner (Eds.), Beliefs: A Hidden Variable in Mathematics Education? 95-113.

2. BELIEFS - DEFINITIONS

In everyday language, the term "belief" is often used loosely and synonymously with terms such as attitude, disposition, opinion, perception, philosophy, and value. Because these various concepts are not directly observable and have to be inferred, and because of their overlapping nature, it is not easy to produce a precise definition of beliefs. In a detailed review of the research literature, McLeod (1992, p. 582) concluded, "it is difficult to separate research on attitudes from research on beliefs". Reference to overlap between different concepts and the need to deduce beliefs from observable activities are both reflected within the psychological literature, as can be seen from the selection of definitions presented in Table 1.

The entries in Table 1 reveal a number of common notions and assumptions as well as differences in emphasis and a subtle increase in complexity in the definition of "beliefs" over time. Thus many of the definitions postulate that beliefs, attitudes and values are intrinsically related, and that beliefs and attitudes have cognitive, affective, and behavioral components. Gopnik and Meltzoff's (1997) use of "perceptions" as a synonym for "beliefs" highlights the filtering role on beliefs of previous experiences, predominant social norms, and cultural expectations. A thorough search of the psychological literature yields a further variety of definitions of beliefs. Rather than listing these, it is more fruitful to argue, as did Halloran (1970) with respect to attitudes, "that despite the complexity and apparent confusion social scientists still seem to understand each other when using the concept" (p. 17). Furthermore, given the common elements evident among many of the definitions, much useful work can be done without full and rigid agreement about the precise definition adopted.

Table 1. Selected definitions of beliefs (chronological order)

Authors	Key elements of the definition
Bem (1970)	"Beliefs, attitudes, and values do seem logically connected" (p. 4).
Cook & Selltiz (1970/1964)	Attitudes involve "an underlying disposition which enters, along with other influences, into the determination of a variety of behaviors toward an object or class of objects, including statements of beliefs and feelings about the object and approach-avoidance actions with respect to it" (p. 23).
Rokeach (1972)	"A belief is any simple proposition, conscious or unconscious, inferred from what a person says or does, capable of being preceded by the phrase 'I believe that …'" (p. 113).

Authors	Key elements of the definition
Rokeach (1972)	An attitude is defined simply as an organization of interrelated beliefs around a common object, with certain aspects of the object being at the focus of attention for some persons, and other aspects for other people (p. 116)...Each belief within an attitude organization is conceived to have three components: a *cognitive* component [it represents a person's knowledge], an *affective component* [the belief can arouse affect], and a *behavioral* component [leads to some action, when suitably activated] (p. 113).
Cooper & McGaugh (1970)	A belief is "an attitude which incorporates a large amount of cognitive restructuring. Operationally, one has an attitude *toward* and a belief *in* or *about* a stimulus object.... Belief connotes an attitude which involves or identifies the subject deeply with the object. The individual uses his[2] [sic] belief as a basis for predicting what will happen in the future" (p. 26, emphasis in the original).
Fishbein & Ajzen (1975)	"Whereas attitude refers to a person's favorable or unfavorable evaluation of an object, beliefs represent the information he has about the object. Specifically, a belief links an object to some attribute.... The object of a belief may be a person, a group of people, an institution, a behavior, a policy, an event, etc., and the associated attribute may be any object, trait, property, quality, characteristic, outcome, or event" (p. 12).
Aiken (1980)	"...attitudes may be conceptualized as learned predispositions to respond positively or negatively to certain objects, situations, concepts, or persons. As such, they possess cognitive (beliefs or knowledge), affective (emotional, motivational) and performance (behavior or action tendencies) components" (p. 2).
Sloman (1987)	"An attitude...is a collection of beliefs, motives, motive generators, and comparators focused on some individual, object or idea... Attitudes are expressed in tendencies to make certain choices *when the opportunity arises*, but need not include continual disturbance of thoughts and decisions" (Italics in original, pp. 228-229).
Gopnik & Meltzoff (1997)	Perception and belief, they contend, share many features: "Both have a mind-to-world direction of fit – they involve changing representations to fit the world. Both are only indirectly reflected in action and behavior. Most significant, both perception and belief may be subject to misrepresentation – with perceptual illusions we see what is not the case, just as with false beliefs we think what is not the case" (p. 118).

In this chapter, the focus is on identifying beliefs about the learning of mathematics and factors which might encourage or discourage continued participation in mathematics and related areas. A selection of typical means for measuring beliefs (and attitudes) is summarized in Table 2. These methods also stem from psychology and have been used in a variety of academic disciplines, including mathematics education, to develop instruments with defensible validity and sound reliability and to explore mathematics-related beliefs and attitudes.

The advantages and disadvantages of the techniques used to measure attitudes and beliefs continue to be debated in the literature. Self-report measures for which the measured constructs are obvious and the implications of responses are apparent to the respondent are susceptible to distortion; for example, in directions of "social desirability" (Cook & Selltiz, 1970/1964; Lemon, 1973). On the other hand cost, ease of administration and of scoring are some of the advantages of many self-report techniques over interviews and observation methods.

Our own approach to the definition and measurement of belief is heavily influenced by two propositions: Rokeach's (1972) assertion that "a belief is any simple proposition, conscious or unconscious, inferred from what a person says or does, capable of being preceded by the phrase 'I believe that ...'" (p. 113) and Schoenfeld's (1992) caution that much research on affect tell us that *something* happens without offering good suggestions as to how or why.

Table 2. Summary of selected methods for measuring attitudes/beliefs

Thurstone's Equal-Appearing Interval scales (Thurstone, 1967/1928). Acceptance or rejection of opinions (verbal expressions of attitude) is the means for measuring and scoring attitudes. Through the method of scale development, each opinion statement is assigned a scale value. People are asked to select a small number of statements with which they agree most closely. A score is obtained by averaging the scaled scores for each item selected. This approach to measuring attitudes and beliefs is now not common.
Likert-scales (Likert, 1967/1932): Summated rating scales. A series of statements about the attitude object comprise a Likert-scale. Items are typically rated from "strongly agree" to "strongly disagree" and five divisions are very common. The divisions are converted to numerical scores. An attitude score is obtained by summing the scores for each item on the scale. Ease of administration and scoring of instruments using such items have ensured that this is a commonly used method for tapping attitudes and beliefs.
Semantic differential scales (Osgood, Suci, & Tannenbaum, 1970/1957). Three factors are typically found which assess the semantic connotations of concepts: evaluation (good-bad), potency (strong-weak), and activity (active-passive). Osgood et al. considered the "evaluative" dimension to be synonymous with a unidimensional conception of attitudes. A series of bipolar adjectives are listed with the same number of divisions (usually 7) between each pair. Each division represents a score on a continuum with the most positive response assigned the highest score. The mean score or the item sum can be used as a "total" scale score. This approach to the measurement of beliefs and attitudes makes explicit the differentiation between the cognitive, affective, and behavioral components of attitudes enumerated in various definitions proposed in the literature.

Guttman scaling (Guttman, 1967/1944) - scalogram analysis. Respondents are asked to agree or disagree with a number of statements (usually about six – Triandis, 1971). On a "true" unidimensional Guttman scale, if the people agree with a particular item, they will also agree with every item with a lower scale value. The reliability of Likert scales can be described in these terms.
Projective techniques. An unstructured or ambiguous stimulus (for example: story, picture) related to the attitude object is presented and respondents are asked to react/explain. Variations include choosing between alternative stories or explanations, and completing sentences or stories. This technique, or variations of it, is often adopted by those who favor qualitative approaches to the measurement of beliefs and by those concerned that respondents to Likert items may not express the beliefs they actually hold but rather those they consider socially acceptable.
Checklists/inventories. A checklist of issues, objects, adjectives etc. related to the attitude object is presented. Respondents mark whether they approve or disapprove of each item. "Unsure" can be included as a response category (see Lemon, 1973). Respondents can be restricted to indicating which of a set of adjectives, for example, are considered applicable to themselves or to a specified group. Inventories typically consist of a list of careers, activities etc. to which respondents indicate which are of interest. Items such as these are often also used in interviews aimed at tapping attitudes and beliefs.
Physiological measures. Physiological reactions are outside the control of respondents. Galvanic skin response has been the most widely used technique associated with attitude measures. Results are not considered reliable. Pupillary size has also been related to attitudes (see Fishbein & Ajzen, 1975; Lemon, 1973). It is unusual to see this approach used outside a laboratory setting.
Repertory grid techniques. Administered on an individual basis, the grid consists of relevant elements (often people) and constructs (usually investigator-generated). Through an investigator-initiated activity (e.g., pen-and paper, see Lemon, 1973) or interview (see Walkerdine, 1998), respondents make connections between the elements and the constructs.
Interviews. Basically an orally administered questionnaire. The 'structured' interview consists of a predetermined list of specific questions to be asked. One advantage of an unstructured interview is that it can uncover views not anticipated in advance. A combination of these approaches – the semi-structured interview – is favored by many engaged in mathematics education research.
Observations. Arguably the best means to gauge attitudes/beliefs; at the same time, the most "expensive" as the researcher cannot predict when pertinent behaviors will be observed. Data gathered in this way can range from structured to unstructured observations collected in a setting that can be natural or highly artificial. Mueller (1986, p. 90) argued that "High-quality observational measurement requires an expert and objective observer, careful selection of behavior situations and of specific behaviors to be observed, a systematic procedure for recording behaviors, and an objective interpretation of the recorded behaviors in making inferences about attitude" (see, for example, Forgasz & Leder, 1996).

3. BELIEFS AND MATHEMATICS

As already noted, mathematics learning and instruction are influenced by cognitive as well as affective issues. Contemporary curricula take for granted "that learning mathematics at school [should be] a *positive* experience in which students develop *confidence* and a *sense of achievement* from what they learn" (Board of Studies, 2000, p. 6, emphasis added). Documents produced by the National Council of Teachers of Mathematics [NCTM] contain similar sentiments (e.g., NCTM, 2000). As a further indication of the perceived interaction between affect about mathematics, large scale international tests of mathematics learning, such as the *Third International Mathematics and Science Study*, include measures of student attitudes and beliefs about different aspects of mathematics (Lokan, Ford, & Greenwood, 1996).

McLeod (1992) described four differentiations commonly used: beliefs about mathematics, about mathematics teaching, about self, and about the contexts in which mathematics education takes place. Pertinent constituent elements for these clusters are identified by Op't Eynde, de Corte, and Verschaffel (chapter 2, this volume) in the framework they put forward for students' mathematics-related beliefs. These chapters, as well as other recent work in mathematics education research reviewed in the next section, illustrate that the variations in the use of the term belief found in the broader psychological literature are mirrored in its application and measurement in mathematics education research:

> data have typically been gathered through observations of students and teachers, as well as through questionnaires and interviews. Researchers have generally not used a consistent framework; instead, the data have been organized in rather different ways in different studies, with each researcher trying to explain the influence of beliefs in each particular context. (McLeod, 1992, p. 579)

This diversity of approaches is also reflected in the data presented by Furinghetti and Pehkonen in an earlier chapter in this book. Their request to a panel of experienced mathematics educators to indicate agreement or disagreement with selected definitions about beliefs revealed substantial individual differences in acceptance of specific components highlighted in the various definitions.

3.1. Recent Research on Beliefs and Mathematics

Examination of recent research published in major, English language, mathematics education journals[3] confirms that a concern with affect is pervasive and that affective issues are incorporated in many explorations of learning and teaching mathematics. As can be seen from the summaries included in Table 3, a broad range of beliefs, measured in a variety of ways, are considered to be relevant to instructional and curriculum issues.

This brief survey of recent articles confirms that the aspects noted by McLeod: beliefs about mathematics, about mathematics teaching, beliefs about self, and about contextual factors relevant to mathematics learning, continue to attract research attention from mathematics educators. Beliefs were gauged using a variety of techniques in the studies reviewed. These included questionnaires, interviews, content analysis of journal entries, reflections, post lesson conferences, and observations. Whether or not inferences could be made about reasons for existing or changing beliefs depended heavily on the method(s) of data gathering used and the length of period over which the researcher and participants interacted.

Table 3: Beliefs in recent mathematics education research (alphabetical)

Author	Theme	Belief	Key measure(s)
Artzt, 1999	Teachers' knowledge, beliefs and goals influence their instructional practice	Beliefs refer to teachers' integrated system of personalized assumptions about the nature of mathematics, of students, about learning, and teaching	Post lesson reflections – through self assessment and exchange with supervisor at conference, and written reflections
Carr, Jessup, & Fuller, 1999	Parent and teacher contributions to gender differences in first-grade mathematics strategy use	Parent and teacher beliefs about best strategies and children's beliefs about others' perceptions of their ability	Interview, using cartoons as stimuli
Forgasz, Leder, & Gardner, 1999	Validity of an instrument for measuring attitudes and beliefs	Beliefs about mathematics as a male domain	Comparison of student responses over time to questionnaire items and behavioral indicators
Frykholm, 1999	Reform and teacher preparation – the interaction between beliefs, content and pedagogical knowledge	Beliefs about mathematics and mathematics instruction in the transition from student to teacher	Observations, post-lesson conferences, interviews, questionnaires, informal discussions
Gellert, 1999	Comprehension of mathematics instruction among pre-service elementary teachers	Beliefs about mathematics and mathematics education	"Structured content analysis" of journal entries

Author	Theme	Belief	Key measure(s)
Perry, Howard, & Tracey, 1999	Comparison of head and classroom teachers' beliefs about learning and teaching mathematics	Beliefs about mathematics, about mathematics learning and teaching	20 item Likert scale
Schuck, 1999	The interaction between mathematical knowledge and facility for teaching mathematics	Beliefs about mathematics and mathematics teaching	Interviews, reflections, open ended questionnaire
Senger, 1999	The recursive nature of teacher change	Personal values and beliefs about traditional teaching and reform	"constant comparative analysis" of variously collected dialogue
Vacc & Bright, 1999	Elementary pre-service teachers beliefs and instructional use of children's thinking	Beliefs about teaching and learning mathematics	48 item Likert scale covering beliefs about the roles of the learner & teacher, the link between skills and understanding, sequencing of topics
Verschaffel, De Corte, Lasure, Van Vaerenbergh, Bogaerts, & Ratinck, 1999	Solving application problems	Beliefs about teaching and learning of mathematical word-problem solving	Likert scale tapping pleasure and persistence, and problem v. process oriented view of problem solving

The representative sampling of pertinent research included in Table 3, illustrative of the wider acceptance of qualitative research designs within mathematics education, indicates that Kiesler, Collins, and Miller's (1969, p. 23) claim that "...social scientists have, almost without exception, settled on pencil and paper or interview techniques for the measurement of attitudes while retaining a theory that specifies behavioral implications for attitudes" is no longer a fair generalization of the mathematics education researchers' methodological repertoire for measuring attitudes and beliefs.

4. MATURE AGE STUDENTS' BELIEFS ABOUT MATHEMATICS[1] - A NEW STUDY

Internationally, concern has been expressed (e.g., Jensen, Niss, & Wedege, 1998) about the drift away from the natural sciences, including mathematics, by senior high school and university students. In an attempt to identify variables critical in influencing students to continue, or discontinue, with mathematics in succeeding years, an extensive survey was administered to students enrolled in first year mathematics courses at five quite diverse universities (see Forgasz, 1998; Forgasz & Leder, 1998a, 1998b; Leder & Forgasz, 1998 for more detailed information). The relatively large number of mature age students among those surveyed was unexpected. Comparisons of school leavers' and mature age students' responses revealed that the latter group, on average, was more highly motivated to study and that their course choices had been more focused and more informed than those of the younger students. We decided to focus on a small group of these older students, who had made a deliberate decision to continue with their formal education and the study of university mathematics, and to explore their beliefs about the tertiary learning environment and about mathematics and mathematics learning.

As in so many of the studies we reviewed, we relied heavily in the first instance on questionnaire responses, interviews, and regular e-mail or snail-mail contacts. We also included a "tag-a-student" period in which time was spent with students on campus to allow us to compare the students' descriptions of the learning environment with our own impressions. However, in addition to these traditional methods which yielded considerable information about students' beliefs, we wished to gather further information which allowed us to understand better the reasons for the beliefs expressed by our sample. Why, for example, did some apparently highly motivated students perceive continued involvement with mathematics as a test of endurance, to be suffered only because further mathematics was a rigid course requirement, while other students in the same class believed mathematics to be enjoyable and stimulating? When did concerns and beliefs about financial responsibilities become so acute that they dominated everything else? Why do so many students believe that repeated practice of problems was the only road to success in mathematics? Why did students in the same class have such differing beliefs about their teachers' level of dedication and commitment?

In the remainder of this chapter we focus on information gathered through the *Experience Sampling Method* (ESM), a novel and rich approach for capturing not only an individual's activities over an extended period, but also that individual's reactions to, and beliefs about, those activities. This additional information, we argue, provides invaluable glimpses of the reasons for students' beliefs. The findings from the study on which we base our discussion are not the emphasis of this chapter. Instead our aim is to document how information elicited through the ESM can offer insights into the reasons for students' espoused feelings, reactions, and beliefs.

4.1. The Experience Sampling Method

The ESM was developed at the University of Chicago almost 30 years ago by Mihaly Csikszentmihalyi. A quarter of a century later, more than 7000 respondents had used this technique to describe the pattern and quality of their daily life, by

> providing a virtual film strip of [their] daily activities and experiences. We can trace a person's activities from morning to night day by day over a week, and we can follow his or her mood swings in relation to what the person does and who he is with. (Csikszentmihalyi, 1997, p. 15)

According to Csikszentmihalyi and his colleagues, American adults and teenagers on average spend between 20 – 45% of their waking time engaged in work or study but are off task (e.g., talking, eating, daydreaming) anywhere between 4 – 15% of this time. Housework and related activities, on average, occupy them for 8 – 22% of their time, eating up to 5%, and traveling on a daily basis from one place to another between 6 – 9% of their time. They are engaged in leisure or social activities anywhere between 20 and 43% of their day. As well as gathering such behavioral data, the ESM is able to capture how respondents feel as they engage in these various activities and thus allows unique insights into motivations, feelings, and beliefs. The ESM has clear parallels with carefully designed observational techniques and self report data gathered over a prolonged period of time.

Csikszentmihalyi (1997) argued that "in everyday life, it is rare for the different contents of experience to be in synchrony with each other" (p. 28). Thus a student in a university mathematics class might experience a number of conflicting emotions as follows:

> In class today my attention was focused, because the instructor gave me a problem to solve that requires intense thinking. But this particular task is not one I ordinarily would want to do, so I am not very motivated intrinsically. At the same time I am distracted by feelings of anxiety about my partner's reactions to the time I spend studying. So while part of my mind is concentrated on the task, I am not completely involved in it. It is not that my mind is in total chaos, but there is quite a bit of entropy in my consciousness – thoughts, emotions, beliefs and intentions come into focus and then disappear, producing contrary impulses, and pulling my attention in different directions. (Based on Csikszentmihalyi, 1997, p. 28)

Such ambiguities, contrary inclinations, and possible anxieties generated by conflicts between personal and peer group beliefs are able to be tracked and described by the ESM more effectively than by other self-report instruments.

The interaction of different professional, academic, and personal concerns, sketched in the extracts above, also characterize the competing influences of variables postulated by Eccles (Parsons) et al. (1985) as critical determinants of academic choice:

> choice is influenced most directly by the students' values (both the utility value of math for attaining future goals and the attainment of interest value of ongoing math activities) and the students' expectancies for success at math. These variables, in turn, are assumed to be influenced by students' goals, and their concepts of both their own academic abilities and the task demands. Individual differences on these attitudinal variables are assumed to result from students' perceptions of the beliefs of major socializers, the

> students' interpretations of their own history of academic performance, and the students' perception of appropriate behaviors and goals. (pp. 97-98)

Thus students' academic choices, including decisions about continued participation in mathematics and related areas, are influenced not only by expectations of attaining success but also by perceptions and beliefs about the value of doing mathematics and the importance attached to this area by themselves, by others in the peer group, and in society. That some individuals, particularly those whose beliefs and ambitions are at variance with those of their peer group or with the cultural group to which they belong, experience a lack of synchrony between these different forces is well documented.

Despite the considerable interest of mathematics educators in measuring affect, using the ESM to chart behaviors and the concomitant affects seems not to have been used before in mathematics education research. The scope, content, and method of administration of the ESM are described in the next section.

4.2. Using the Experience Sampling Method

As already indicated, the ESM allows insights into the motivations, attitudes, and beliefs associated with an individual's behaviors, through extensive monitoring of activities over an extended period of time. In response to signals from an electronic pager, participants chart the course of their daily life and experiences by filling out a detailed report of their current activities, thoughts, companions, and feelings on the specially designed *Experience Sampling Form* or ESF. When paged, participants can choose whether or not to complete an ESF. It might be argued that the data are thus readily biased in the direction of what participants decide that the researchers should learn about their activities. Whether this has occurred in a particular study can be checked, provided researchers using the ESM report participants' response rates. It should also be remembered that interviewees and questionnaire respondents can similarly refuse to respond to particular questions and/or items. Incorporating multiple data gathering approaches, whether in research using the ESM or other approaches, will enhance the reliability of data interpretation.

Excerpts from the *Experience Sampling Form* we used are shown in Figure 1. As can be seen, respondents have the opportunity to describe and comment on the activities being undertaken as well as on the attitudes, beliefs, emotions, and moods elicited by those activities.

4.3. The Sample and Methods

Twenty volunteer mature age students, eight women and 12 men were part of the "beeper" component of our project[4]. They were asked to carry an electronic pager for six consecutive days which included a weekend. Six signals were sent between the hours of 7am and 10pm on week days and between 10am and 10pm on weekend days. Participants were asked to complete the ESFs within 30 minutes of receipt of each signal.

Name: ________
Date: ________ **Time Beeped:** ________ **Time filled out:** _______
As you were beeped:
(You may have been engaged in an academic, household, employment related or leisure pursuit. Any of these are relevant for the purpose of this study)
Where were you?
Who were you with?
....
What were you thinking about?
Tick the column which best describes your response to each of the following questions

	Not at all	A little	Moderately	Quite a bit	Completely
Were you concentrating?					
Was it hard to concentrate?					
Were you living up to your own expectations?					
....					

In the table below you are asked to describe your mood as you were beeped ....
Tick the appropriate column which best describes your mood along (the) continuum.

	Very	Quite	NEITHER	*Quite*	*Very*	
happy						*sad*
lonely						*sociable*
distracted						*focussed*
... ...						... ...

If you had the choice when you were beeped:
What would you prefer to have been doing?
....
Has anything happened, or have you done anything, since you were last beeped which has affected the way you now feel?
Not applicable ( ) No ( ) Yes ( )
If "yes", please elaborate:
Any other comments?

Figure 1: Excerpts from the Experience Sampling Form

The potential number of responses from each participant was 36 ESFs. All paging occurred during the second semester of the academic year. Some students participated mid semester; others towards the end of semester when the pressure of examinations began to build.

4.4. Relevant Data

Out of the approximately 720 ESFs distributed[5] to the 20 students, 582 forms were returned. The wealth of cumulative, sustained, and detailed information and insights about this number of students could not have been obtained, with the limited resources available to us, through more commonly used instruments. As appropriate, data offered on the ESFs were cross referenced with facts already available to us. This confirmed students' honesty and candor in filling out the ESFs.

The response rate generally fell over the six day period. Typically it was high for the first two days, fell away on the third day and remained steady over the remainder of the signaling period. Yet the overall response rate (81%) exceeded our expectations, since we had informed students that completion of at least four of the six sheets each day would be quite acceptable. We interpreted the much higher response rate as indicative of the group's strong commitment to the research project.

The activities in which our respondents were engaged as they were paged were divided into eight broad categories: study, paid work, relaxation, family, chores, transit, eating, and sleeping. When coding was not clear cut, for example, when the respondent was eating with other family members, the dimension emphasized elsewhere on the same ESF determined the category. Summary data are shown in Figure 2.

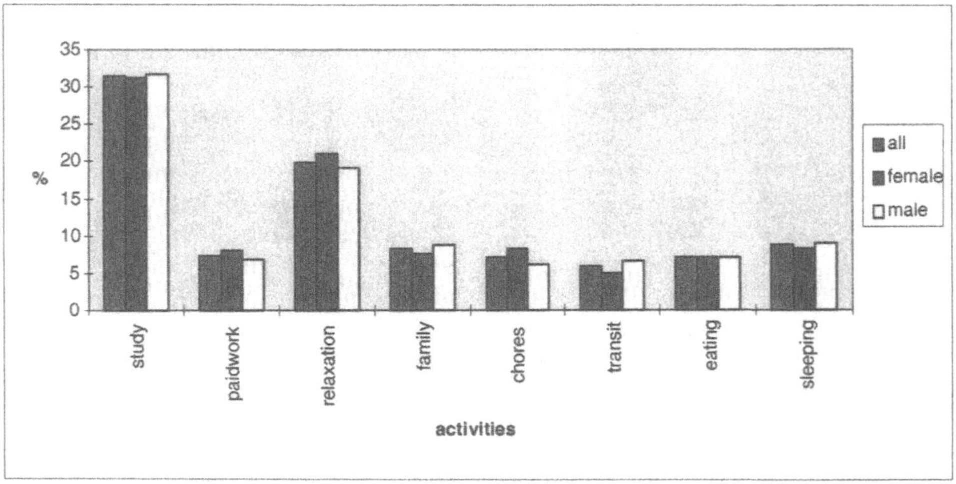

Figure 2: The spread of activities in which students in our study were engaged at the time they were signaled, overall and by gender.

Given the undoubted differences in the category definitions used, comparisons between our volunteer Australian sample and those described for American adults and teenagers by Csikszentmihalyi and his colleagues, and summarized earlier in the chapter, are necessarily crude. Nevertheless, the overview of daily activities for the two groups shows some startling similarities. For instance, our sample indicated that they were engaged in study or paid work approximately 40% of the times they were beeped, compared with the 20 to 45% each day apparently spent on such activities by American samples. When paged, our sample was engaged in activities best described as relaxation 20% of the time, compared with the between 20 and 43% given for American groups. Times spent eating, in transit, and on household chores by our group are also comparable with American groups. Using these crude measures, our sample, on average, seems much like the Americans whose activities have been charted with the ESM.

4.5. Snap Shots of Daily Activities

To indicate more fully the nature and scope of the information provided by the ESM, snap shots of daily activities described by two students, Caitlin and Boyd[6], are shown in Figure 3. These two students were selected because they were conveniently enrolled in the same institution and in the same second year mathematics subject. The data are again collated into the eight main categories already described.

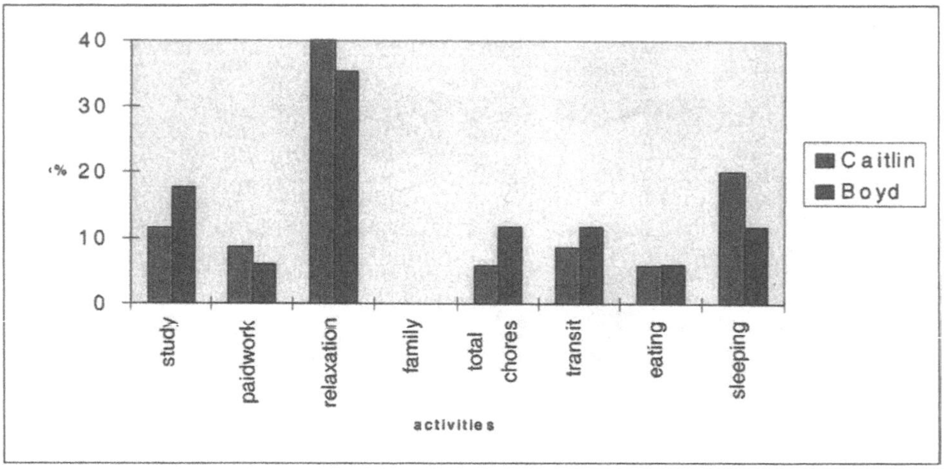

Figure 3: Overview of Caitlin and Boyd's activities when beeped

Comparison of Caitlin's and Boyd's ESM profiles showed overlap with, as well as differences from, the ESM profile recorded for the whole group. The number of times the signal found them to be engaged in paid work, in transit, or eating was similar to that recorded by the full group. In contrast to the responses for the full group, no involvement with family was recorded by the two students, neither of whom lived with family – as we knew from their survey responses. Both seemed to be spending less time studying and more time relaxing, than was recorded by the larger group: Boyd was involved in study activities almost 20% of the times he was beeped. For Caitlin this figure was just over 10%. Annotations on the ESFs revealed the most likely reasons for the discrepancies recorded between the two students and the collective responses of the larger group. For example, Caitlin was in bed with a heavy cold during the week she was monitored. Boyd seemed to be finding the demands of his part time (predominantly night) job physically exhausting. His ESF responses indicated that he spent considerable time recovering from his nightly labors. His more frequent than expected involvement in domestic chores appeared to be an attempt to avoid studying.

The profile of Boyd's thoughts and beliefs when paged during class or studying times is shown in Figure 4, and contrasts with Caitlin's more positive reflections.

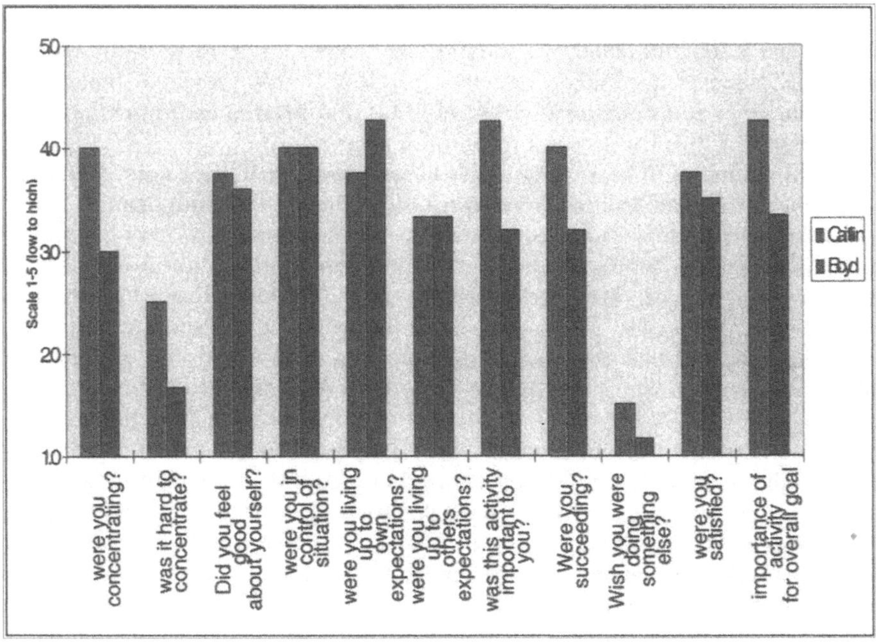

Figure 4. Results from selected ESFs for two case studies – data represent the average result for ESFs completed during study activity (N=4 for Caitlin and N=6 for Boyd).

The activities recorded on Boyd's ESFs clarify the apparent contradictions found in his email and interview responses: his fervent interest in mathematics but acknowledged difficulties in remaining motivated to tackle the course work.

We often caught Boyd engaged in mathematical activities, but these were frequently not, or only peripherally, related to the requirements of his course. For example, time spent in transit was used to read *The French mathematician* by Tom Petsinis, a book he enjoyed thoroughly, although he became "... concerned that Everiste Galois may not make it to his entrance exam for the Polytechnic on time". While in bed, getting ready to get up, Boyd "was casually wondering about ways of looking for patterns in the set of prime numbers ... to derive a formula for generating them". When sitting alone in his lounge room he was "playing with some ideas on prime numbers". Instead of working on problems involving the Fourier Series he was reflecting on a conversation he had had earlier that day with a post-graduate student about "motivational problems with studying". Boyd's lack of motivation for the course contrasts sharply with his beliefs about mathematics:

> I like playing with ideas in mathematics and setting myself little experiments to find out what happens. It's kind of like doodling for me even though these mathematical doodles have little to do with the course material. I see different tricks and techniques in mathematics as being rather similar to the different materials and mediums available to an artist. The more tricks I learn, the larger my palette becomes and the more art it is possible to create through mathematics. I see mathematics being as much an art as it is a science. Maybe this can be a motivational inspiration in itself.

His further reflection that "by studying the course material I can expand my technique" seems a creative attempt to transfer his positive beliefs about mathematics *per se* to a course in which he is disappointed and thus find renewed motivation.

Caitlin, in her ESF responses, interviews, and email messages revealed her enjoyment in the mathematics covered in the course. Her commitment to her work and academic confidence were confirmed as she was monitored over the week. The tutorial sessions were a high point of the course: "Math tutorials. I love them, working through questions on the blackboard ..., either alone or working with someone else". She was consistently enthusiastic about her studies, believed she would do well, and that completion of the course would help her ultimate career path. The mathematical activities she recorded when signaled were on task and centered on review of her work and completion of set assignments. Distressed by her inability to keep up with work requirements because of her severe cold and enforced bed rest, she nevertheless remained focused and confident: "I plan to work pretty much full-time next week to make up for this week. I'm stressed about being behind with work and study, but feel confident that I can make it up". Several days after that entry she wrote that she was feeling really good about herself: "completion of the assignment and improving health has lifted my mood". These excerpts capture well her consistently positive feelings towards her studies and beliefs about herself and reveal a consistency between her behaviors and her expressed beliefs.

5. A FINAL COMMENT

The similarities in the range of activities reported by our mature age students and the findings from many studies in the US are clear. This finding provides support for the validity of the ESM as an instrument, appropriately used, for tracking people's daily activities.

The prompts to elicit affective responses included in the ESFs used in this study produced some very strong and emotional responses from participants. It was not difficult to infer patterns of beliefs. Referring back to the definitions of beliefs summarized in Table 1, it is clear that the ESM with appropriately designed ESFs has the potential to tap the belief systems of individuals. In our case, the focus was on beliefs associated with factors affecting and relevant to their engagement in tertiary mathematics study.

From the perspective of research methods, we would argue that the ESM is a valid and reliable approximation to gathering observational data, with three key advantages over traditional observation methods: it is cost effective; it is not open to the criticism that the "natural setting" has been "disturbed" by the presence of the researcher; and participants' written comments are less susceptible to errors of interpretation than are the inferences drawn from behaviors. The precise form of the data gathered using the ESM has the potential to allow for unexpected and unanticipated findings to be more easily detected during analysis.

In our study, the more detailed, consistent, and prolonged monitoring of students via the ESM proved an economic yet extremely informative measure of activities, feelings and reflections generated by those activities. As already noted, the successful use of the ESM is critically dependent on the commitment and honesty of respondents; this was clearly not an issue in our study.

Based on our experiences, we would advocate the use of the ESM in conjunction with other forms of data collection in which the resultant pool of data is mutually complementary. As for others in our sample, Caitlin's and Boyd's responses to the ESF prompts confirmed and fleshed out the impressions gained from other data we had gathered through interviews and regular e-mail interactions. Because of the repeated application of the measure and breadth of responses elicited, the ESM seems less prone to distortions – unintentional or intentional – so often associated with self report data. The procedures followed enabled a larger group to be monitored more intensively than would be possible, with the same resources, through regular observations or interviews. By comparing students' activities at and away from study a useful context was provided against which students' beliefs about mathematics and related activities could be assessed.

6. NOTES

[1] Students who are 21 or over on March 1 of the year in which University entry is sought.

[2] The masculine pronoun has been used generically to represent both males and females. Although this remains an acceptable grammatical construction, it is now considered inappropriate. Similar generic pronouns or nouns found elsewhere in this chapter have not been individually identified; the conventional use of "sic" should be assumed.

[3] *Educational Studies in Mathematics*; The *Journal of Mathematics Teacher Education*; The *Journal for Research in Mathematics Education*; *Mathematical Thinking and Learning*; and The *Mathematics Education Research Journal*

[4] As already described, the following additional data gathering instruments were used in the study (sample sizes varied): a survey questionnaire which covered: biographical and background details, enrolment

issues, affective dimensions, and perceptions of the learning environment; interviews; regular e-mail or snail-mail contacts, and a "tag-a-student" period in which time was spent with students on campus.
[5] Human error led to the sending of five rather than six signals on a small number of days. One participant had returned to his home in the country for the weekend and was out of reach of our signals. The total number of ESFs that could have been completed was thus less than 720.
[6] Pseudonyms

7. REFERENCES

Aiken, L. R. (1980). Attitude measurement and research. In D.A. Payne (Ed.), *Recent developments in affective measurement* (pp. 1-24). San Francisco, Jossey-Bass.

Artzt, A. F. (1999). A structure to enable preservice teachers of mathematics to reflect on their teaching. *Journal of Mathematics Teacher Education*, 2(2), 143-166.

Bem, D. J. (1970). *Beliefs, attitudes, and human affairs*. Belmont, California. Brooks/Cole Publishing Co.

Board of Studies. (2000). *Mathematics. Curriculum and Standards Framework II*. Carlton, Victoria: Author.

Carr, M., Jessup, D. L., & Fuller, D. (1999). Gender differences in first-grade mathematics strategy use: Parent and teacher contributions. *Journal for Research in Mathematics Education, 30*(1), 20-46.

Csikszentmihalyi, M. (1997). *Finding flaws: The psychology of engagement with everyday life*. New York: Basic Books.

Cook, S. W., & Selltiz, C. (1970). A multiple-indicator approach to attitude measurement. In G. F. Summers (Ed.), *Attitude Measurement* (pp. 23-41). Chicago: Rand McNally. (Reprinted from *Psychological Bulletin*, 1964, *62*, 36-55.)

Cooper, J. B,. & McGaugh, J. L. (1970). Attitude and related concepts. In M. Jahoda & N. Warren (Eds.), *Attitudes. Selected readings* (pp. 26-31). Harmondsworth, England: Penguin.

Eccles (Parsons), J., Adler, T. F., Futterman, R., Goff, S. B., Kaczala, C. M., Meece, J. L., & Midgley, C. (1985). Self-perceptions, task perceptions, socializing influences, and the decision to enroll in mathematics. In S. Chipman, L. R. Brush, & D. M. Wilson (Eds.), *Women and mathematics: Balancing the equation* (pp. 95-121). Hillsdale, NJ: Lawrence Erlbaum.

Fishbein, M., & Ajzen, I. (1975). *Belief, attitude, intention and behavior: An introduction to theory and research*. Menlo Park, California: Addison-Wesley Publishing Company.

Forgasz, H. J. (1998). The typical Australian university mathematics student: Challenging myths and stereotypes? *Higher Education, 36*(1), 87-108.

Forgasz, H. J., & Leder, G. C. (1996). Mathematics classrooms, gender and affect. *Mathematics Education Research Journal*, *8*(2), 153-173.

Forgasz, H. J., & Leder, G. C. (1998a). Tertiary mathematics students: Why are they here? *Nordisk Matematik Didaktik (Nordic Studies in Mathematics Education)*, *6*(2), 7-27.

Forgasz, H. J., & Leder, G. C. (1998b). Affective dimensions and tertiary mathematics students. In A. Olivier & K. Newstead (Eds.), *Proceedings of the 22nd conference of the International Group for the Psychology of Mathematics Education* (pp. 2-296 - 2-303). Program committee of the 22nd PME conference: Stellenbosch, S.A.

Forgasz, H. J., Leder, G. C., & Gardner, P. L. (1999). The Fennema-Sherman 'Mathematics as a male domain' scale re-examined. *Journal for Research in Mathematics Education, 30*(3), 342-348.

Frykholm, J. A. (1999). The impact of reform: challenges for mathematics teacher preparation. *Journal of Mathematics Teacher Education*, 2(1), 79-105.

Gellert, U. W. E. (1999). Prospective elementary teachers' comprehension of mathematics instruction. *Educational Studies in Mathematics, 37*(1), 23-43.

Gopnik, A., & Meltzoff, A. N. (1997). *Words, thoughts and theories*. Cambridge, Massachusetts: MIT Press.

Guttman, L. (1967). A basis for scaling qualitative data. In M. Fishbein (Ed.), *Readings in attitude theory and measurement* (pp. 96-107). New York: John Wiley & Sons. (Reprinted from *American sociological Review*, 1944, *9*, 139-150.)

Halloran, J. D. (1970). *Attitude formation and change*. Leicester: Leicester University Press.

Jensen, J. H., Niss, M., & Wedege, T. (Eds.), (1998). *Justification and enrolment problems in education involving mathematics or physics*. Roskilde, Denmark: Roskilde University Press.

Kiesler, C. A., Collins, B. E., & Miller, N. (1969). *Attitude change: A critical analysis of theoretical approaches*. New York: John Wiley & Sons.

Leder, G., & Forgasz, H. (1998). Tertiary mathematics: Perceptions of school leavers and mature-age students. In C. Kanes, M. Goos, & E. Warren (Eds.), *Teaching mathematics in new times* (pp. 311-318). Mathematics education Research Group of Australasia Incorporated: Griffith University, Brisbane.

Lemon, N. (1973). *Attitudes and their measurement*. London: B. T. Batsford.

Likert, R. (1967). The method of constructing an attitude scale. In M. Fishbein (Ed.), *Readings in attitude theory and measurement* (pp. 90-95). New York: John Wiley & Sons. (Excerpted from the Appendix of 'A technique for the measurement of attitudes', *Archives of Psychology*, 1932, No.140, pp. 44-53.)

Lokan, J., Ford, P., & Greenwood, L. (1996). Maths & science on the line: Australian junior secondary students' performance in the Third International Mathematics and Science Study. Melbourne: Australian Council for Educational Research.

McLeod, D. B. (1992). Research on affect in mathematics education: A reconceptualization. In D. A. Grouws (Ed.), *Handbook of research on mathematics teaching and learning* (pp. 575-596). New York: MacMillan.

Mueller, D. J. (1986). *Measuring social attitudes: A handbook for researchers and practitioners*. New York: Teachers College Press.

National Council of Teachers of Mathematics. (2000). *Principles and standards for school mathematics*. Reston, VA: Author

Osgood, C. E., Suci, G. J., & Tannenbaum, P. H. (1970). Attitude measurement. In G. F. Summers (Ed.), *Attitude measurement* (pp. 227-234). Chicago: Rand McNally. (Reprinted from C. E. Osgood, G. J. Suci, & P. H. Tannenbaum, *The measurement of meaning*, 1957, (pp. 189-199). University of Illinois Press.)

Perry, B., Howard, P., & Tracey, D. (1999). Head teachers' beliefs abut the learning and teaching of mathematics. *Mathematics Education Research Journal, 11*(1), 39-53.

Rokeach, M. (1972). *Beliefs, attitudes and values*. San Fransisco: Jossey-Bass Inc.

Schoenfeld, A. H. (1992). Learning to think mathematically: Problem solving, metacognition, and sense making in mathematics. In D. A. Grouws (Ed.), *Handbook of research on mathematics teaching and learning* (pp. 334-370). New York: MacMillan.

Schuck, S. (1999). Teaching mathematics: A brightly wrapped but empty box. *Mathematics Education Research Journal, 11*(2), 109-123.

Senger, E. S. (1999). Reflective reform in mathematics: The recursive nature of teacher change. *Educational Studies in Mathematics, 37*(3), 199-221.

Sloman, A. (1987). Motives, mechanisms and emotions. *Cognition and Emotion, 1*(3), 209-216.

Thurstone, L. L. (1967). Attitudes can be measured. In M. Fishbein (Ed.), *Readings in attitude theory and measurement* (pp. 77-89). New York: John Wiley & Sons. (Reprinted from *Journal of Sociology*, 1928, *33*, 529-554.)

Vacc, N. N., & Bright, G. W. (1999). Elementary preservice teachers' changing beliefs and instructional use of children's mathematical thinking. *Journal for Research in Mathematics Education, 30*(1), 89-110.

Verschaffel, L., De Corte, E., Lasure, S., Van Vaerenbergh, G., Bogaerts, H., & Ratinck, E. (1999). Learning to solve mathematical applications problems: A design experiment with fifth graders. *Mathematical Thinking and Learning, 1*(3), 195-229.

Walkerdine, V. (1998). *Counting girls out* (new edition). London: Falmer Press.

CHAPTER 7

DOUGLAS B. MCLEOD AND SUSAN H. MCLEOD

SYNTHESIS - BELIEFS AND MATHEMATICS EDUCATION: IMPLICATIONS FOR LEARNING, TEACHING, AND RESEARCH

Abstract. Recent research on beliefs has made considerable progress in explaining the processes involved in mathematics learning and teaching. Beliefs have strong relationships to both affective and cognitive processes that are important in mathematics education. The chapters in this section are reviewed with regard to the varied definitions of the term "belief," the difficulty of reaching a consensus on one definition and the general agreement on the core commonalties of the construct. A framework for classifying and using different types of definitions is proposed, and methods and implications for research are discussed.

1. INTRODUCTION

Research on the learning and teaching of mathematics continues to make progress in illuminating the factors that influence students and teachers. As this field of research develops, certain key ideas regularly play a crucial role in our understanding; the importance of beliefs related to mathematics education is one such key idea. Pehkonen and Törner (1996) provide a useful overview of how beliefs came to be seen as a central topic in the field.

During the 1980s, research on problem solving in mathematics made substantial progress in identifying crucial aspects that limited student performance on complex mathematical tasks (Schoenfeld, 1985). Student beliefs were identified as one of those crucial aspects, and much of the current research in the area stems from that research focus on problem solving that developed out of the work of Schoenfeld (1985), Silver (1985), and others.

But other approaches to beliefs had stimulated significant research on more traditional mathematical tasks. As early as the 1970s Fennema (1989), in her remarkably successful and sustained research program on gender-related differences, investigated the role of beliefs about mathematics and their influence on student achievement. Although Fennema's research program focused on the affective domain, the instruments that she developed gathered useful data on beliefs about mathematics and learning that are still part of our focus today.

Researchers examining mathematics teaching during the 1970s and 1980s were also trying to identify factors that constrained the performance of teachers. As

G. C. Leder, E. Pehkonen, & G. Törner (Eds.), Beliefs: A Hidden Variable in Mathematics Education? 115-123.

Thompson (1992) tried to sort out the connection between teacher knowledge and teacher performance, she was led to a focus on teacher beliefs, an approach that has had a significant impact on the direction of research on mathematics teaching. As Thompson pointed out, the difficulties that are involved in changing teacher performance are intimately connected with what teachers believe, as well as with what they know.

When the influence of cognitive science on educational research was near its peak during the 1980s, the application of information processing theories of cognition to issues in problem solving was a popular approach. The relation of the affective domain to problem solving performance was an interesting topic for both of us, and we each pursued it in our own way (McLeod & Adams, 1989; S.H. McLeod, 1997). For McLeod and Adams, the focus was affective influences on mathematical problem solving; for S. H. McLeod, the focus was the relation of affect to writing as a problem solving process. Again, as we tried to understand how affect influenced student performance on tasks in mathematics and in written composition, we ran into the importance of beliefs in student learning. For a summary of how beliefs became central in both lines of research, see D. B. McLeod (1992) and S. H. McLeod (1997).

Our interest in beliefs has centered on their role in linking affective and cognitive processes. In mathematics, we are fortunate that much has happened in this area since the summary included in McLeod (1992). First of all, we have the chapters that are included in this volume, which are discussed in more detail below. Goldin's chapter, in particular, has expanded in significant ways our understanding of the linkages between the affective and cognitive domains. We also have a number of other significant efforts that expand on the connections among beliefs, affective issues, and mathematical performance. These contributions come from a number of different countries, an encouraging sign of international interest in deepening and strengthening the research on beliefs, affect, and cognition in mathematics education.

From England, a new book by Evans (2000) emphasizes social aspects of emotion and cognition related to mathematics learning and beliefs about mathematics. He uses a variety of methods and perspectives to examine adults' mathematical thinking and emotional responses in school and in everyday contexts. His research suggests that we should broaden the notion of "context" to include the two fundamental meanings of the term suggested by Wedge (1999): *task-context* (the linguistic features or wording of the task and the assumption a student must make in order to solve the problem mathematically) and *situation-context* (the social, historical, psychological and other circumstances in which problem solving and learning takes place). His discussion of discourse practices and the variety of subject-positions (with regard to authority or institutional authority, social difference, social relations, power) that students can take enlarges our understanding of the importance of emotion as related to mathematical problem solving.

In Finland, Malmivuori (2001) has developed an extensive theoretical framework to guide research on beliefs, affect and mathematics learning. Her central concern is to explain the dynamic interaction of affect and cognition in relation to the learning processes involved in mathematics education. In her analysis, she considers all major areas of research related to the affective domain, including causal attributions, self-efficacy, and self-confidence, and how the research in these areas is related to mathematics learning. Some of these topics were also summarized briefly by D. B. McLeod (1992), but the treatment by Malmivuori (2001) is much deeper, more satisfying, and more useful to the field.

Researchers in several other countries have also produced significant advances in the study of beliefs and affect in mathematics learning. From Spain, Gómez Chacón (2000) considered the ways in which affect influences student performance. She proposes a model for studying the interaction between cognition and affect in mathematics education, focusing on both the local and the global affect structures that students possess. Japanese researchers have also recognized that the study of cognition can be enriched through connections to the study of affect (Hatano, Okada, & Tanabe, 2000). In relation to mathematics learning, Isoda and Nakagoshi (2000) demonstrated how researchers can use measures of physiological change (e.g., heart rate) as indicators of emotional change during problem solving lessons in mathematics. In the United States Goldin (2000) has continued his insightful research on how affect influences mathematical problem solving, and his latest work is included in the chapters we discuss below. From Australia we note the influential plenary address by Leder (1993), with its timely focus on reconciling affective and cognitive approaches to research on mathematics learning. No doubt the presentation of this address in an international conference in Japan helped to stimulate the attention that has been devoted to research on beliefs, affect, and mathematics learning around the world in recent years.

These contributions are all useful in understanding the impact of beliefs on mathematics education. More general treatments of these issues are also of interest in the research community. For example, Goleman (1998) continues his emphasis on emotional intelligence, and Damasio in his recent book (1999) speculates about the biology and neurology of knowing and feeling. Although these contributors come from outside of mathematics education, their work contributes to our general understanding of beliefs, affect, and the teaching and learning of mathematics.

The chapters in this section of the present volume contribute more specifically to our understanding of beliefs and their influence in mathematics education. These five chapters have a number of common themes, including three that we address now: the difficulty of establishing a clear definition for the term "belief" (illustrated by the varied definitions in the literature) and the related need for a common understanding of the term, key issues involved in research studying beliefs and belief systems (e.g., methodology), and implications for further research.

2. DEFINITIONS OF "BELIEF"

Let us first examine the issue of definition of terms. The chapters in this section list and discuss the more prevalent definitions of the term "belief" found in the literature, pointing out the variations and commonalties; several attempt a single definition themselves in an attempt to bring about consensus on how we might define the term. One might ask why consensus on a definition is necessary, especially if, as Leder and Forgasz argue in their chapter, useful work can be done without full agreement about the precise definition of the construct under discussion. In his chapter, Günter Törner points to one important reason: clarification of terminology helps determine research focus. In scientific contexts terms play a functional role - they help us to define areas of needed research and to pose pertinent research questions. Research studies can build on other research and help the field come to cumulative understanding only if there is agreement about what the research object is. If, as Törner tells us Pajares claims (1992, p. 309), the terms beliefs, values, attitudes, judgments, opinions, ideologies, perceptions, conceptions, conceptual systems, preconceptions, dispositions, implicit theories, and perspectives have been used almost interchangeably in the literature, then research studies based on such differing definitions will not have much to contribute to one another. Törner proposes that we approach the issue of definition by examining the core characteristics of beliefs. Much the same idea is put forward in the chapter by Furinghetti and Pehkonen; the authors selected nine characteristics related to belief (taken from the literature) and asked a panel of experts to respond, agreeing or disagreeing, to arrive at some shared understanding of the term. What they found was that there was agreement on certain identifiable commonalties in the construct called "belief". With this sort of general agreement on the core characteristics of the construct, it would seem that forming a clear, concise definition of "belief" would not be difficult. Yet so far at least, there is no real agreement on a single definition of the term in the literature, not even in the present volume.

We propose that one of the difficulties involved in coming to agreement about a definition has been that the *types* of definitions as well as the definitions themselves differ. Experts in the field of technical and professional communication distinguish among several separate types of definitions, each appropriate for a particular kind of audience. Three are appropriate for the present discussion:

1. Informal definitions - these are "rule of thumb" definitions (often used parenthetically) for a general audience.
2. Formal definitions - these consist of three parts: the term to be defined, the class of objects or concepts to which it belongs, and the distinguishing characteristics that separate it from all other objects or concepts in the class. (The class should be small, just large enough to include all members of the term, but no larger. Hence it is not helpful in a formal definition to use the broad term "conception" as the class to which beliefs belong.) The intended audience is more sophisticated in its understanding of technical terms than a general audience, but is still a broad one.
3. Extended definitions - these start with a formal definition, but continue in more technical language to include more complete characteristics and instantiations of the term. The intended audience is specialists in a particular field (e.g., mathematicians writing for other mathematicians). This is perhaps the sort of definition that Furinghetti and Pehkonen have in mind when they say that they dropped the idea of a multipurpose characterization suitable for all fields and instead referred to a given context and specific population.

With this classification in mind, we can see that most of the definitions proposed for the term "belief" (here and elsewhere) are in fact useful for different audiences and purposes, and that one may choose from a range of types, depending on the situation - for, say, an explanation to beginning students, to a group of faculty colleagues from different disciplines, or to researchers in the field of Mathematics Education. Let us demonstrate by examining several of the definitions quoted or proposed in this section of the book.

An early definition of belief systems, quoted by Furinghetti and Pehkonen, comes from Schoenfeld (1985): "Belief systems are one's mathematical world view". Technically speaking, this is not a definition at all, but a metaphor. The comparison gives us an image of what belief systems are and how globally they operate. In that way, the metaphor functions much as an informal definition would - to clarify the concept for a general audience. Furinghetti and Pehkonen's characterization of belief as "subjective knowledge" is a true informal definition, one that could be used to explain the construct to a lay audience using terms the audience would find familiar.

Goldin's chapter provides us with an example of a formal definition of beliefs: "multiply-encoded cognitive/affective configurations, usually including (but not limited to) propositional encoding, to which the holder attributes some kind of *truth value*". The class and distinguishing characteristics are clear, and the terminology familiar enough so that researchers in the sciences and social sciences would find it useful. (A similar formal definition was worked out collectively by the attendees at the 1999 Oberwolfach workshop on Mathematical Beliefs and their Impact on Teaching and Learning of Mathematics, adding "including associated warrants" after "truth value".) Such a definition might easily be used to guide research studies in several fields and promote useful discussion across disciplines. Finally, Törner proposes the following definition in his chapter: "A belief B constitutes itself by a quadruple $B = (O, C_0, \mu_i, \Sigma j)$, whereby O is the debatable belief object, with C_0 representing the content set of mental associations (what traditionally and shortened is called a belief), μ_i modeling the membership degree function(s) of a belief and Σ_j denoting the evaluation maps". Törner goes on to list five characteristics that B should fulfill in a probabilistic sense. To understand the definition completely (e.g., to understand what is meant by "membership degree functions" or "evaluation maps") one must read the entire chapter, which is itself an extended definition. It is written in language that specialists - mathematicians and mathematics educators - are familiar with, and would be useful to researchers in those domains.

These chapters are not the last word on attempts to define beliefs. Indeed, other chapters in this volume will also clarify in their own ways the meaning of the term. Just as Thompson (1992) used the term "conceptions" to broaden the thinking of researchers and to extend the domain of research interest, writers in the future will shape the research using terminology that meets their needs. Philipp (in preparation), for example, discusses beliefs as theories at some points in his analysis, and elsewhere as a sort of lens through which students view mathematics and learning. We believe such flexibility is not only acceptable, but even advisable.

To summarize, we propose that there is no single definition of the term "belief" that is correct and true, but several types of definitions that are illuminative in different situations (for different audiences and purposes). If the purpose is explanation to an audience of non-specialists, an informal definition is appropriate; if the purpose is research, then a formal or extended definition is more useful.

3. KEY ISSUES AND IMPLICATIONS FOR RESEARCH

The chapters in this section focus on a number of relevant issues having to do with research in the field of Mathematics Education. We consider two of the most important: methodological approaches and proposed models of students' beliefs with regard to mathematics. Leder and Forgasz summarize and discuss various qualitative and quantitative methods for measuring attitudes and beliefs, some of which use questionnaires (Likert-scales or checklists asking respondents to rate or to agree/disagree with statements or respond to a stimulus), some involving physiological measures (like galvanic skin response), some involving interviews and/or observations of subjects. They suggest another approach, using the Experience Sampling Method (ESM), developed some 30 years ago by Mihaly Csikszentmihalyi at the University of Chicago. This method involves extensive monitoring of a person's activity over an extended period of time, relying on periodic reports prompted by an electronic pager. Leder and Forgasz demonstrate the usefulness of this approach by reporting data from a study of mature age students of mathematics. These students were asked to record their beliefs associated with factors relevant to their engagement in university work. The approach, as Leder and Forgasz show, can be used in conjunction with other forms of data collection to examine students' belief systems in impressive detail. The ESM has been used by other researchers to examine students' emotional states while they responded to school writing assignments (Larson, 1985), and to study relationships among interest, achievement motivation, mathematical ability, the quality of experience when doing mathematics, and mathematics achievement (Schiefele & Csikszentmihalyi, 1995). As Leder and Forgasz suggest, this approach is useful in examining students' beliefs as well.

Op 't Eynde, de Corte, and Verschaffel discuss the lack of clarity with regard to the available models for students' beliefs. These authors point out that, as with the issue of definitions, there is no agreement about the structure and content of the relevant categories of student beliefs. After an analysis of the nature and structure of beliefs, they propose a theoretical framework that integrates the major components of other discussions. Specifically, they propose a model that differentiates among the students' beliefs about mathematics education, their beliefs about the self, and their beliefs about the social context, including particularly the context of the mathematics classroom. This analysis is very comprehensive and extremely helpful. It provides a useful foundation on which the field can build and make progress, a particularly useful advance in our thinking.

Leder and Forgasz provide a useful overview of a variety of research methods that can be used to investigate beliefs and related topics. For an extensive treatment of qualitative and quantitative research methods relevant to mathematics education, see also the work of Kelly and Lesh (2000). In particular, the Kelly and Lesh volume discusses strategies for using clinical methods with new technologies (e.g., video) in ways that should be very useful in research on beliefs.

Although research on beliefs continues to make substantial progress in explaining the difficulties that students have in understanding mathematics, there is still much that needs to be done. We still have colleagues who look narrowly at cognitive processes, neglecting to observe and account for those dimensions of learning that go beyond the purely cognitive. In a recent publication from the National Academy of Sciences in Washington, DC, the authors note how the research has focused mainly on cognitive issues related to school mathematics learning (Kilpatrick, Swafford, & Findell, 2001). Research in mathematics education has often lagged behind the reform efforts in the schools, and this report is a reminder of how much work is left to be done to make sure that beliefs and related issues receive sufficient attention from the research community.

One area where research has made substantial progress is in recognizing the important role that the social context plays in shaping student beliefs and student learning. In fact, the pendulum has moved so far from a focus on the individual learner to the learner in a social context that it appears that some researchers think learning by an individual is unlikely, if not impossible. A moment's reflection on our own learning should remind us that a good deal of learning occurs individually. As we examine the current interest in distance learning and innovative uses of new technology, we see perhaps new ways in which individual learners acting independently can learn effectively. Clearly the social context plays an important role in learning, but also it seems worthwhile to investigate how students learn individually, as well as in groups.

Our research methods often focus on an analytical approach where we take beliefs and break them down into smaller categories in order to investigate the specific aspects that seem to be the most important to the learning and teaching of mathematics. Generally this approach is quite satisfactory and serves us well. However, there are times when a more holistic approach seems quite effective as well. For example, Ivey (1996) approached her study of an algebra classroom by taking students who had different approaches to mathematics, and describing them in terms of their worldview. Students whose view of the world was mechanical were often dealing with only surface aspects of the mathematical concepts that their teacher was presenting. However, Ivey's case study of one student showed a pattern of understanding that she termed "organic"; this student's approach to mathematics led him to a deeper understanding of the content through a more integrated view of the subject. Again, we recommend that a multitude of approaches to research can be useful, including those that are more holistic, as well as those that are more analytic.

Lubienski (2001) has taken a different approach to research on beliefs that seems particularly promising. In her analysis of data from the National Assessment of Educational Progress, Lubienski examined how students' ethnicity and social class were related to disparities in achievement and beliefs related to mathematics. Her insightful analyses suggest how new approaches to research on beliefs may help us understand the persistent achievement gap between advantaged and disadvantaged students in different ethnic groups.

Finally, we have been impressed by the usefulness of Goldin's notion of meta-affect, and more generally by the emphasis that he puts on the linkages between the cognitive and affective domains. His discussion of how the affective reactions of students are linked to beliefs, and how these belief structures are stabilized by meta-affect, provides a significant advance in our thinking on these topics. Moreover, these advances undertaken by Goldin are significantly enhanced by the recent work of Malmivuori (2001). In her discussion of self-regulation and the general research on motivational factors, Malmivuori provides a broad theoretical framework that links research on beliefs, affect, and mathematics learning to the larger context of educational research, including new aspects of personal agency and social learning environments.

These ideas advance our thinking in terms of theory, but they have practical applications as well. The ways that we teach students to be aware of their beliefs may have more implications for changing those beliefs and improving student learning than anything else that we do. Certainly we need more research on strategies to help students monitor the way that beliefs and affect may influence their performance in mathematics classes, but these ideas already provide a reasonable framework for improving how we think about pedagogical matters.

As we consider the implications of this research area for improving instruction in mathematics classrooms, we believe it is time for more extensive studies that evaluate the theories outlined by Evans (2000), Malmivuori (2001), and others. We know of only one major attempt to help teachers improve their understanding of how beliefs and affect might influence student performance. Ellerton and Clements (1994), with the support of the Australian Association of Mathematics Teachers, developed materials on beliefs and affect that were designed to improve the knowledge of teachers in mathematics classrooms. These professional development materials were based on the research and theory available at the time, and provided a reasonable way to test certain hypotheses about how teachers might help students change their beliefs and affective responses with regard to mathematics. Further efforts to improve classroom instruction are surely needed, and we hope that the recent advances in theory can be tested and refined in classroom studies, as well as in more carefully controlled laboratory studies.

In closing, we note that the chapters in this section have made an important contribution to our understanding of beliefs and their role in the teaching and learning of mathematics. Later sections of this volume will expand on these topics. We look forward to the continuation of this research and to seeing its impact in mathematics classrooms.

4. REFERENCES

Damasio, A. (1999). *The feeling of what happens: Body and emotion in the making of consciousness.* New York: Harcourt.

Ellerton, N., & Clements, K. (1994). *Fostering mathematical attitudes and appreciation.* Adelaide: Australian Association of Mathematics Teachers.

Evans, J. (2000). *Adults' mathematical thinking and emotions: A study of numerate practices.* London: Routledge.

Fennema, E. (1989). The study of affect and mathematics: A proposed generic model for research. In D.B. McLeod & V.M. Adams (Eds.), *Affect and mathematical problem solving: A new perspective* (pp. 205-219). New York: Springer-Verlag.

Goleman, D. (1998). *Working with emotional intelligence.* New York: Bantam.

Goldin, G. A. (2000). Affective pathways and representation in mathematical problem solving. *Mathematical Teaching and Learning* 2, 209-219.

Gómez Chacón, I. M. (2000). Affective influences in the knowledge of mathematics. *Educational Studies in Mathematics, 43,* 149-168.

Hatano, G., Okada, N., & Tanabe, H. (Eds.) (2000). *Affective minds.* New York: Elsevier.

Isoda, M., & Nakagoshi, A. (2000). A case study of students' emotional change using changing heart rate in problem posing and solving Japan (sic) classrooms in mathematics. In T. Nakahara & M. Koyama (Eds.), *Proceedings of the 24th International Conference for the Psychology of Mathematics Education* Vol. 3 (pp. 87-94). Hiroshima, Japan: Hiroshima University.

Ivey, K.M.C. (1996). Mechanistic and organic: Dual cultures in a mathematics classroom. *Journal of Research and Development in Education, 29,* 141-151.

Kelly, A. E., & Lesh, R.A. (Eds.) (2000). *Handbook of research design in mathematics and science education.* Mahwah, NJ: Lawrence Erlbaum Associates.

Kilpatrick, J., Swafford, J., & Findell, B. (Eds.) (2001). *Adding it up: Helping children learn mathematics.* Washington, DC: National Academy Press.

Larson, R. (1985). Emotional scenarios in the writing process: An examination of young writers' affective experiences. In M. Rose (Ed.), *When a writer can't write: Studies in writer's block and other composing-process problems* (pp. 19-42). New York: Guilford.

Leder, G. (1993). Reconciling affective and cognitive aspects of mathematics learning: Reality or a pious hope? In I. Hirabayashi et al. (Eds.), *Proceedings of the 17th Annual Conference for the Psychology of Mathematics Education* Vol. 1 (pp. 46-65). Tsukuba, Japan: University of Tsukuba.

Lubienski, S. T. (2001, April). *Are the NCTM Standards reaching all students? An examination of race, class, and instructional practices.* Paper presented at the Annual Meeting of the American Educational Research Association, Seattle.

Malmivuori, M. L. (2001). *The dynamics of affect, cognition, and social environment in the regulation of personal learning processes: The case of mathematics.* University of Helsinki, Department of Education. Research Report 172.

McLeod, D B. (1992). Research on affect in mathematics education: A reconceptualization. In D.A Grouws (Ed.), *Handbook of research on mathematics teaching and learning: A project of the National Council of Teachers of Mathematics* (pp. 575-596). New York: Macmillan.

McLeod, D. B. & Adams, V.M. (Eds.) (1989). *Affect and mathematical problem solving: A new perspective.* New York: Springer-Verlag.

McLeod, S. H. (1997). *Notes on the heart: Affective issues in the writing classroom.* Carbondale, IL: Southern Illinois University Press.

Pajares, M.F. (1992). Teachers' beliefs and educational research: Cleaning up a messy construct. *Review of Educational Research 62,* 307-332.

Pehkonen, E. & Törner, G. (1996). Mathematical beliefs and different aspects of their meaning. *International Reviews on Mathematical Education (ZDM) 28* (4), 101 - 108.

Philipp, R. (in preparation). *Preservice teachers' beliefs for teaching mathematics.* Center for Research in Mathematics and Science Education, San Diego State University.

Schiefele, U. and Csikszentmihalyi, M. (1995). Motivation and ability as factors in mathematics experience and achievement. *Journal of Research in Mathematics Education, 26,* 163-181.

Schoenfeld, A.H. (1985). *Mathematical problem solving.* Orlando, FL: Academic Press.

Silver, E. A. (1985). Research on teaching mathematical problem solving: Some underrepresented themes and needed directions. In E.A. Silver (Ed.), *Teaching and learning mathematical problem solving: Multiple research perspectives* (pp. 247-266). Hillsdale, NJ: Erlbaum.

Thompson, A.G. (1992). Teachers' beliefs and conceptions: A synthesis of the research. In D.A. Grouws (Ed.), *Handbook of research on mathematics teaching and learning: A project of the National Council of Teachers of Mathematics* (pp. 127-146). New York: Macmillan.

Wedge, T. (1999). To know or not to know—mathematics, that is a question of context. *Educational Studies in Mathematics, 39,* 205-227.

PART 2

TEACHERS' BELIEFS

CHAPTER 8

MELVIN (SKIP) WILSON AND THOMAS COONEY

MATHEMATICS TEACHER CHANGE AND DEVELOPMENT

The Role of Beliefs

Abstract. The chapter focuses on impacts of mathematics teachers' beliefs on their ability and tendency to change. A discussion of common definitions of belief precedes a review of reports about elementary, middle, and high school mathematics teachers' beliefs from three international research journals (1995-1999). Implications of the reports' often implicit definitions of beliefs are considered, as well as other implications for teacher change. For example, an assumption that beliefs are dispositions to act in certain ways, rather than simply verbal proclamations, leads to data collection methods that involve both discussions with and observations of teachers. The reports also point out the need to reconsider a tendency to separate teachers' mathematical and pedagogical beliefs

1. INTRODUCTION

Research on the teaching and learning of mathematics has taken many twists and turns throughout the years. During the late 1960s and early 1970s the primary emphasis of research was on students' learning, particularly that of young children, and often evoked a strong Piagetian flavor. Little emphasis was given to how teachers influence learning through their instructional programs or to how instructional programs were influenced by what teachers believed about mathematics or its teaching. Most research on teaching mathematics that was conducted in the 1970s used a behavioristic framework that shaped both questions and methodologies. In general, various characteristics of teaching were quantified and correlated with specific learning outcomes. Although this line of research led to a few interesting results in terms of revealing high inference variables that were significantly correlated to learning, e.g., clarity and flexibility, in general, this line of research was limited. There was concern that the mathematics being assessed was too narrow and that the notions of clarity, flexibility, and the like defied definition in a mathematics classroom.

In the late 1970s there was an emerging methodological crisis that suggested researchers were not focusing on what was really significant about the teaching of mathematics. There was, for example, a brief foray into investigations about teachers' attitudes toward mathematics and teaching. This research was generally

G. C. Leder, E. Pehkonen, & G. Törner (Eds.), Beliefs: A Hidden Variable in Mathematics Education? 127-147.

well received but it lacked a cognitive component and ultimately failed to be free of the shackles of the dominant behaviorist paradigm. Still, there was a sense that in order to understand the teaching and learning of mathematics one must contend with something other than well-defined variables and a tightly quantifiable notion of teaching behavior.

Perhaps spurred by Kuhn's (1962) classic work *The Structure of Scientific Revolution*, combined with the movement toward teaching experiments borne out of Russian methodologies in mathematics education, sense-making became a central theme for much of the research during the 1980s. Some researchers shifted away from viewing teaching and learning from an ontological perspective toward one in which an individual's construction of meaning was paramount. Given the diverse forces that were influencing research in mathematics education, there evolved a greater sense that the context in which teaching occurred influenced what was being taught and learned in the classroom. The notion of context became recognized not just as the physical arrangement of classrooms but also of the teachers' beliefs about mathematics and its teaching. Over the past 15 years, there has been a considerable amount of research on teachers' beliefs based on the assumption that what teachers believe is a significant determiner of what gets taught, how it gets taught, and what gets learned in the classroom.

Thompson (1992) provided an extensive review of research on teachers' beliefs and conceptions. One thrust of her review was about the impact of teachers' beliefs on change. Thompson understood, as many authors point out in this volume, that understanding teachers' beliefs is vital to reform. However, Thompson's review was a decade ago and a considerable amount of research has occurred since that review. Additionally, recent research has focused even more extensively on the implications of teachers' beliefs for change. Like many of the chapters in this section, we focus on the impact of teachers' beliefs on their ability to grow, change, and develop teaching practices consistent with reform recommendations (e.g., National Council of Teachers of Mathematics [NCTM], 1989, 2000).

The review that follows involves articles from three mathematics education journals: *Educational Studies in Mathematics (ESM)*, the *Journal for Research in Mathematics Education (JRME)*, and the *Journal of Mathematics Teacher Education (JMTE)*. Articles about mathematics teachers' beliefs or teacher change that were published from 1995 through 1999 constituted the data set for our analysis. We have chosen to review articles from these three journals, given their international readership and the fact that they provide important outlets for researchers concerned with teaching or teacher education.

To set a context for our review, we present a characterization of beliefs and link this notion to the concept of teacher change. We then present a summary and analysis of several sets of recent empirical studies that focus on elementary and secondary mathematics teachers' beliefs or changes in their teaching of mathematics. We conclude our discussion with an analysis of how this research can impact future research in mathematics teacher education.

Previous research on teachers' beliefs has focused primarily on the empirical question of what teachers believe about mathematics, the teaching of mathematics, and the learning of mathematics. Often unnoticed in this research, however, is what is meant by an individual's belief or how one's beliefs relate to teacher change. This neglect can be attributed to the fact that the primary emphasis has usually been on trying to understand why teachers make the decisions they do without getting mired down on disputing various constructs of teachers' beliefs. Nevertheless, at this point in time, it may be helpful to consider just what has been examined under the umbrella of research on teachers' beliefs and consider the implications for what different, often implicit, definitions have for our subsequent research and development activities. Consequently, one of our intentions here is to review and analyze different research reports in an effort to understand what researchers seem to be investigating. Our intent in doing so is to provide a context in which we can examine the notion of teacher change, a notion seldom addressed in the 1980s and early 1990s in mathematics education. We maintain that researchers' definitions, often implicit, influence what they investigate and consequently what they find.

2. DEFINITIONS AND COMMON CHARACTERIZATIONS OF BELIEFS

The words *belief* or *believe* have many meanings in common usage. "I believe I'll go the movies." is more a statement of intent than a claim. On the other hand, "I believe it is cold outside." is a claim about weather conditions. This claim could be based on considerable evidence, e.g., the thermometer outside our window or a recent weather forecast. If a person states that he/she believes there is life in other galaxies, we might consider this claim based on evidence that is not verifiable. In contrast, a claim that you know something is to make a claim based on certain logic or evidence that appears irrefutable. If I say, "I know that lead is heavier than wood." the assumption is that this claim has been verified by me or others. On the other hand, if I state, "I know I can break the record." I am stating a claim that has not yet been verified. In some sense, I am misusing the word "know" since my claim is not based on irrefutable evidence.

All of this is to say that a claim that we know something is a stronger claim than to say that we believe something. Belief is usually seen as a construct that has a cognitive component but is a weaker condition than knowing. Scheffler (1965) claimed that X knows Q if and only if

i. X believes Q

ii. X has the right be to sure Q

iii. Q

Accordingly, X believing Q is a necessary condition for X knowing Q. Indeed, it would be strange to say that I know that wood floats but I don't believe it. On the other hand, it would not be strange to say that I believe wood is lighter than plastic, but I don't know it.

In order for X to have the right to be sure that Q is the case, there must be reasonable evidence to support the existence of Q, that is, criterion ii. The evidence may, of course, point us in the wrong direction but some evidence must exist. We can wrongly believe that the universe is geo-centered or that "bleeding" is an appropriate medical treatment. But to know that these propositions are true, evidence must exist - as historically we know was the case. The last criterion, the actual existence of Q, is a tricky criterion. How can we be sure that Q actually exists? Science has revealed that the universe is not geocentered and that bleeding is a dangerous medical treatment. But can we be so sure that what now appears to be the case is, in fact, the case? The famous book *Flatland* (Abbott, 1963) cautions us about thinking that our perspective constitutes reality. The very essence of constructivism is that we can never know reality as such but rather we construct models that have viability (Von Glasersfeld, 1991) for describing the world in which we exist. For the constructivist, we can never determine that condition iii holds. Rather, we can speak with confidence, but not certainty, that Q holds. This perspective suggests that criteria ii and iii ought to be combined into a single condition that might be stated as follows.

iiR (revised). X has reasonable evidence to support Q

Even here, we are not out of deep water because we have the question of what constitutes evidence. What is evidence for some is not for others. Nevertheless, condition iiR avoids the trap of trying to determine what an individual knows or believes based on some perceived absolute reality.

The logic of the above explication suggests that belief is a necessary but not sufficient condition for knowing. What does it mean for me to say, "I believe life is good." or that "I believe the essence of mathematics is problem solving." or that "I believe it is cold outside". Most of us would agree that replacing the word "believe" with "know" would result in a questionable proclamation. Further, it would be strange to act in a way that seemed counter to one's stated belief. For example, if a person really believed it was cold outside, it seems unlikely that he or she would venture outside for an extended period of time wearing only shorts and a tee shirt. We might say, "He says he believes it is cold outside but I don't believe he means it." which suggests a certain insincerity or inconsistency. Scheffler (1965) addressed this point when he wrote,

> A belief is a cluster of dispositions to do various things under various associated circumstances. The things done include responses and actions of many sorts and are not restricted to verbal affirmations. None of these dispositions is strictly necessary, or sufficient, for the belief in question; what is required is that a sufficient number of these clustered dispositions be present. Thus verbal dispositions, in particular, occupy no privileged position vis-à-vis belief. (p. 85)

Based on this definition we encounter difficulties with our present research on beliefs. According to Scheffler, the determination of one's beliefs requires a variety of types of evidence including not only what a person says but also what the person does. What do we conclude, for example, when a teacher steadfastly maintains that the essence of mathematics is problem solving, yet we see only procedural knowledge being emphasized in the classroom? Typically, the researcher claims that there exists an inconsistency between the teacher's belief and his or her practice. But other interpretations exist. Consider the following possibilities:

1. We do not have a viable interpretation of what the teacher means by problem-solving.

2. The teacher cannot act according to his or her belief because of practical or logistical circumstances.

3. The teacher holds the belief about problem solving subservient (or peripheral in Green's (1971) terms), to the belief that the teaching of mathematics is about certainty and procedural knowledge.

Acceptance of any one of these three alternative interpretations would not lead us to a conclusion about inconsistency but rather would require us to understand in a deeper sense how the teacher constructs meaning.

We adopt the view that believing is a weaker condition than knowing. Thus, the locution "I believe the sum of the angles of a triangle is 180 but I don't know it." is a perfectly normal way of speaking. But if one says that "I just proved that the sum of the measures of a triangle is 180 but I don't believe it", phrases sometimes heard in classrooms by students who "distrust" proofs, we take this to be a colloquial form of speech rather than a logical consequence. For the purpose of the review that follows, we take authors' claims about teachers' beliefs to be based on evidence and made in good faith based on that evidence. Nevertheless, we do not take those statements to be based on irrefutable evidence.

3. CONNECTING BELIEFS AND TEACHER CHANGE

When the emphasis of research shifts towards a sense-making perspective, boundary lines between knowing and believing become blurred as we seek to understand the phenomena of teacher change and what drives that change. One of the consequences of this blurring is the use of terms such as "conception" or "cognition" by many researchers in place of "belief." Many of the reports we analyze do this, perhaps, in part, to avoid the truth conditions that tend to separate knowledge and beliefs. Research that quantifies and correlates beliefs with changes in classroom behavior can be very useful. Indeed, as some of the studies we describe illustrate, this type of research can provide significant insights into what teachers value and the relative importance they assign to different aspects of mathematics or the teaching of mathematics. Nevertheless, in-depth studies of individuals emphasize the value of telling stories about teachers' professional lives and what shapes those lives. From a research perspective, good stories are not simply descriptions but are grounded in theoretical constructs that have the power to explain what is described. We will now offer a glimpse as to the kinds of theoretical orientations that we consider important and helpful in the telling of those stories.

Given the uncertainty that exists between criteria ii and iii in Scheffler's definition of knowledge, we are faced with the condition that knowing is a relative condition necessarily dependent on the context in which one lives. This is not to say that such words as *true* and *false* have no meaning. Rather these terms, when applied to the human condition, must be tempered with the realization that we all see the world differently. Because of this, knowing is a construct that depends on circumstances and context. Accordingly, the professional development of teachers can be envisioned as progressing from seeing the doing of mathematics and the teaching of mathematics in dualistic terms to seeing these activities as contextual.

To say it another way, we might say that teacher change consists not solely of changes in classroom behavior but of conceptualizing in a relativistic way the very act of teaching. We contend that this perspective can enable us to conceptualize how teachers come to understand and appreciate the relative flexibility of their knowing. For many teachers, a dualistic orientation toward mathematics leads to an emphasis on product, such as the acquisition of procedures, without accompanying meaning. Similarly, a dualistic orientation toward teaching mathematics leads to an instructional style that is determined a priori and is dominated by telling to help insure the certainty of classroom events. In short, preconceived instructional styles are necessarily insensitive to the contexts in which they are used - contexts that are determined, in part, by what students know and believe about mathematics. In contrast, a relativistic view of mathematics has an emphasis on process and leads to a dynamic view of mathematics. A relativistic view of mathematics teaching is based on the context of teaching, the most important of which is student understanding. From a relativistic perspective, the question becomes less of whether an activity or teaching strategy is good or bad per se and more a question of the context in which an activity facilitates student learning.

This line of reasoning has led to the use of theoretical perspectives that attend to ways of knowing, i.e., that consider an individual's personal journey from a dualistic orientation to a relativistic one. This includes schemes developed by Perry (1970), Belenky, Clinchy, Goldberger, and Tarule (1986), Baxter-Magolda (1992), and King and Kitchener (1994). Each of these schemes provides a basis for conceptualizing one's beliefs and how those beliefs are related to the way in which the individual comes to know. Cooney, Shealy, and Arvold (1998) and Wilson and Goldenberg (1998) have posited similar schemes that are specific to mathematics education. Consistent with these schemes are the perspectives offered by Green (1971) and Schön (1983). These theoretical orientations offer a means for describing and shaping our ways of thinking about teacher development. If we characterize reform-oriented teaching as that teaching which attends to context, including basing instruction on what students' know, then teaching becomes a matter of being adaptive (Cooney, 1994) rather than a matter of using a particular sequence of instructional strategies.

The development of a reform-oriented teacher so characterized, is rooted in the ability of the individual to doubt, to reflect, and to reconstruct. Teacher education and mathematics teaching in general then become a matter of focusing on reflection and on the inculcation of doubt in order to promote attention to context. This opens new vistas for creating situations in teacher education in which teachers can develop a reflective posture toward their teaching. This requires considerable maturity with respect to the knowledge domains of mathematics, pedagogy of mathematics, and student learning. (See Lappan & Theule-Lubienski, 1994.) A limited view of mathematics, preconceived teaching strategies, and a reductionist view of learning clearly work against the development of a reflective teacher. Consequently, our teacher education programs should model the kind of knowledge development that we expect teachers to promote for their students. For otherwise, we will find ourselves mired in a significant moral dilemma as our medium and our message are inconsistent if not incoherent. Many of the studies we analyze in the next section support this view.

4. RECENT REPORTS OF RELATIONSHIPS BETWEEN MATHEMATICS TEACHER CHANGE AND BELIEFS

Our review consists of articles about the teaching of mathematics at the elementary, middle school, and high school levels. We found that the most prominent issue in the recent literature about elementary teachers' beliefs involved their views about mathematics learning, particularly about how their students learn. There was very little attention to teachers' mathematical knowledge per se. Perhaps this is because there had already been a great deal of research revealing elementary teachers' weak understandings of mathematics. On the other hand, literature related to secondary teachers' beliefs has placed little emphasis on the interplay between students' and teachers' beliefs or teachers' beliefs about pedagogical principles. In summary, we found different emphases in research with different grade level teachers. We examine the literature to consider the researchers' emphases in promoting reform-oriented teaching and what factors seem to facilitate the development of that kind of teaching.

4.1. Research Involving Elementary Teachers

We begin with a set of investigations by Terry Wood, Erna Yackel, Paul Cobb, and others during the 1990s at Purdue University. Although we focus on two specific articles from *JRME* (Wood & Sellers, 1997; Wood, 1999), these reports are but a small part of the extensive research that has been reported and the emerging theoretical orientations based on that work.

Wood and Sellers (1997) focused on third-grade students' mathematical understandings and explored the pedagogical beliefs communicated by teachers in what the authors referred to as "problem-centered classes" (p.163). Wood (1999) discussed a second-grade teacher's actions during classroom discussions involving student argumentation. (For a report on students' conceptions in classrooms where this kind of teaching occurs see Wood (1996) and Wood and Sellers (1996).) This research communicates the importance of teachers having a child-centered or constructivist view of learning. Included in the authors' view of constructivism is the belief that students learn mathematics most effectively when they construct meanings for themselves, rather than simply being told. One important factor in teachers' successful use of child-centered strategies is an awareness of the originality of children's thinking. Additionally, Wood (1999) contended that "teachers need to understand and to some extent accept the fundamental tenets that distinguish a constructivist theory of learning from other theories" (p.171).

The primary tenet discussed by Wood (1999) is Piaget's notion of cognitive confusion and contradiction, a tenet many teachers have difficulty accepting and thus implementing. Wood emphasized the importance of both individual constructions and the social context created in classrooms that promote such constructions. In addition to expecting disagreement and helping students to deal with these situations in appropriate ways, e.g., by listening and responding politely, Wood illustrated how the teacher was able to establish mathematical norms that enabled students to not only disagree, but discuss significant mathematical ideas which subsequently led to a deeper understanding of mathematics. Wood referred to this as an atmosphere of "inquiry" (p. 187). Wood and Sellers (1997) referred to a similar environment as "problem-centered" (p.163) and illustrated how such an atmosphere had positive effects on student attitudes and learning.

These reports provide evidence that understanding by students, defined as the ability to construct rational, meaningful solutions to meaningful problems, is enhanced by such an inquiry approach by mathematics teachers. A teacher's emphasis on the nature of children's cognitive development, with special attention to ways in which children experience and deal with cognitive conflict, is the primary focus of this research. A teacher's inquiry orientation to student conceptions contrasts with a more traditional view of children's understanding as being right or wrong. Wood does not focus on the issue of correctness but rather on the importance of teachers listening to students and encouraging debate among them. This emphasis on students' understanding of mathematics requires that teachers decenter from their own activities and base their instruction on others' (students') thinking. The notion of decentering is rooted in one's ability to think contextually or relativistically. Teachers whose beliefs are more dualistically oriented would likely experience difficulty in attending to students' mathematical understanding.

Wood and Sellers (1997) found that teachers who volunteered to participate in a problem-centered program had initial pedagogical beliefs more consistent with a constructivist philosophy than with a philosophy typically associated with traditional teaching. However, after participating in project activities for one year, these teachers' pedagogical beliefs, determined by a written questionnaire, were even more substantially different than those of traditional teachers. Wood and Sellers (1997) suggested that, "teachers shift their beliefs, not only about teaching but also about what mathematics is and what it means to do mathematics, in a direction proposed by reform initiatives" (p.180). Together, these reports indicate that teachers' beliefs about reform are enhanced by the teachers' participation in reform-oriented instructional settings. In short, teachers' beliefs can change when they are provided opportunities to consider and challenge those beliefs.

A strand of research commonly referred to as Cognitively Guided Instruction (CGI) focused on the teaching of elementary students by teachers trained to attend to their students' mathematical understandings. This project began in the mid 1980s at the University of Wisconsin under the direction of Tom Carpenter and Elizabeth Fennema (see, e.g., Carpenter, Fennema, Peterson, & Carey, 1988). CGI has subsequently spread to many other locations. For example, Vacc and Bright (1999) reported on elementary preservice teachers' changing beliefs and their use of children's mathematical thinking in their instruction based on data collected in North Carolina. Like Wood and Sellers (1997), Vacc and Bright (1999) administered a questionnaire that assessed teachers' beliefs about mathematics, teaching, and learning at four times over a 2-year period during the interns' teacher education program. This survey was the primary means of assessing participants' beliefs, although journal entries, interviews with, and observations of selected participants were also analyzed. The survey measured individuals' beliefs about the role of the mathematics learner, the relationship between skills and understanding, the sequencing of topics, and the role of the teacher. Responses of the 34 preservice teachers indicated very little change during the first year of the study, but substantial change during the second year when the teachers participated in a field-based methods course and student teaching. This suggests the importance of time in teachers changing their beliefs. In the methods course interns also participated in CGI sessions in which specific elementary problems, typical solutions, and solution strategies of children were studied. Survey responses reflected changes in the following areas: increased consensus that children are able to construct their own knowledge (rather than being recipients), a stronger belief that skills should be taught in relationship to understanding and problem solving rather than in isolation, a stronger belief that sequencing of topics should be based on children's understanding and not on the structure of formal mathematics, and an increased belief that mathematics instruction should facilitate children's construction of knowledge rather than teachers simply presenting information. The authors characterized this shift as a more firmly held "constructivist orientation about the learning of mathematics" (p. 103). They attributed this shift to strong attention to mathematics pedagogy and to a research-based model (CGI) in the methods class (before student teaching).

Such research is valuable in helping us understand what preconceptions teachers hold prior to interacting with students and how their professed beliefs evolve as they engage students. But, again, we must consider Scheffler's (1965) analysis that responses, written or oral, tell only part of the story about what one believes. Recall that Scheffler considered beliefs as propensities to act in certain ways. This includes both reactions to surveys and interviews, and actions within the classroom. We have no quarrel with the assumption underlying the studies reported here that beliefs can be assessed through written surveys so long as those author-interpreted beliefs are seen as tentative interpretations of beliefs and are inherently limited in terms of predicting action in the classroom. One can describe the kinds of beliefs found by Vacc and Bright (1999) as stated, claimed, or espoused beliefs. As such they can provide us important clues as to what the teacher might do in the classroom but they do not capture *what* the teacher does in the classroom. Vacc and Bright (1999) studied two of the interns more closely during their student teaching, observing and commenting on the interns' teaching activities. But it was clear that beliefs were seen as professed and separate from actions. This is not to disparage research that focuses on espoused beliefs. As Hart explains in this volume, espoused beliefs often provide a valuable window into understanding an individual's experiences. However, stories only about espoused beliefs are largely unfinished stories.

Schifter (1998) explored the experiences of a sixth grade teacher as that teacher explored mathematical content from the elementary curriculum. Based on her study, Schifter offered suggestions for the teaching of that curriculum. Specifically, she considered the impact on teachers teaching mathematical inquiry in grades K-6 and of the teachers' analyses of children's mathematical thinking. Shifter provided an analysis of a set of events in a sixth-grade class that dealt with a problem students agreed to solve using fraction division:

> Nancy has 6 2/3 meters of material. It takes 5/6 of a meter to make her fabulous fancy hair ribbons. How many fabulous fancy hair ribbons can she make? (p. 59)

The ensuing episode demonstrated the nature of the espoused beliefs among students and the complexity of students' understanding of fraction division, particularly with respect to identifying the unit involved. Schifter pointed out that teachers often ignore the complexity of the situation as they are unaware of its significance for assessing student understanding. As Schifter's work illustrates, teachers who strive to adopt reform by encouraging discussion and even debate among students are forced to confront their own mathematical understandings. Had the teacher in question not understood that the unit changes, depending upon one's perspective and interpretation of the situation, students' comments might simply "derail" the lesson.

Schifter's (1998) study illustrates the importance of attending to actions in addition to descriptions. Schifter's work is about the beliefs teachers hold about learning. But what it really points out is the importance of seeing the implications of those beliefs in the classroom, namely the importance of decentering and thinking contextually. Thus we see two kinds of information being provided by Vacc and Bright (1999) and by Schifter (1998). The former informs us about teachers' beliefs as revealed through written surveys substantiated in part by observation. The latter provides a more detailed analysis of a single teacher's beliefs and the potential naivety of teachers' interpretation of complex learning situations, a naivety not easily detected from a single data source.

4.2. Research Involving Middle Grade Teachers

Sowder et al. (1998) provide an extensive analysis of the understandings of multiplicative structures teachers should have to teach middle grade mathematics. The authors develop four sets of recommendations for the preparation and professional development of middle school teachers. One group of recommendations relates to the role of problem solving. The authors contend that teachers should be given opportunities to solve complex and difficult problems. They argue that unlike simple problems, whose solutions can often be obtained by identifying key words in the problem statement, complex problems require teachers to understand the mathematical procedures and concepts involved in problem solutions. Some of the same arguments have been made about appropriate mathematical activity for students. Sowder and her colleagues make a strong case about the appropriateness of such activities for teachers.

In describing another of their recommendations, Sowder and her colleagues (1998) pointed out that moving from additive to multiplicative structures is cognitively demanding for both teachers and students. This demand is due partly to the fact that multiplicative reasoning, like fractions, requires reconceptualization of the idea of unit. The complexity of this and other middle grade mathematical concepts supports the authors' contention about the importance of conceptual understanding. Briefly stated, the authors contend that: (1) like students, teachers need to explore a wide range of realistic problems and make sense of quantitative relationships and other mathematical ideas for themselves, and (2) conceptual understanding and sense-making should be emphasized for teachers as well as for students. Like the Schifter article (1998), Sowder et al. (1998) illustrate the complexity of a seemingly simple idea and the importance of flexible understandings by teachers so that they can make sense of, and deal appropriately with, a variety of student reactions. What is important about the Sowder et al. work is that it provides a model to facilitate teachers understanding students' mathematics. It provides a context for teachers to analyze students' work beyond the simplicity of right versus wrong, should teachers intend to engage in such analysis.

Another strand of research related to middle school mathematics teachers' beliefs is by Raffaella Borasi and others at the University of Rochester (e.g., Borasi, Siegel, Fonzi, & Smith, 1998; Siegel, Borasi, & Fonzi,1998; Borasi, Fonzi, Smith, & Rose, 1999). In particular, Borasi et al. (1999) illustrate the importance of inquiry in teachers' attempts to implement innovative practices. As Wood and Sellers (1997) demonstrated for elementary teachers, and as was emphasized in recent U.S. reform recommendations (e.g., NCTM, 1989, 2000), it is important for middle school mathematics teachers to provide student-centered instruction for their students. Borasi and her colleagues described a professional development program that introduced participating teachers to and engaged them in an inquiry approach to mathematics instruction.

Borasi's program, like CGI and other programs that focus on students' thinking, was designed to help teachers align their beliefs and practices with reform recommendations that emphasize problem solving, conceptual understanding, and appreciation for mathematics. This involved helping teachers from grades 5-8 plan instructional activities related to tessellations and area during a week-long summer institute. Teachers implemented their summer plans during the following academic year. During the implementation period, teachers were supported by peers at their own or at nearby schools with whom they met weekly and by university personnel. Borasi, et al. (1999) claimed that the experiences were successful in "*initiating* the process of rethinking beliefs and practices" (p. 63, emphasis in the original). Most of the 39 teachers who completed the program not only used the 2 common "required" inquiry units in their classes, but also designed and delivered other similar lessons. Although the authors pointed out that there was considerable variation among participating teachers, the observed and reported experiences "represented a substantial step forward toward implementing the vision for school mathematics articulated in the NCTM (1989) Standards" (Borasi et al., 1999, p. 63)..

Borasi et al. (1999) indicated that this project confirmed their belief about the importance of teachers experiencing innovative activities as learners before implementing them as teachers. Nevertheless, the success experienced by teachers who implemented activities similar to the ones they had completed as students the summer before did not eliminate the need for classroom support during implementation. The authors concluded that providing teachers with the time and opportunity for reflection both as students and as teachers of inquiry-based mathematics was not sufficient to ensure the successful implementation of ideas. Rather, the authors argued, it is important to provide specific reflection prompts for participants. During the summer workshops the teachers were asked to (a) reconstruct key components of the activities they experienced, (b) write down things that helped or hindered their learning, and (c) discuss what they wrote with other teachers in their group. These specific reflections required teachers to critically examine their experiences, moving beyond mere recollection. What seemed critical in the Borasi et al. program was the continual contact with the teachers and the reinforcing nature of the program, a context that likely promoted further reflective thinking on the part of the teachers. Another critical factor in the Borasi et al. program was the participation of 3 or 4 teachers from the same school. The resulting socialization process not only provided increased opportunities for peer support, but it provided a certain resilience against factors that might otherwise undermine using inquiry-based methods.

What becomes obvious from these studies involving middle grade teachers is that the mathematical and pedagogical ideas with which they are dealing go beyond the simplistic ideas that often characterize teachers' beliefs. Further, in order to internalize these more complex phenomena, teachers need some sort of networking (Cooney & Krainer, 2000) in order to solidify the ideas. Without a support group of some type or significant planned opportunities to reflect, the ideas will likely vanish or render themselves impotent in the classroom. This perspective has significant implications for teacher education. Too often we see individual teachers as the units of analysis. In fact, perhaps the unit of analysis should be the group of teachers as they participate in their professional development programs. It is in this sense that the idea of networking and mutual support seems so central to success in implementing reform. This is consistent with claims by Hart in this volume - her teachers identified collaboration and colleagues as major factors in their abilities to change.

The studies involving middle grade teachers also illustrate the risk of inferring beliefs based on snapshots of professed beliefs. The research reveals the complexity associated with changes in classroom behaviors and with the beliefs that accompany, if not drive, those behaviors. Without the benefit of multiple data sources, it would be difficult to ascertain how beliefs, including both professed and classroom-oriented aspects of beliefs, changed. We have here a potentially significant research agenda. How is it that we could describe middle school teachers' professed beliefs when juxtaposed against their actions? Would we conclude that one provides a foundation for the other? Or would we address these two components of beliefs as being either consistent or inconsistent? If the latter, what rationale could we provide to account for this apparent inconsistency?

4.3. Research Involving High School Teachers

There have been many studies involving high school teachers that focus on the interplay between teachers' beliefs and their instructional decisions. For example, Even and Tirosh (1995) illustrated how a preservice teacher in Israel explained that 4/0 is undefined because division by zero is not allowed. When asked how she would explain why this is so to her students, she responded, "In mathematics we have rules . . . These rules often do not seem reasonable. When studying mathematics, one has to adopt these rules and operate accordingly" (p. 10). As Even and Tirosh pointed out, this teacher viewed mathematics as a "bag of unexplainable rules" (p. 10) and she expected her students to unquestioningly memorize those rules. Like many of the other articles dealing with mathematics teachers' beliefs, this article illustrates how a weak mathematical understanding can severely limit teachers' abilities to use innovative and student-centered instructional strategies. Even and Tirosh also asked teachers to react to student responses to specific mathematical tasks. These reactions focused on anticipating common student mistakes. The authors pointed out how identifying the correctness of student responses is dependent on teachers' content knowledge, but that strong mathematical knowledge by itself is not enough to guarantee that teachers envision the complexity of ideas with which students must grapple.

Like the CGI study discussed previously, the issue of correctness dominated Even and Tirosh's (1995) analysis. Not only did examples of teachers' mathematical knowledge focus on whether or not teachers correctly understood procedures and definitions (e.g., the univalence requirement for functions), but the authors' portrayal of teachers' reactions to student interpretations also concentrated on the accuracy of teachers' interpretations. For example, teachers commonly (and incorrectly) speculated that a student's mistake involving the relationship between slope and angle of the graph of a linear function was due to an error in estimation. Although such an interpretation did not accurately describe the student's thinking, and initial accurate guesses by teachers are certainly preferred, we believe that a critical factor in mathematics teaching also involves teachers' flexibility in considering multiple possibilities. Unlike some of the previous authors cited whose focus was on the importance of student constructions of less well-defined concepts such as debate and problem solving, Even and Tirosh (1995) focused on the match between student constructions and accepted mathematical definitions and procedures.

In another study Even (1999) emphasized the dialectical nature of high school mathematics teachers' beliefs and practices and illustrated how changing teachers' beliefs can lead to changes in teaching practices. This report also pointed out the negative feelings that secondary mathematics teachers sometimes feel toward researchers and mathematics education research because it is not practical enough. Even discussed how this perceived gap can be bridged. As in the Even and Tirosh (1995) article, Even's article focused on the importance of helping teachers learn to base their instructional decisions on student thinking. Perhaps because teachers often seek "reliable and relevant rules . . . that can be put to immediate use" (Even, 1999, p. 236), strong pedagogical content knowledge was characterized primarily in terms of teachers' ability to accurately interpret students' errors.

Lloyd and others focused on high school teachers in the United States who attempted to implement reform-oriented curriculum materials (Lloyd 1999; Lloyd & Frykholm, 2000; Lloyd & Wilson, 1998; Wilson & Lloyd, 2000). In this body of work, Lloyd and her colleagues explored relationships between teachers' beliefs and their abilities to successfully implement reform curricula. Lloyd's chapter in this volume further elaborates many of her claims about the potential impact of curriculum materials and teachers' beliefs about curriculum. Lloyd and Wilson (1998) described the "content conceptions" of secondary mathematics teachers. For example, one teacher communicated deep and integrated understandings that supported his successful implementation of curriculum materials that emphasized multiple representation, application, and covariation in its presentation of functional relationships. The authors' presentation of the teacher's mathematical understandings (initially assessed through an interview) and his classroom implementation of a unit about functions, illustrated many of the claims made by other researchers about the importance of a deep and flexible understanding of mathematical ideas. The teacher's own preference for thinking about functions and functional relationships was more consistent with traditional portrayals of functions as formal objects, e.g., sets of ordered pairs that satisfy the univalence condition. However, his ability to also see functions as covariations enabled him to adopt an unfamiliar classroom approach that emphasized patterns and relationships, real-world applications, and multiple representation in instruction about functions (as opposed to determining the functionality of relations). It also allowed him to implement more student-centered activities, such as small-group discussions, than he had previously used. In short, the teacher's understanding of mathematics facilitated his implementation efforts.

The Lloyd (1999) study illustrated more extensively how teachers' beliefs about pedagogical principles impact their teaching. Lloyd described how curriculum materials designed to encourage student cooperation and exploration were used by two high school teachers. One of the teachers viewed the curriculum's problems as very open-ended and challenging for students; the other teacher claimed that the same problems were overly structured. The two teachers' classroom decisions reflected these contrasting perspectives. As Lloyd stated, "curriculum implementation consists of a dynamic relation between teachers and particular curricular features. This notion is consistent with warnings that reform recommendations and associated curriculum materials cannot and do not bring about change alone - educational change is a complex human endeavor" (p. 244). This article goes beyond the claim that traditional beliefs often "interfere" with teachers' abilities to implement novel curricula. In fact, Lloyd concluded that, "When a reform-minded teacher uses traditional materials in the classroom, he or she may be afforded more room for personalization because the goals of the materials are so different from his or her own goals" (p. 246).

A third set of studies about high school mathematics teachers by Artzt (1999) and by Artzt and Armour-Thomas (1999) focused on both preservice and inservice teachers' abilities to reflect on their teaching, a prominent concept in many recent theoretical and empirical discussions of mathematics teacher change. In these articles, beliefs were defined as "assumptions regarding the nature of mathematics, of students, and of ways of learning and teaching" (Artzt, 1999, p. 145). Knowledge involved an awareness of the content, pedagogical strategies, and students' understanding of mathematics. In both studies, participants' beliefs and knowledge were determined using written and interview data, both before and after teaching, as well as multiple observations of their teaching. To avoid the possibility of participants "giving lip-service to their beliefs about student-centered instruction"

(p. 148), information about beliefs was based on post-lesson interviews in which teachers discussed video-taped segments of lessons they had taught. Such a methodological choice is consistent with Scheffler's (1965) definition of beliefs as dispositions to act in certain ways. Artzt's (1999) examples illustrate the power of written and verbal reflection on teaching events. For example, one preservice teacher recorded after a lesson that she was not satisfied with her students' procedural knowledge. This teacher sensed an underlying conceptual weakness, which she successfully addressed in subsequent lessons. Artzt claimed that this teacher's ability to meet her students' needs was facilitated by her written reflection. Verbal reflections consisted primarily of conferences with teachers in which they were asked questions about their lessons.

The Artzt and Armour-Thomas (1999) article described a framework for conceptualizing relationships between teachers' cognitions (knowledge, beliefs, and goals) and their instructional practice. The authors referred to this as the *Phase-Dimension Framework (PDF)* because the two main categories in it are *Lesson Phases and Lesson Dimensions*. This model enabled the researchers to break observed lessons into segments and to analyze systematically those lesson segments. Not surprisingly, teacher behaviors that promoted students' conceptual understandings, across all phases and dimensions, were consistent with teacher cognitions that valued student understanding. In contrast, the cognitions of a second group of teachers, who were much less likely to stress conceptual understanding with their students, were more extensively focused on their own practices rather than on student learning. The implications of this result are interesting given the emphasis in the Artzt (1999) article on reflection about teaching practice. Taken together, these studies suggest that reflection on teaching practice can be an extremely positive thing, especially when it is focused on student understanding. Although Artzt and her colleagues do not explicitly address the notion of beliefs, their studies suggest that they see teachers' beliefs as a broad phenomenon encompassing many kinds of behavior.

5. DISCUSSION AND IMPLICATIONS

In addition to those previously discussed, there are several themes that emerged from our review that deserve attention particularly as they relate to the general topic of teachers' beliefs and their impact on change. The themes that are addressed in this section are: (1) reflection, (2) teachers' abilities to attend to students' understanding, and (3) content versus pedagogy.

5.1. The Importance of Reflection

A prominent theme in nearly every article we reviewed, as well as in many of the chapters in this volume about teachers' beliefs, was that of reflection. The term even appeared in the titles of the Artzt (1999) and the Artzt and Armour-Thomas (1999) articles. The Schifter (1998), Vacc and Bright (1999), and the Borasi et al. (1999) studies focused on teachers reflecting on their own beliefs and teaching behaviors.

These studies illustrate how reflection on beliefs allows teachers to connect their thoughts and actions, to recognize and perhaps confront contradictory or otherwise problematic beliefs, and, particularly, to change their teaching behavior. What is especially striking about these studies is that the importance attributed to teachers' ability to be reflective is unquestionably linked to teacher change. This circumstance suggests that reflective teachers' ways of knowing are relativistic in nature. Reflective thinking about teaching requires that teaching be viewed as contextual. Reform then becomes a matter of altering the context of instruction.

As many of the articles included in this review pointed out, context includes a different mode of teaching mathematics and a different instructional environment, e.g., one that involves inquiry-based instruction or the use of cooperative learning groups. Sometimes teacher change is temporary and superficial and fails to realize the intended reform, at least from the perspective of researchers. What is often missing from these failed efforts is a basic shift in beliefs about what constitutes an appropriate role for the teacher of mathematics. The second teacher in the Lloyd (1999) study exemplifies this failure. Although her teaching practices illustrated many things commonly characterized as innovative, Ms. Faye maintained a relatively traditional view about curriculum materials and textbooks.

Many of the CGI studies deal with the issue of teachers' beliefs in a more implicit way. The investigators probably impact on their teachers' beliefs if we assume that beliefs have something to do with propensities to act in certain ways. It seems unlikely that fundamental changes in teaching could occur without fundamental changes in accompanying beliefs. Although it might be more difficult upon reading CGI studies to conclude that beliefs have a fundamental role in effecting teacher change, it does seem reasonable to conclude that the infusion of information housed in a reflective environment influences teacher change and likely teachers' beliefs.

Another theme communicated in some of the studies is that focused, specific reflection is necessary to avoid merely recalling past events and experiences. Both Borasi et al. (1999) and Lloyd (1999) emphasized that in order to accommodate change, teachers need first-hand experiences working on specific innovative investigations and activities that they are attempting to use in their classrooms. These experiences as both students and teachers influence what teachers ultimately think and do. But in order for these actions to be internalized, they must be made the objects of explicit actions that promote reflection about the materials and their use. This suggests that attention to beliefs about teaching in general may result in certain peripheral beliefs (Green, 1971) but lack connections to more centrally held beliefs. It is through the act of reflecting on specific events that those centrally held beliefs can be affected in fundamental ways.

5.2. Teachers' Ability to Attend to Students' Understanding

One of the primary tenets of reform-oriented teaching is that teachers base their instructional approaches on what and how students are thinking about the mathematics being taught. As shown explicitly by Artzt and Armour-Thomas

(1999), teacher reflection about student understanding is particularly powerful in terms of helping teachers connect their instruction to the work students are doing. Even (1999) similarly communicated the importance of helping teachers learn to base their instructional decisions on student thinking. Both Schifter (1998) and Sowder et al. (1998) emphasized how teachers' understandings can enable them to deal effectively with a variety of potential student reactions.

As previously pointed out, some of the articles reviewed in this chapter deal with the correctness of teachers' thinking and actions. For example, Vacc and Bright (1999), operating in a CGI context, concentrated on whether teachers' decisions and actions were research-based, thus indicating some sort of fundamental rationale for teachers' interpretations of students' work. Even and Tirosh (1995) analyzed the implications of the accuracy of teachers' interpretations about students' understanding of mathematics. We certainly do not want students to do incorrect mathematics or teachers to use ineffective techniques. However, if correctness is judged against some absolute (mathematical or pedagogical) standard, such comparisons run the risk of being dualistic.

Mathematical correctness has the potential for blurring students' thinking that lead to "mistakes." Much can be learned by taking into consideration those thought processes and circumstances that lead to common "misconceptions". As illustrated by Wood (1999), and Wood and Sellers (1997), the resolution of cognitive conflict helps students gain deeper mathematical understandings. It is important for teachers to develop teaching strategies that support such a belief. For example, as these authors point out, encouraging debate among students not only helps teachers become more sensitive to student understandings, it emphasizes the importance of inquiry in the development of understanding.

5.3. Content Versus Pedagogy

The reports summarized here about elementary teachers focused on teaching and learning processes (e.g., inquiry, problem solving, student construction, argumentation), while the secondary reports dealt more extensively with issues involving mathematical content. One oddity of this trend is that it seems reasonable to assume that secondary teachers have stronger and more flexible mathematical understandings and that elementary teachers have more constructivist-oriented views of learning. During the 1970 and 1980s there was a considerable amount of research that measured elementary teachers' mathematical understandings (or lack of understanding). Although it is helpful for researchers to focus on only one thing (content or pedagogy), it might be time to revisit elementary teachers' mathematical knowledge as it relates to important pedagogical processes. Similarly, an increased focus on secondary teachers' pedagogical beliefs and strategies might provide valuable information. A limited focus on a single set of teachers' cognitions, be it mathematical or pedagogical, can be misleading.

This point is illustrated by the two articles described in this chapter that were both about the same teacher. An examination of only Mr. Allen's mathematical conceptions (Lloyd & Wilson, 1998) did not provide a complete picture of Mr.

Allen. Although his deep and flexible mathematical understanding helped him successfully implement student-centered activities, his beliefs about exploration and cooperation inhibited his innovation (Lloyd, 1999). Although Mr. Allen's understanding of mathematics and his beliefs about mathematics were not isolated from his beliefs about the teaching of mathematics, some teachers, either implicitly or explicitly, hold clusters of beliefs apart from one another. If teachers do hold separated, clustered beliefs, it is likely that these clusters of beliefs will remain isolated unless explicitly brought into focus. For Mr. Allen there were connections between two sets of beliefs, but his concerns for how to run a classroom, e.g., management issues, often overpowered his willingness to engage in more reform-oriented kinds of teaching. Alternative explanations such as these require empirical evidence for substantiation but they do highlight the direction in which such an investigation might be directed.

Ernest (1991) claimed that teachers' knowledge of mathematics is important but does not account for every difference observed in the classroom. His perspective helps explain why many researchers focus on teachers' beliefs in an effort to account for these differences. But the separation of knowledge and beliefs is a tricky issue, as was pointed out earlier. In addition, such a separation often results in incomplete descriptions. To learn what a person knows without knowing something about the structure of that knowledge leaves much to question. Shulman's (1986) characterization of the terms content knowledge and pedagogical content knowledge addresses this issue. A framework that builds on Shulman's work and other theoretical ideas was designed specifically to help describe mathematics teachers' knowledge and beliefs, that is, what Wilson and Lloyd (2000) call teachers' conceptions. Teachers' conceptions, according to these authors, include the categories mathematical authority and pedagogical authority. Both categories emphasize the importance of teachers sharing authority regarding intellectual issues with their students and developing teaching strategies that promote this sharing. This construct of teachers' conceptions forces researchers to attend to both mathematical and pedagogical issues.

5.4. Some Observations

There does not appear to be a consensus about what constitutes beliefs or whether they include or simply reflect behavior. Generally speaking, neither does there seem to be agreement about the notion of teachers' conceptions or teachers' cognitions. However, regardless of whether one calls teacher thinking beliefs, knowledge, conceptions, cognitions, views, or orientations, with all the subtlety these terms imply, or how they are assessed, e.g., by questionnaires (or other written means), interviews, or observations, the evidence is clear that teacher thinking influences what happens in classrooms, what teachers communicate to students, and what students ultimately learn. Each of the articles discussed in this chapter focused on understanding some aspect of teacher thinking, broadly defined. If one accepts Scheffler's (1965) definition of beliefs as dispositions to act in certain ways, it seems that both observing and interviewing teachers are necessary if one is interested in comprehending how teachers make sense of their worlds. Questionnaire responses represent dispositions to respond to a written stimulus, but they do not constitute strong evidence of what an individual might do when interacting in the classroom. As we have also pointed out, understanding context and developing alternative explanations for phenomena require researchers to dig deeply. What questionnaire data can provide is information across a wide array of teachers, thus offering the possibility of revealing different beliefs among teachers from different grade levels, different geographical settings, or different teaching environments. Given this diversity, it behooves teacher education programs to make explicit those thoughts and actions that are central to teachers' daily practices.

In short, although we agree with Sheffler's characterization of beliefs, we conclude that concern over a precise definition of belief pales in importance compared with the issue of understanding the nature of teachers' thinking and what provides the foundation for teacher change. It might be argued that a more precise definition of beliefs could better shape our research on teacher change and provide a framework in which our research could advance more rapidly. But we question the value of this perspective. We also conclude that attempts to understand teacher change in the absence of understanding the beliefs that encase that change is lacking as well. The studies reviewed in this chapter provide valuable information regarding teacher change. It seems highly speculative that these studies would further enhance our understanding if the authors had provided sharp delineations between knowledge and beliefs. Nevertheless, it might move the field forward to view research on teacher change as one kind of manifestation of beliefs. The literature seems to suggest that the "acting out" of reform-oriented teaching strategies is but one facet of a propensity to act in a certain way. The human condition is always beset with a strange mixture of rationality and irrationality that defy sharp lines of demarcation. If we reject the deterministic methodologies of the past we should hesitate to rush to a methodology that is lacking as well in accounting for teachers' beliefs as they constitute propensities to act in a variety of circumstances.

6. CONCLUSION

There is much to be appreciated about the artistry of teaching and what contributes to that artistry. Certainly, quantification with well-defined constructs can enable us to better understand the teaching and learning of mathematics. But to the extent that our research is about individual's sense-making activities, we must recognize that the individual is driven by many considerations, some of which are amenable to a form of Aristotelian logic but some of which are not. We suggest that neither situation, logical or not, should dissuade us from a certain empiricism that demands the kinds of analysis that reveal insights about how we come to know. We believe the studies reviewed in this chapter contribute to that understanding and through the careful analysis that these authors have provided, we can chart a clearer path of productive research. This task is surely a challenge. But it is a challenge we must face if we are to provide the kind of evidence that informs our education of teachers.

Acceptance of this challenge signals a certain professionalism for both our practice and our research.

7. REFERENCES

Abbott, E. A. (1963). *Flatland: A romance of many dimensions*. New York: Barnes & Noble.

Artzt, A. F. (1999). A structure to enable preservice teachers of mathematics to reflect on their teaching. *Journal of Mathematics Teacher Education, 2*(2), 143-166.

Artzt, A. F., & Armour-Thomas, E. (1999). A cognitive model for examining teachers' instructional practice in mathematics: A guide for facilitating teacher reflection. *Educational Studies in Mathematics, 40*(3), 211-235.

Baxter-Magolda, M. (1992). *Knowing and reasoning in college*. San Francisco: Jossey-Bass, Inc.

Belenky, M., Clinchy, B., Goldberger, N., & Tarule, J. (1986). *Women's' way of knowing: The development of self, voice, and mind.* New York: Basic Books.

Borasi, R, Siegel, M, Fonzi, J., & Smith, C. (1998). Using transactional reading strategies to support sense-making and discussion in mathematics classrooms: An exploratory study. *Journal for Research in Mathematics Education, 29*(3), 275-305.

Borasi, R, Fonzi, J., Smith, C., & Rose, B. J. (1999). Beginning the process of rethinking mathematics instruction: A professional development program. *Journal of Mathematics Teacher Education, 2*, 49-78.

Carpenter, T. P., Fennema, E., Peterson, P. L., & Carey, D. A. (1988). Teachers' pedagogical content knowledge of students' problem solving in elementary arithmetic. *Journal for Research in Mathematics Education, 19*(5), 385-401.

Cooney, T. (1994). Teacher education as an exercise in adaptation. In D. Aichele & A. Coxford (Eds.), *Professional development for teachers of mathematics. 1994 Yearbook of the National Council of Teachers of Mathematics* (pp. 9-22). Reston, VA: National Council of Teachers of Mathematics.

Cooney, T., & Krainer, K. (2000). The notion of networking: The fusion of the pubic and private components of teachers' professional development. In Fernández, M. (Ed.), *Proceedings of the 22nd Annual Meeting of the North American Chapter of the International Group for the Psychology of Mathematics Education* Vol. 2 (pp. 531-537). Columbus, OH: ERIC Clearinghouse for Science, Mathematics, and Environmental Education.

Cooney, T., Shealy, B., & Arvold, B. (1998). Conceptualizing belief structures of preservice secondary mathematics teachers. *Journal for Research in Mathematics Education, 29*(3), 306-333.

Ernest, P. (1991). *The philosophy of mathematics education*. London: Falmer Press.

Even, R. (1999). Integrating academic and practical knowledge in a teacher leaders' development program. *Educational Studies in Mathematics, 38*(3), 235-252.

Even, R., & Tirosh, D. (1995). Subject-matter knowledge and knowledge about students as sources of teacher presentations of the subject-matter. *Educational Studies in Mathematics, 29*(1), 1-20.

Green, T. (1971). *The activities of teaching*. New York: McGraw-Hill.

King, P., & Kitchener, K. (1994). *Developing reflective judgment*. San Francisco: Jossey Bass, Inc.

Kuhn, T. (1962). *The structure of scientific revolution*. Chicago: University of Chicago Press.

Lappan, G., & Theule-Lubienski, S. (1994). Training teachers or educating professionals? What are the issues and how are they being resolved? In D. Robitaille, D. Wheeler, & C. Kieran (Eds.), *Selected lectures from the 7th International Congress on Mathematical Education* (pp. 249-261). Les Presses de L'Université: Quebec.

Lloyd, G. (1999). Two teachers' conceptions of a reform-oriented curriculum: Implications for mathematics teacher development. *Journal of Mathematics Teacher Education, 2*, 227-252.

Lloyd, G. M., & Frykholm, J. A. (2000). How innovative middle school mathematics materials can change prospective elementary teachers' conceptions. *Education, 21*, 575-580.

Lloyd, G. M., & Wilson, M. (1998). Supporting innovation: The impact of a teacher's conceptions of functions on his implementation of a reform curriculum. *Journal for Research in Mathematics Education, 29*, 248-274.

National Council of Teachers of Mathematics (1989). *Curriculum and evaluation standards for school mathematics*. Reston, VA: Author.

National Council of Teachers of Mathematics (2000). *Principles and standards for school mathematics.* Reston, VA: Author.

Perry, W. (1970). *Forms of intellectual and ethical development in the college years.* New York: Holt, Rinehart, & Winston.

Scheffler, I. (1965). *Conditions of knowledge*. Chicago: Scott Foresman and Company.

Schifter, D. (1998). Learning mathematics for teaching: From a teacher's seminar to the classroom. *Journal of Mathematics Teacher Education, 1*(1), 55-87.

Schön, D. (1983). *The reflective practitioner: How professionals think in action.* New York: Basic Books, Inc.

Shulman, L. (1986). Those who understand: Knowledge growth in teaching. *Educational Researcher, 15*(2), 4-14.

Siegel, M, Borasi, R., & Fonzi, J. (1998). Supporting students' mathematical inquiries through reading. *Journal for Research in Mathematics Education, 29*(4), 378-413.

Sowder, J, Armstrong, B, Lamon, S, Simon, M, Sowder, L, & Thompson, A. (1998). Educating teachers to teach multiplicative structures in the middle grades. *Journal of Mathematics Teacher Education, 1*(2), 127-155.

Thompson, A. G. (1992).Teachers' beliefs and conceptions: A synthesis of the research. In D. A. Grouws (Ed.), *Handbook of research on mathematics teaching and learning* (pp. 127-146). New York: Macmillan.

Vacc, N. B., & Bright, G. W. (1999). Elementary preservice teachers' changing beliefs and instructional use of children's mathematical thinking. *Journal for Research in Mathematics Education, 30*(1), 89-110.

Von Glasersfeld, E. (1991). Abstraction, re-presentation, and reflection: An interpretation of experience and Piaget's approach. In L. Steffe (Ed.), *Epistemological foundations of mathematical experience* (pp. 45-67). New York: Springer-Verlag.

Wilson, M. R., & Goldenberg, M. P. (1998). Some conceptions are difficult to change: One middle school mathematics teacher's struggle. *Journal of Mathematics Teacher Education, 1*, 269-293.

Wilson, M., & Lloyd, G. M. (2000). The challenge to share mathematical authority with students: High school teachers reforming classroom roles. *Journal of Curriculum and Supervision, 15,* 146-169.

Wood, T., & Sellers, P. (1996). Assessment of a problem-centered mathematics program: Third grade. *Journal for Research in Mathematics Education, 27*(3), 337-353.

Wood, T., & Sellers, P. (1997). Deepening the analysis: Longitudinal assessment of a problem-centered mathematics program. *Journal for Research in Mathematics Education, 28*(2), 163-186.

Wood, T. (1996). Events in learning mathematics: Insights from research in classrooms. *Educational Studies in Mathematics, 30*(1), 85-105.

Wood, T. (1999) Creating a context for argument in mathematics class. *Journal for Research in Mathematics Education, 30*(2), 171-191.

CHAPTER 9

GWENDOLYN LLOYD

MATHEMATICS TEACHERS' BELIEFS AND EXPERIENCES WITH INNOVATIVE CURRICULUM MATERIALS

The Role of Curriculum in Teacher Development

Abstract. This chapter draws attention to the educative potential of teachers' experiences with the curriculum materials that have been developed in the context of recent efforts to improve K-12 mathematics education in the United States. Examples from two different professional development settings illustrate how teachers' beliefs can change on the basis of experiences with these innovative curriculum materials. Discussion of these examples suggests the need for greater attention to teachers' beliefs about mathematics curriculum.

1. INTRODUCTION

In the United States and many other parts of the world, ongoing curriculum initiatives aim to revise the conventional view of mathematics learning as the mastery of a fixed set of facts and procedures to more centrally locate the processes of investigation, sense-making, and communication in classroom activities. At the heart of such reforms is the distinction between *inquiry mathematics* and *school mathematics* (Cobb, Wood, Yackel, & McNeal, 1992). In contrast with traditional classroom activities that emphasize repetition, practice, and routinized means to some focused endpoint, inquiry mathematics instruction emphasizes student engagement in problem-solving and theory-building about important mathematical situations and concepts.

Bringing about such dramatic changes in mathematics instruction demands that teachers possess beliefs about mathematics, learning, and teaching that depart significantly from school mathematics traditions (Battista, 1994; Thompson, 1992). Perhaps the greatest obstacle for teachers is a lack of personal familiarity with mathematical problem-solving and sense-making – processes that most have never experienced themselves, as students or teachers. Even when teachers' efforts to emphasize inquiry mathematics are supported by specially designed curriculum materials, teachers often struggle to bring about significant changes in classroom practice (Cohen, 1990; Grant, Peterson, & Shojgreen-Downer, 1996; Wilson &

G. C. Leder, E. Pehkonen, & G. Törner (Eds.), Beliefs: A Hidden Variable in Mathematics Education? 149-159.

Lloyd, 2000). Because many teachers' beliefs and practices are deeply tied to school mathematics traditions, the success of current mathematics education initiatives depends on our identification of viable ways to encourage and enable teachers to make significant shifts in their beliefs.

The purpose of this chapter is to draw attention to the educative potential of teachers' experiences with the curriculum materials that have been developed in the context of recent efforts to improve K-12 mathematics education in the United States. Of particular interest is how teachers' beliefs can *change* on the basis of experiences with these innovative curriculum materials. The chapter begins by outlining how teachers' beliefs relate to the visions of instruction advocated in recent years. A subsequent section distinguishes the new curriculum materials from traditional texts. Next, two sections offer discussions and illustrations of two viable contexts for using innovative curriculum materials to help teachers develop reform-oriented beliefs about mathematics pedagogy. A final section calls for greater attention to teachers' beliefs about mathematics curriculum.

2. TEACHERS' BELIEFS AND INQUIRY MATHEMATICS INSTRUCTION

The responsibilities of mathematics teachers are extensive. In one of their influential reform documents, the National Council of Teachers of Mathematics [NCTM] (1991) delineates four categories of mathematics teachers' work: setting goals and selecting or creating mathematical tasks to help students achieve these goals; stimulating and managing mathematical discourse; creating a classroom environment to support teaching and learning mathematics; and analyzing student learning, the mathematical tasks, and the environment in order to make ongoing instructional actions. Teachers' beliefs about and knowledge of mathematical subject matter are critical influences on how they cope with the challenges of classroom instruction (Brophy, 1991; Chapman, this volume; Fennema & Franke, 1992; Thompson, 1992). The review by Wilson and Cooney (this volume) suggests that there is a complex relationship between beliefs and classroom practices. Whereas many teachers learned mathematics by memorizing rules, they must now learn to view rich mathematical understanding as the capacity to use mathematics to reason, to communicate, and to pose and solve meaningful problems (Hiebert et al., 1996; NCTM, 1989, 1991, 2000). Doing so involves learning that a mathematical idea or solution should be judged to be appropriate or correct because it is meaningful and works, not just because the teacher, textbook, or some other outside authority says it is so (Cooney, 1994; Wilson & Goldenberg, 1998; Wilson & Lloyd, 2000).

Mathematics is learned through an active, social process of construction (Cobb, 1995; Davis & Maher, 1990; Mathematical Sciences Education Board [MSEB] and National Research Council [NRC], 1989; von Glasersfeld, 1984). Helping teachers to make sense of constructivist learning theories is a major challenge for those involved with the professional development of teachers. A teacher's beliefs about how students engage in mathematical activity and learn are critical factors in the ability and tendency to design and carry out inquiry-based instruction. Researchers on the Cognitively Guided Instruction (CGI) project have promoted the theory that teacher development involves a fundamental change in the content and organization of teachers' knowledge about children's mathematical thought (Fennema et al., 1996). Teachers need to have a sense of how student understanding develops so that they can anticipate what sorts of mathematical activities will help specific students' learning (Even & Tirosh, 1995). For most teachers, development of this sense will involve a shift in how they conceptualize the mathematical learning process. Understanding learners and subject matter as interactive is one of the most important beliefs about teaching. Ball (1993) describes this conception as a "bifocal perspective – perceiving the mathematics through the mind of the learner while perceiving the mind of the learner through the mathematics" (p. 159). For teachers to appreciate and strive for this relationship is a major challenge of reform-oriented teacher development programs.

Inquiry about the nature of teachers' learning from experiences with innovative curriculum materials relies upon analysis of teachers' mathematical and pedagogical beliefs. Entwined within these beliefs are teachers' beliefs about mathematics curriculum. A better understanding of teachers' beliefs about mathematics curriculum is vital to the success of current reform efforts. Although textbooks have long held prominent roles in guiding practice in American classrooms (Tyson-Bernstein & Woodward, 1991), we know surprisingly little about how teachers' beliefs about curriculum materials relate to their beliefs about mathematics, teaching, and learning, and how they develop during teacher education and school-based experiences.

3. REFORM-ORIENTED MATHEMATICS CURRICULUM MATERIALS

To support teachers in reforming mathematics classroom activity, numerous sets of reform-oriented curriculum materials have been developed in the United States (e.g., Investigations of Number, Data, and Space; Connected Mathematics Project; Core-Plus Mathematics Project; Interactive Mathematics Program; Mathematics in Context; etc.). Although the curriculum materials of these programs incorporate specific aspects of reform recommendations in diverse ways (emphasizing different themes or activities), the materials share certain qualities that distinguish them from traditional mathematics textbooks.

First, reform curricula explicitly incorporate reform ideas about mathematics and pedagogy by emphasizing inquiry mathematics: student explorations of real-world mathematical situations and discussions of problem-centered activities. Furthermore, the materials are formatted to support these mathematical and pedagogical differences. American texts are typically divided into chapters outlining self-contained daily lessons for the teacher to present (composed primarily of definitions and examples of the lesson's content) followed by practice exercises for the student. In contrast, most reform-oriented curriculum materials are published in unit booklets (offering greater flexibility in ordering) that pose large-scale problems and situations, centered on particular mathematical themes and content areas, for students to investigate and debate.

A second substantive difference is that reform-oriented materials generally offer more extensive information for teachers than do traditional texts. In addition to providing problem solutions, the teachers' guides for most of these new materials offer details about different representations of content, historical information about mathematical and pedagogical ideas, examples of what students might believe or understand about particular activities and content, potentially fruitful questions for eliciting discussion, and so on. The inclusion of these details has been motivated in part by the failure of the "teacher-proof" curriculum materials of the 1950s and 1960s to facilitate substantial educational change. After all, it is *teachers* who determine how the innovations envisioned by reformers and curriculum designers become implemented in mathematics classrooms (Cooney, 1988; Freudenthal, 1983).

3.1. Contexts for Teacher Learning with Reform-Oriented Curriculum Materials

There exist many potential ways that teachers can learn from engagement with innovative K-12 mathematics curricula. During inservice workshops or preservice courses at the university, teachers may work collaboratively "as students" on the mathematical lessons outlined in the materials. Doing so can offer teachers critical opportunities as learners because they can personally experience unfamiliar mathematics in novel ways. Another rich context for educative experiences with curricula is teachers' own classrooms. As teachers implement curriculum materials in their classrooms (or as student teachers), they may develop new mathematical and pedagogical beliefs and skills on the basis of their design of lessons, interactions with students, use of technology, and so on. As Llinares (this volume) explains, these contexts (university and school practice) offer quite different situations for learning. The next two sections illustrate how these two contexts offer unique and useful ways for teachers to learn with curriculum materials.

4. TEACHER LEARNING WITH CURRICULUM MATERIALS IN THE MATHEMATICS CLASSROOM

When implementing innovative curriculum materials in K-12 classrooms, teachers are afforded frequent and extensive learning opportunities. Because the act of teaching (regardless of the context - reform or otherwise) is a learning process, instruction necessarily impacts teachers' beliefs. As Ball (1994) describes, teachers continually construct new knowledge from classroom experiences:

> Teachers must figure things out as they teach. They are constantly faced with the data of their own experience. They must develop knowledge of particular children, of the material they are teaching, and of ways to engage students in the content. (p. 9)

The potential for learning from classroom experiences increases as teachers attempt to enact reform visions. Existing beliefs and practices may be directly challenged by the process of interpreting and implementing reform recommendations and curricula. In other words, teachers' sense of efficacy may be brought into question when confronted with novel curricular materials (Philipou & Christou, this volume).

Implementation of innovative materials offers one potentially powerful site for teachers to learn about themselves, their students, mathematics, and the teaching and learning of mathematics (Ball & Cohen, 1996; Lloyd, 1996; Russell et al., 1995).

Consider, for example, the experiences of Mr. Allen, a high school mathematics teacher in the United States who was studied for several years as he taught about functions with reform-oriented curriculum materials (Lloyd, 1996; Lloyd & Wilson, 1998; Wilson & Lloyd, 2000). The Core-Plus curriculum unit that he implemented required students to create and analyze multiple representations (tables, graphs, sentence descriptions, equations) for a variety of real-world situations. In the first year that he used the new materials, Mr. Allen demonstrated a strong preference for graphical representations of functions. His discussions with students reflected this tendency. For example, as he said to his class on one occasion, "The table gives you times and heights, but the graph gives you *the relationship* between time and height". After teaching with this curriculum unit, Mr. Allen's beliefs that graphs offer the optimal display of a relationship *changed.* At the beginning of his second year, he articulated much richer views of different representations. As he explained,

> With a graph it's visual and you can see the pattern, but you don't necessarily have right in front of you the actual pairings of the data to be able to look critically at "This is an x with a y" and "This is a new x with a y." . . . If there is some constant increase, with a graph you can see that it's a line, but you don't necessarily see the data that might be able to tell you exactly a specific rate of change. Having the equation allows you to generate values of the function quickly and maybe interpolate what might be happening down the road for some piece of data x where you might not see all of that in just a small portion of the graph or just in five or six pieces of data.

These comments are consistent with his classroom instruction in year 2, when he treated the different representations more equitably in his conversations with students. For instance, as he explained to his class, "There are many ways to take a look at a function: there's a graph, there's a table, there's a word expression, or you can get an equation or a rule".

What was the reason for this change in Mr. Allen's beliefs? There are several: While interacting with students, he came to appreciate that they did not all share his belief that graphs provide the optimal display of a relationship. He had also learned himself about mathematical situations in which a non-graphical representation is more useful. Moreover, he had become familiar with this curricular approach which emphasizes multiple representations of functional relationships. From Mr. Allen's perspective, his change was directly related to his experiences teaching with the new curriculum materials. As he explained during an interview:

> Traditional textbooks tended to emphasize the equation and the graph, but not so much the table. You get away from doing the table and you try to learn quick ways to take an equation to a graph. I mean they always tell you, "Don't make a table. This is the better way, there's a slope and a y-intercept." . . . Using the graphic calculator allows you to take a look at the change in the variables as you look at the table. It's right there in front of you and you don't have to go through the messy computation of creating the table. You're focusing on trying to take a look at how x is changing and how y is changing and if there's some pattern there.

As these examples suggest, curriculum implementation offered a context in which Mr. Allen was challenged to change his beliefs as he learned to teach with the new materials.

Mr. Allen's case illustrates how engagement with innovative curriculum materials allows teachers to learn or re-learn mathematical subject matter currently recommended for school mathematics. Whether it occurs in the classroom while teaching with novel curriculum materials or in university settings while working on curricular activities as students (as discussed below), for most teachers, this learning involves revisiting mathematical ideas that they learned as students to extend their knowledge to include more conceptual or relational understandings. Teachers' learning may also involve exploring entirely new mathematics. For example, many teachers have never learned about probability, statistics, and discrete mathematics – areas now viewed as "big ideas" in the K-12 curriculum. Because reform-oriented curriculum materials have been designed to include these topics, and emphasize conceptual connections, they are an excellent source of mathematical activities that can give these teachers first-hand experiences with the types of mathematics they are expected to teach. Furthermore, reform-oriented curriculum materials portray mathematics as a vibrant and useful subject to be explored and understood. When teachers learn personally from their work with these materials, or share in their students' engagement with these materials, they are better prepared to make, and more personally invested in making, important changes in their views of appropriate mathematical activity for the classroom.

5. TEACHER LEARNING WITH CURRICULUM MATERIALS IN UNIVERSITY AND INSERVICE SETTINGS

A useful preservice and inservice activity involves inviting teachers to work carefully through the mathematical activities presented in innovative K-12 curriculum materials. Engagement with curriculum *as learners* invites teachers to think about challenging mathematics and the nature of mathematical activity, reflect on the process of learning mathematics to develop empathy for future students, and contemplate teaching mathematics to create new personal visions of classroom practice. Such experiences are critical:

> Teachers themselves need experiences in doing mathematics – in exploring, guessing, testing, estimating, arguing, and proving ... they should learn mathematics in a manner that encourages active engagement with mathematical ideas. (MSEB & NRC, 1989, p. 65)

As teachers revisit mathematical content from new perspectives, they can begin to translate the knowledge developed as learners into pedagogical content knowledge – knowledge of mathematics *for teaching* (Shulman, 1987). Teachers can also begin to revise their views of the types of activities that give rise to rich mathematical understanding, and their views of what constitutes evidence of student understanding. Davenport and Sassi (1995) report that the veteran teachers in their study were profoundly affected by reading detailed classroom narratives in curriculum materials and other print resources. Such detailed descriptions of lessons, including important images of students engaging in meaningful mathematics, helped the teachers to develop visions for their own classrooms. It is likely that preservice and beginning teachers would also benefit in multiple ways by reading and reflecting on the variety of information provided in innovative curriculum materials.

Consider an example from a university mathematics class for preservice elementary teachers in which middle school materials were used (Lloyd & Frykholm, 2000). In this class, without exception, every student (preservice teachers) came to recognize that there was mathematical content in the curriculum materials (intended for grades 5-8) with which they were unfamiliar or uncomfortable. Some students indicated that they preferred middle school mathematics the first time (when they were in middle school) and wished they were not asked to relearn it. However, most students appreciated the novelty of the activities in which they engaged. For instance, Stella, one of the preservice teachers, expressed,

> I am learning about how to look for reasons and explanations as opposed to simply believing "the rules" that some really ancient dead guy came up with. I prefer being able to use my own mind in solving problems. This class seems to use more common sense instead of book smarts.

This comment suggests that Stella was experiencing a different kind of mathematics than she had known before, and was developing new beliefs about what counts as "doing mathematics".

Experiences of learning and teaching with innovative curriculum materials may compel teachers to recognize that the nature of mathematics communicated in the classroom is intimately linked to the way it is shared with students. If teachers wish to communicate vibrant and useful images of mathematics, they must incorporate a range of pedagogical strategies that engage students in genuine problem-solving and problem-posing. Preservice teachers, whose process of learning to teach in reformed ways is compounded by the pressures of teaching for the first time, may greatly benefit from explicit attention to the development of models of practice during teacher education experiences. Reform-oriented curriculum materials offer useful images of what reformed mathematics teaching can look like (Davenport & Sassi, 1995).

As a field we know very little about how teachers can learn to center their instructional plans on student development. How do teachers come to view student learning as both the goal and guide of their mathematics classroom practices? A worthy area for research is the potential for teachers' own experiences doing mathematics in reform-oriented ways to support their development of instructional practices that honor and build on students' understandings. Teachers' reflection on experiences learning mathematics with reform-oriented curriculum materials can allow teachers to extract important theories about the nature of the mathematical learning process. Doing so may help them to recognize the significance of the learning that can occur during inquiry and student-centered activities. Further, when they work with teachers' editions (which include descriptions of possible student responses or work) and use innovative curricula to plan instructional activities, teachers may learn about the processes through which students develop understandings during particular classroom activities.

6. BELIEFS ABOUT MATHEMATICS CURRICULUM

Curriculum-based professional development has the potential to involve and impact teachers' beliefs about mathematics, student learning, and mathematics pedagogy, as well as their beliefs about mathematics curriculum. This branch of teachers' beliefs is intimately related to beliefs about mathematics, learning, and teaching. Teacher learning *about* curriculum is one goal of learning *with* innovative curriculum materials. For instance, reflection on the design and uses of curriculum may help teachers to develop rich beliefs about the role of curriculum in mathematics teaching.

Let us consider again the cases of Mr. Allen and Stella. After teaching with new curriculum materials, Mr Allen was able to look at the traditional curriculum from a distance. He described the traditional curriculum as "just the way we've always done things". Experiencing something different enabled Mr. Allen to see qualities of the "old way" that had been invisible to him before. What did Stella learn about mathematics curriculum? Recall her suggestion that, when working with reform-oriented curriculum materials, she uses "[her] own mind in solving problems . . . more common sense instead of book smarts". It is noteworthy that Stella refers to "book smarts": in contrast to other experiences where the textbook is a primary source of mathematical authority, her work with these materials allowed her to recognize herself as a sense-making authority. Understanding mathematics demands much more than reproducing rules found in textbooks, and as Stella indicates, curriculum materials can provide the basis for conceptual exploration and knowledge.

The notion that teachers possess beliefs about curriculum is certainly not new. For instance, Shulman (1987) identifies "curriculum knowledge, with particular grasp of the materials and programs that serve as 'tools of the trade' for teachers" (p. 8). However, in the present climate of reform in mathematics education, teachers' beliefs about curriculum have seldom been discussed. Because textbooks and curriculum materials are often teachers' sole contact with reform visions (Ball & Cohen, 1996), this neglect is particularly alarming. Given the prominence of textbooks in teachers' classroom decision-making (Bush, 1986; Tyson-Bernstein & Woodward, 1991), we would be wise to attend more fully to teachers' beliefs about mathematics curriculum and their role in the teacher change process. Reform recommendations and associated curriculum materials cannot and do not bring about change alone – educational change is a complex human endeavor involving teachers *and* texts (Cooney, 1988; Freudenthal, 1983).

Teachers' beliefs about mathematics curriculum include more than a familiarity with the currently available materials for designing mathematics instruction. Beliefs about curriculum encompass understandings of the role of curricular materials in the teaching and learning process, the philosophies of teaching and learning that underlie diverse curriculum materials, knowledge of the appropriateness of particular materials for certain classes and individuals, and the practical and intellectual understandings necessary for making adjustments to curricular approaches. The notion that a textbook outlines *one* of many possible mathematical and pedagogical approaches is central to teachers viewing curriculum as adaptable. Teachers are often dissatisfied with features of textbooks and curriculum materials but tend not to change or adapt those features (e.g., Lloyd, 1999). Teachers' treatment of curriculum as fixed suggests that teachers may struggle to conceive of curriculum as a flexible guide that permits and encourages alterations with respect to the changing needs and demands of particular students. As Prawat (1992) explains, a static view of curriculum is one impediment to significant teacher change:

> Instead of viewing students and curriculum interactively . . . teachers tend to regard them as similar factors that somehow must be reconciled. . . . Teachers focus on the packaging and delivery of content, instead of on more substantive issues of knowledge selection and construction. (p. 389)

If teachers are to view students and curriculum dynamically, they need to learn to make classroom-based developments within the curriculum implementation process. After all, curriculum developers may wish to create certain learning experiences for students, but they cannot fully anticipate how particular students will interact with the mathematical activities. Teachers require support not only in coming to recognize the need to adapt curriculum, but also in learning *how* to adapt it.

Experiences with innovative curriculum materials can directly challenge teachers' beliefs about curriculum. The distinctions between reform-oriented and traditional curricula provide immediate opportunities for teachers to explore, and possibly experience, multiple approaches to mathematical subject matter and mathematics pedagogy. Teachers' recognition of the multiplicity of curricular approaches is critical to their movement toward adopting more innovative instructional practices. As teachers identify and weigh the value of specific characteristics of curriculum, they may be pressed to recognize the need to make contextual, classroom-based decisions about instructional design. When teachers' beliefs about curriculum include an inquiry perspective toward *their own* development of pedagogy, we may see a corresponding increase in teachers' ability and inclination to honor and capitalize on students' processes of mathematical sense-making in the classroom. In other words, just as the quality of students' learning hinges upon their ability to make sense of mathematical problems and situations, teachers' development hinges upon educative opportunities to engage in sense-making and problem-solving about mathematics curriculum.

We should take more seriously the powerful role that curriculum materials can play in the learning of teachers throughout their careers. Most teachers rely upon one or two primary textbooks to guide their classroom instruction. If teachers can learn to use their textbooks for their own personal development, then they will be better prepared to learn from and deal productively with the types of materials that will continue to emerge in school settings in the future. Teachers need guidance in learning to make reasoned pedagogical decisions about how to judiciously incorporate the recommendations of curriculum materials into their own instruction. Such learning must extend beyond making choices among particular practices or activities to the broader development of sensible and useable theories of teaching and learning. Explicit emphasis in professional development activities on the role of curriculum materials in students' learning will support teachers in more effectively using textbooks and other resource materials to teach themselves and their students in the future.

7. REFERENCES

Ball, D. L. (1993). With an eye on the mathematical horizon: Dilemmas of teaching elementary school mathematics. *The Elementary School Journal, 93*, 373-397.

Ball, D. L. (1994, November). *Developing Mathematics Reform: What Don't We Know About Teacher Learning – But Would Make Good Hypotheses?* Paper presented at the conference Teacher Enhancement in Mathematics K-6, National Science Foundation, Washington D. C.

Ball, D. L., & Cohen, D. K. (1996). Reform by the book: What is - or might be - the role of curriculum materials in teacher learning and instructional reform? *Educational Researcher, 25(9)*, 6-8, 14.

Battista, M. T. (1994). Teacher beliefs and the reform movement in mathematics education. *Phi Delta Kappan*, 75, 462-470.

Brophy, J. E. (Ed). (1991). *Advances in research on teaching: Teachers' knowledge of subject matter as it relates to their teaching practice* Vol. 2. Greenwich, CT: JAI.

Bush, W. S. (1986). Preservice teachers' sources of decisions in teaching secondary mathematics. *Journal for Research in Mathematics Education, 17(1)*, 21-30.

Cobb, P. (1995). Where is the mind? Constructivist and sociocultural perspectives on mathematical development. *Educational Researcher, 23(7)*, 13-20.

Cobb, P., Wood, T., Yackel, E., & McNeal, B. (1992). Characteristics of classroom mathematics traditions: An interactional analysis. *American Educational Research Journal, 29*, 573-604.

Cohen, D. K. (1990). A revolution in one classroom: The case of Mrs. Oublier. *Educational Evaluation and Policy Analysis, 12*(3), 327-345.

Cooney, T. J. (1988). The issue of reform: What have we learned from yesteryear. *Mathematics Teacher, 81*, 352-363.

Cooney, T. J. (1994). Research and teacher education. *Journal for Research in Mathematics Education, 25*, 608-636.

Davenport, L. R., & Sassi, A. (1995). Transforming mathematics teaching in grades k-8: How narrative structures in resource materials help support teacher change. In B. S. Nelson (Ed.), *Inquiry and the development of teaching* (pp. 37-46). Newton, MA: Education Development Center

Davis, R. B., & Maher, C. A. (1990). The nature of mathematics: What do we do when we 'do' mathematics? In R. B. Davis, C. A. Maher, and N. Noddings (Eds.), *Constructivist views on the teaching and learning of mathematics*. Journal for Research in Mathematics Education, Monograph no. 4, 65-78.

Even, R., & Tirosh, D. (1995). Subject-matter knowledge and knowledge about students as sources of teacher presentations of the subject-matter." *Educational Studies in Mathematics, 29*, 1-20.

Fennema, E., & Megan L. F. (1992). Teachers' knowledge and its impact. In D. Grouws (Ed.), *Handbook of research on mathematics teaching and learning* (pp. 147-164). New York: Macmillan.

Fennema, E., Carpenter, T. P., Franke, M. L., Levi, L., Jacobs, V. R., & Empson, S. B. (1996). A Longitudinal study of learning to use children's thinking in mathematics instruction. *Journal for Research in Mathematics Education, 27*(4), 403-434.

Freudenthal, H. (1983). Major problems of mathematics education. In M. Zweng, T. Green, J. Kilpatrick, H. Pollak, and M. Suydam (Eds.), *Proceedings of the Fourth International Congress on Mathematical Education* (pp. 1-7). Boston: Birkhauser.

Grant, S. G., Peterson, P. L., & Shojgreen-Downer, A. (1996). Learning to teach mathematics in the context of systemic reform. *American Educational Research Journal, 33*(2), 509-541.

Hiebert, J., Carpenter, T. P., Fennema, E., Fuson, K., Human, P., Murray, H., Olivier, A., & Wearne, D. (1996). Problem solving as a basis for reform in curriculum and instruction: The case of mathematics. *Educational Researcher, 25*(4), 12-21.

Lloyd, G. M. (1996). Change in teaching about functions: Content conceptions and curriculum reform. In E. Jakubowski, D. Watkins, and H. Biske (Eds.), *Proceedings of the Annual Meeting of the North*

American Chapter of the International Group for Psychology of Mathematics Education Vol. 2 (pp. 200-206). Columbus, OH: ERIC CSMEE.

Lloyd, G. M. (1999). Two teachers' conceptions of a reform curriculum: Implications for mathematics teacher development. *Journal of Mathematics Teacher Education, 2*, 227-252.

Lloyd, G. M., & Frykholm, J. A. (2000). Middle school mathematics curricula at the university? Using innovative materials to challenge the conceptions of preservice elementary teachers. *Education, 21*, 575-580.

Lloyd, G. M., & Wilson, M. (1998). Supporting innovation: The impact of a teacher's conceptions of functions on his implementation of a reform curriculum. *Journal for Research in Mathematics Education, 29*, 248-274.

Mathematical Sciences Education Board and National Research Council. (1989). *Everybody counts: A report to the nation on the future of mathematics education*. Washington DC: National Academy Press.

National Council of Teachers of Mathematics. (1989). *Curriculum and Evaluation Standards for School Mathematics*. Reston, VA: Author.

National Council of Teachers of Mathematics. (1991). *Professional Standards for Teaching Mathematics*. Reston, VA: Author.

National Council of Teachers of Mathematics. (2000). *Principles and Standards for School Mathematics*. Reston, VA: Author.

Prawat, R. (1992). Teachers' beliefs about teaching and learning: A constructivist perspective. *American Journal of Education, 100*(3), 354-395.

Russell, S. J., Schifter, D., Bastable, V., Yaffee, L., Lester, J. B., & Cohen, S. (1995). Learning mathematics while teaching. In B. S. Nelson (Ed.), *Inquiry and the development of teaching: Issues in the transformation of mathematics teaching* (pp. 9-16). Newton, MA: Center for the Development of Teaching.

Shulman, L S. (1987). Knowledge and teaching: Foundations of the new reform. *Harvard Educational Review, 57*(1), 1-22.

Thompson, A. G. (1992). Teachers' beliefs and conceptions: A synthesis of the research. In D. Grouws (Ed.), *Handbook of research on mathematics teaching and learning* (pp. 127-146) New York: Macmillan.

Tyson-Bernstein, H., & Woodward, A. (1991). Nineteenth century policies for twenty-first century practice: The textbook reform dilemma. In P. Altbach, G. Kelly, H. Petrie, & L. Weis (Eds.), *Textbooks in American society* (pp. 91-104). Albany: SUNY Press.

von Glasersfeld, E. (1984). An introduction to radical constructivism. In P. Watzlawick (Ed.), *The invented reality* (pp. 17-40). New York: Norton, 1984.

Wilson, M., & Goldenberg, M. P. (1998). Some conceptions are difficult to change: One middle school mathematics teacher's struggle. *Journal of Mathematics Teacher Education, 29*(1), 269-293.

Wilson, M R., & Lloyd, G. M. (2000) The challenge to share mathematical authority with students: high school teachers reforming classroom roles. *Journal of Curriculum and Supervision, 15*, 146-169.

CHAPTER 10

LYNN C. HART

A FOUR YEAR FOLLOW-UP STUDY OF TEACHERS' BELIEFS AFTER PARTICIPATING IN A TEACHER ENHANCEMENT PROJECT

Abstract. Beliefs as a practical indicator offer insight into a person's previous experience and a "method for indirectly evaluating the instruction he has received" (Pehkonen & Torner, 1999). Using this meaning of beliefs, research was carried out to examine the beliefs teachers hold about their own change process. Four years after participating in a teacher change project designed to provide teachers in grades three through nine with experiences that encourage reflection on teaching and learning and to highlight methods that are consistent with a constructivist theory of learning, 37 teachers participated in a research study that asked them to share their beliefs about the experience. Twenty-nine of the 37 teachers were active in a teacher group that continued the mission of the project. Eight were no longer participating. In both groups, the teachers' believed three factors were instrumental in their change process: collaboration, colleagues in the project, and modeling of thinking and behaviors advocated. Three factors were believed by both groups to have had very little effect or actually hindered their change: the principal or school administration, colleagues in their school, and their day-to-day working conditions. These beliefs provide useful information in "understanding the nature of teachers' professional development" (Cooney, 1999) and can help us frame future professional development models.

1. INTRODUCTION

For more than a decade the mathematics education community has made recommendations for reform in curriculum and pedagogy based on theoretical work coming out of cognitive science. Through this reform lens the teacher in the mathematics classroom has a different charge. Students are not seen as empty vessels waiting to be filled up, and teaching is not perceived as a set of routines and scripts to be followed. The teacher is asked to be a facilitator of learning and to engage students in discussions that require explaining and justifying their thinking. This type of mathematics classroom looks and sounds very different from a conventional lecture-oriented classroom. Teaching from this perspective feels very different and requires fundamental change in how teachers behave and perhaps in what they believe.

In the United States, this movement was spearheaded by the National Council of Teachers of Mathematics [NCTM] (NCTM, 1995, 1991, 1989). Curriculum developers built programs that, to various degrees, supported the recommendations of reform. Individual teachers, schools, and entire school systems sought teacher

G. C. Leder, E. Pehkonen, & G. Törner (Eds.), Beliefs: A Hidden Variable in Mathematics Education? 161-176.

development models and methods to assist them as they attempted to respond to these recommendations; and numerous projects were initiated to support teacher change consistent with the recommendations (Annenberg Foundation, Guide, 1996; Grouws & Schultz, 1996; Ferrini-Mundy, 1997). Yet change is difficult and far from universal (Cooney & Shealy, 1997; Pehkonen, 1999).

Why are some teachers reluctant to change and hold fast to their traditional methods while others are embracing reform practices and changing the environment of their mathematics classrooms? Why are some projects able to impact classroom teaching in significant ways? There is a large body of literature that points to teachers' beliefs about teaching and learning mathematics as driving pedagogical behavior. In order to change teaching, teachers often must examine and change their beliefs (Richardson, 1996; Thompson, 1992). If we continue the logic, however, we must ask another question. Why are some projects, courses, experiences, etc., able to impact teachers' beliefs about teaching and learning mathematics and others are not? And, to continue, what beliefs do teachers hold about their change process? What do they believe has impacted change in their teaching? To understand teacher development and change (and eventually to impact that change) a critical piece of the puzzle is teachers' understandings of their own change process.

A variety of definitions exist around beliefs (Furinghetti & Pehkonen, 1999), even among researchers within the field. To situate my research let me begin by sharing my understandings. I interpret beliefs to be part of our subjective knowledge with a strong affective component. This is different from knowledge which can at some level be socially agreed upon as true or false. For example, I may have a belief that addition of fractions is hard (based on my unsuccessful and stressful experience as a student standing at the front of the room), but I can say I know that ½ plus ½ is 1 with some certainty. From a constructivist perspective this is as close as we come to reality since reality at its best is the creation of a viable model (von Glasersfeld, 1983; Wilson & Cooney, this volume). Whether I am able to articulate my belief (that fractions are hard) or not, it drives my behavior and I may avoid situations where I have to perform operations on fractions. In this case my belief acts as a regulating system (Pehkonen & Törner, 1999) influencing my behavior. But taken from the perspective of the researcher looking at the individual who holds the belief, knowledge of that belief informs our interpretation of the actions of the subject and provides a window of understanding into a person's experiences. If we observe behavior (avoidance of fractions) we may intuit the related belief. If the individual further confirms this, it provides a window of understanding into past experiences and helps frame our understanding of a person's mathematical development.

At another level, suppose the fraction avoider is a teacher of fifth grade students and limits her instruction of fractions to a minimum. If at some later time the teacher is seen teaching fractions with confidence and enthusiasm, one must ask questions. Have the teacher's beliefs about fractions changed? What impacted that change? And, equally importantly we must ask: *What does the teacher believe impacted the change?* To describe fully the story of a teacher's change we must understand the beliefs she holds about her change process - about why she has (or has not) changed her practice. This deeply held subjective knowledge is an indicator of the experiences teachers have had (Pehkonen & Torner, 1999) and can be used to inform practical decisions about teacher change and teacher development in the future.

In this chapter those questions are explored with a group of teachers who had a common professional development experience that was motivated by reform recommendations. In Part One, I will describe briefly the professional development model the teachers experienced (The Reflective Teaching Model) and how it was implemented with them in the Atlanta Math Project (AMP). In Part Two, I will describe the follow-up research on that group of teachers who, four years after the end of AMP, were asked to share their beliefs about their own change process.

2. THE TEACHER DEVELOPMENT MODEL

Crucial to the success of any reform effort is the professional development model chosen to implement the effort (Clarke, 1997). The model used in this research was the Reflective Teaching Model (RTM) (Schultz, Hart, & Najee-ullah, 1995; Hart, 1996; Hart & Najee-ullah, 1997). The model is grounded in the theories of constructivism and metacognition. It also is based on beliefs in the value of shared authority and modeling. These constructs are fundamental to all activities and experiences in the RTM.

2.1. Philosophy and Assumptions

2.1.1. Constructivism

In the RTM, teachers and mathematics educators are seen as *learners* of new thinking and pedagogy associated with mathematics education. As such, teaching in the RTM mirrors teaching within a mathematics classroom framed by a constructivist concept of learning. Within the model the teachers engage in both individual and group experiences that challenge existing beliefs and that provide opportunities to examine, share and rethink their ideas; to evaluate, argue and justify their thinking; to construct new knowledge; and to reflect on their experiences. For example, teachers in the RTM regularly plan lessons collaboratively. After teaching the lesson, they individually and collaboratively reflect on the experience. Through the dialogue about the experience teachers make public their pedagogical content knowledge, content knowledge, and beliefs about both. As teachers wrestle with their own concepts and beliefs these are challenged by the concepts and beliefs of others, forcing them to reexamine, publicly or privately, their thinking and perhaps construct new knowledge. These experiences blend the radical constructivist view of learning as an individual cognitive process (von Glasersfeld, 1983), and the social constructivist view of learning as a social process (Bauersfeld, 1992; Ernest, 1992). The Reflective Teaching Model takes the position that both perspectives are helpful in nurturing and understanding knowledge development (Simon, 1995).

2.1.2. Metacognition

The theory of metacognition (Flavell, 1979) refers in part to our ability to monitor and regulate what we are doing and thinking while we are experiencing it and our ability to reflect on experiences and to learn from them. Increasing metacognitive activity through private reflection and shared conversations increases teachers' awareness of their subjective knowledge. As mentioned above, beliefs are often challenged through this process, which lays the groundwork for the construction of new knowledge and for real change in teaching practice. For example, Teacher A is engaged in collaboratively planning a lesson. The topic is area of rectangular regions. Teacher B (who will be teaching the lesson) is proposing some ideas for integrating the concepts of area and perimeter. Teacher A becomes quite uncomfortable, realizing she has *always* taught the concepts separately to avoid confusion. Teacher A is now aware of her belief about the appropriate pedagogy for teaching these concepts – they should be taught separately. Her belief is challenged through the discussion about the lesson. If the integrated lesson is successful, she may reflect on the experience and choose to try a similar lesson with her students. If that too works, her belief may eventually change. Through carefully crafted experiences, teachers in the RTM are given opportunities that encourage the monitoring of their knowledge and beliefs and the subsequent regulation of their behaviors. For teachers in the RTM, developing the ability to monitor their thinking *while they are teaching* is enhanced through the process of planning *before* the lesson *and* reflecting *after* the lesson. These are fundamental parts of the experience and critical to the process of change (Hart, Schultz, Najee-ullah, & Nash, 1992).

2.1.3. Shared Authority

The RTM is also based on the assumption that sharing authority (Cooney, 1993; Hart, 1993; Wilson & Lloyd, 2000) is critical in the interaction of teachers and teacher educators. The ability of a teacher or teacher educator to relinquish intellectual control and allow others to share in the generation of mathematical or pedagogical ideas is a subtle but significant shift in roles from the traditional teacher educator as teller. It builds trust, ownership, and cohesion among those involved. In the RTM, teacher educator and teacher sit down at the table together and *collaboratively* plan lessons with the teacher educator/researcher acting as audience and critical friend (Clarke, 1997). Through this process of sharing between teacher educators and teachers as well as between teachers, participants in the RTM build collegial and collaborative relationships that sustain and support them as they examine and consider their beliefs and behaviors.

2.1.4. Modeling

Goldsmith and Schifter claim, "Teachers seeking to change their practice may not have useful images from their personal experience to guide the creation of a focused and productive classroom culture" (Goldsmith & Schifter, 1997, p. 25). Since much human behavior is learned through observation and modeling (Bower & Hilgard, 1981; Rotter, 1992) and most teachers were students in traditional classrooms, they do not have a picture of what a reform classroom looks like or feels like. Given the very complex nature of teaching within a constructivist framework of learning, modeling is a particularly effective way to teach abstract behaviors such as examination of values and standards of conduct. It is reasonable to assume that modeling can be an effective component in facilitating teacher change. The more complex the skill or thinking process the greater the need for an opportunity to observe performance of the skill or thinking process. Hermann (1990) found that mental modeling of reasoning processes enhanced the effectiveness of instruction. It follows that the mental modeling of the thinking processes engaged in planning lessons, teaching lessons or of reflecting on the act of teaching assists teachers in developing their own mental models of the thinking involved. By modeling their thinking as they plan a lesson, teacher educators demonstrate the reasoning processes they engage in to develop a plan. By modeling the teaching of the lesson, teacher educators in the RTM give teachers an opportunity to see what their role in the classroom might look like. By modeling the act of reflecting on their teaching, teacher educators can explain the decisions they made during the act of teaching and make public the process of productive self-assessment.

2.2. The Model/Experience/Reflect Framework

All activities in the RTM follow a model/experience/reflect framework. This includes all aspects of summer or school year staff development. The facilitators first model an activity (e.g., planning, teaching, problem solving); the teachers then experience this activity; and each activity concludes with a reflection on the experience. As is evident, this framework exemplifies the constructs of the model: constructivism, metacognition, modeling and shared authority.

2.3. Essential Activities in the RTM

Using the model/experience/reflect framework, teachers engage in several essential activities during a project using the RTM. They participate in an initial inservice that introduces the philosophy and language of reform and lays the groundwork for sharing authority and thus building relationships necessary to carry on long-term support. They are introduced to think-aloud paired problem-solving sessions where they learn to listen to the thinking of others as they solve problems and learn to communicate their own thinking. This process mimics classroom interaction. During the school year they engage in monthly plan/teach/debrief cycles where pairs work together to think-aloud through the process of planning a lesson, teaching the lesson and debriefing on the lesson. They are involved in various formats for reflecting on their teaching: videotaping, oral reflections, reflection logs, etc. Each of these experiences is modeled first by RTM teacher educators, then experienced by the teachers. Each experience is reflected on by all. Finally, they meet regularly as a group to exchange and share ideas and concerns and to nurture their professional growth and collegial relationships.

2.4. The Atlanta Math Project

The research described in this chapter involves teachers who participated in the RTM (described above) through the Atlanta Math Project (AMP), a National

Science Foundation Teacher Enhancement Project conducted from 1990-1994 in a large urban community in the United States. AMP worked with teachers from several school systems in a large metropolitan area. Although designated a middle-school project, AMP involved teachers in grades three through nine. Beginning with 13 teachers in 1990 and concluding with a database of 98 in 1994, the project engaged teachers in on-going, on-site support.

At the conclusion of the formal funding period, the AMP teachers independently formed a professional group (the AMP Council) dedicated to continuing the professional development and collegial relations begun in the Atlanta Math Project. Four years later (1998), the Council had a membership of 53 from the original 98. Forty-five were not participating in the Council. Of those 45, nine were working in a school system but not as classroom teachers; eight teachers had retired; four had moved with no known address; eight had left teaching; two were deceased; and 14 were still teaching in the Atlanta area. In the remainder of this chapter the follow-up research on the classroom teachers participating in the AMP Council and those not participating in the Council is described.

3. THE STUDY

3.1. Rationale for the Study

In 1998 the funded AMP project had been over for 4 years, yet over half the project teachers were still members of an independently formed teacher group whose purpose was to continue the efforts and vision of the project. These teachers saw themselves as reform teachers who were actively continuing their professional development as well as sharing their expertise with others in their school systems through various staff development activities. They offered a unique opportunity to study teachers' beliefs about their own change process. In addition, 14 of the original 98 were still teaching but were not associated with the AMP Council. They might offer a slightly different perspective. All had had several years to reflect on their experiences and situate those related to AMP within a larger framework of their change. What beliefs do those teachers hold about their own change? What do they believe were the factors that most influenced them? Do teachers who were no longer members of the council have different beliefs? The purpose of this research was to ask the teachers to look back on their experience and attempt to address these questions.

3.2. Conceptual Framework

This research is about teachers' beliefs about their change as they attempted to reform their teaching. Following is a discussion of the conceptual framework used to guide my thinking as I conducted the research.

3.2.1. Teacher Change

Too often in the reform literature we talk about *teacher change* as if it were some clearly defined process. But, as Cooney and Shealy (1997) discuss, change is merely doing things differently. They further observe that there are many behaviors we might discuss in terms of change, e.g., the novice to expert teacher, increase in the effective use of time, etc. So when we claim that teachers have *changed*, it seems important to identify the dimensions we are looking at that may be *different*. Once identified, we can look for how a teacher has changed along those dimensions. How much has she changed? In what ways? How do we measure or describe this change?

Why is teacher change within the reform not well defined? Perhaps because those of us who are conducting research on teacher change within the reform paradigm are only just beginning to define what change means within this arena. Research on teacher change is in its infancy. Teachers and researchers are, in fact, only now inventing the "changed" practice that we speak of (Schifter, 1996). With so much to be learned about how teaching from a reform perspective looks and feels, it is premature to claim "change" as some absolute. We can discuss change - how it feels to the teacher and the student and how it looks to the observer looking in. We can ask teachers to describe their experiences and share their beliefs about that process (Schifter, 1996). As a research community we can share our findings, but the researcher provides only one perspective. A critical component in developing this body of knowledge is the teachers' voice. Do they believe they have changed? Can they describe how they have *changed?* What do they believe impacted or deterred that change?

3.2.2. Teachers' Beliefs

Studies of the origins of teachers' beliefs indicate that life experiences are a major contributor to the formation of beliefs (Richardson, 1996). If teachers experience change in their beliefs about teaching and learning in a way that is consistent with the philosophy of a particular model of change, then it is imperative not only to examine the nature of that change, but also to examine the model that motivated the change (including teachers' beliefs about that model) and to describe and identify factors that facilitated change. In order to develop effective teacher education programs, we must not only identify the presence of change, but teachers' beliefs about their change. Beliefs from this perspective are a "practical indicator" providing a good estimation of teachers' experiences (Pehkonen & Torner, 1999) and lay the groundwork for future professional development.

3.3. Methods of Inquiry and Data Sources

To begin to explore the questions mentioned above, a 16-item survey was developed (Appendix A) listing factors that may have impacted teacher change. Several generic factors were used (e.g., the reform movement in general, innovative curricular materials, etc.). These factors had been identified by Clarke in his research on innovative curricular materials to study teacher change. Of his list he comments, "few items on the list would come as a surprise to those experienced in the professional development of mathematics teachers" (Clarke, 1997), suggesting a broadly defined and accepted set of factors. Some of the items from Clarke's list were used verbatim and some of the factors were adapted to match the specifics of the AMP experience. Five items were unique to the AMP experience (items number 9, 10, 11, 12, & 13). Space was allowed at the end of the survey to add additional remarks or comments.

On the survey, AMP teachers were asked to indicate which factors they believed were very helpful in their change, somewhat helpful in their change, somewhat hindered their change, or significantly hindered their change over the last few years. After the surveys were returned and analyzed a sample of teachers was interviewed to confirm and expand the survey data. In the interview, teachers were asked to answer three questions. They were asked if they saw themselves having a "before" and "after" with respect to the reform in mathematics education and if so, to describe how their teaching had changed. They were asked to identify factors that contributed to that change and finally they were asked to identify factors that may have hindered the change. Descriptive statistics were used to analyze the survey items. For the interview questions, qualitative methods were used to analyze responses looking for themes that confirmed the survey or for new themes that were not identified on the survey.

The survey was mailed to the 53 members on the AMP Council database current in the fall, 1998 and to the 14 members who were still teaching but no longer participating in the AMP Council. Where possible, follow-up phone calls were made to those who did not respond. Thirty-three surveys were eventually returned from the AMP Council. Four were unopened with no forwarding address, resulting in 29 surveys available for analysis. Eight surveys were returned from the teachers who were not participating in the AMP Council. Ten of the AMP Council members were interviewed. Four of those not participating in the Council were interviewed.

3.4. Results

3.4.1. The Survey

The results from the survey for both groups are reported in Table 1. The percent of Council members responding to each item is listed first and data for non-Council members are in parentheses immediately following. The results were not glaringly different across the two groups on the survey. The majority of both groups indicated that most of the 16 factors were very or somewhat helpful in their change. However, some trends were apparent. The table has been organized to highlight the trends.

Among the Council members, three factors specific to AMP were identified by 90% of the respondents as being very helpful in supporting their change: item #5, colleagues in AMP; item #12, modeling in AMP; and item #14, collaboration in AMP. Among the non-Council members the same three very helpful AMP factors mentioned above were identified by 75% of the respondents. In addition, two other generic factors were equally identified by non-Council members: item #1, the reform in general; and item #6, innovative curricular materials. These items also received relatively high responses from Council members, although not as high as the previous three.

Three generic factors were identified by a majority of the Council teachers as only somewhat helpful, somewhat hindered or absolutely hindered: item #2, the principal/school administration (76% total of the three categories); item #3, colleagues in my school (55% total of the three categories); and item #15, my day-to-day working conditions (69% total of the three categories). The non-Council members identified the same three factors that were only somewhat helpful or actually hindered their change.

The remaining items from the survey were more in the moderate range. They did not demonstrate as strong a consistency within the teacher groups.

Table 1. Percent of Teachers Responding by Item on Factors They Believe Influenced Their Change

Item	Item Description	Very Helpful	Somewhat Helpful	Somewhat Hindered	Absolutely Hindered	Omit
Most Helpful						
5	Colleagues in AMP	90(75)	10(25)			
12	Modeling of strategies in AMP	90(75)	7(25)			3
14	Collaboration in AMP	90(75)	7(25)			3
1	The reform movement	79(75)	21(25)			
6	Innovative curriculum materials	69(75)	28(25)			3
Least helpful or Hindrance						
2	The principal/administration	24(13)	66(50)	7(37)	3	
3	Colleagues in my school	45(37)	42(38)	3(25)	10	
15	Day-to-day working conditions	31(13)	52(37)	14(50)	3	
Interim Ratings						
4	Support personnel at my school		28(50)	3(13)		3
7	External support personnel		28(50)		6	
8	AMP in general	80(63)	10(37)			10
9	The on-site support of AMP	72(63)	21(37)			7
10	The long-term support of AMP	69(63)	28(37)			3
11	AMP plan/teach/debrief cycle	80(63)	17(37)			3
13	Observing other APM teachers	79(63)	17(37)			3
16	My math content knowledge	73(37)	24(63)			3

N = 29 AMP Council members
(N) = 8 non-Council members

3.4.2. Interview

All of the teachers interviewed (Council and non-Council members) saw themselves as having changed their teaching over the last eight years. The most common theme that emerged had to do with discourse in the classroom. One AMP Council member reported:

> The major differences I see in my teaching deal with the open communication in the classroom. I elicit student responses in a much more open ended way than I did when I first began teaching. I accept a wide range of responses and deliberately ask questions that elicit varying responses in order to facilitate conversation.

Another Council member said:

> I try very hard to teach the students to communicate with one another and explain their reasoning. During the mathematics lesson more time is spent with students talking, thinking and listening now and less time in paper and pencil tasks.

One teacher spoke of how the tasks she is presenting are different and often encourage multiple solution strategies.

> I have always taught with manipulatives, hands-on, but more teacher-directed. Now my questioning strategies have become better. My questions are more open ended.

Another indicated:

> I realize math is not cut and dry . . . before I thought there was one and only one answer or method of solving a problem. Now I realize that one problem could have 10-plus different ways of finding the solution!

Some of the teachers newer to the profession suggested that they had changed, but that it was less dramatic.

> I had heard in school that I should be asking questions instead of telling, allowing my students to work in groups and think for themselves, and encouraging the girls to speak up in class. As I began participating in the Atlanta Math Project this began to make sense to me. My teaching matured and developed into a style I was quite proud of!

The teachers also spoke of changes in their students.

> I believe my students understand math better and enjoy it more because of my experience in AMP.

When asked specifically to discuss what they believed as having most influenced change in their teaching, the AMP Council members were remarkably consistent with their survey responses. They spoke frequently of collaboration and the value of their AMP colleagues.

> The time with other colleagues was invaluable.

> The collegiality allowed me to see outside the loneliness of the 'one teacher' classroom.

> Through collaboration with other AMP teachers I was able to enhance my teaching.

> The modeling of strategies and working with other teachers was rejuvenating.

They also spoke repeatedly of the importance of modeling and actually "seeing" others teach and talk about the thinking involved in planning lessons.

> The 'modeling of the model' helped me appreciate and understand the experience.

> Seeing my mentor teach before I tried it gave me confidence and understanding.
>
> I think it was important for me to watch other teachers and be able to talk to them about what they were thinking.
>
> As a beginning teacher, the early sessions were instrumental in helping me develop a picture of teaching this way.
>
> The modeling of strategies and working with other teachers was rejuvenating.
>
> I learned when topics were discussed and demonstrated in my classroom.

Additionally, Council members spoke vividly of two additional factors that did not appear as high in the survey, but surfaced repeatedly in the interviews. They spoke of the value of reflection.

> The model helped me develop the habit of mind of reflection.
>
> I was encouraged to reflect and analyze my teaching in a non-threatening way.
>
> The debriefing was critical to my success.
>
> Videotaping helped me think about my teaching more carefully.
>
> I learned when topics were discussed and demonstrated in my classroom.

And, related to reflection, they also spoke specifically of the plan/teach/debrief cycle.

> The plan/teach/debrief cycle was important in the success of the project.
>
> Planning and debriefing with others was very helpful.
>
> My teaching has been influenced greatly by having had the opportunity to go through the steps of planning, teaching and then debriefing a lesson with someone else.
>
> To be able to sit down and plan with someone, then actually watch them teaching and then talk to them about what they were thinking as they were teaching – was invaluable. It made it come alive.

Not surprisingly the non-council members spoke positively in general about AMP but were less focused on the AMP experience. They seemed to incorporate that experience with their overall professional growth. They also mentioned the importance of colleagues in AMP, but mentioned modeling much less frequently. Other factors that they believe impacted their change included:

(1) learning from *their* students,

> I saw changes in my students; student behavior actually improved; and my students understand math better and enjoy it more;

(2) becoming a student themselves through inservice and college courses,

> being in the student position helped me understand how important teaching this way is;

(3) and, the NCTM standards,

> learning about the standards was an important part of what happened to me.

During the interviews other factors not identified on the survey were mentioned as hindrances to change. Both groups mentioned parents who were not supportive. Several AMP Council members mentioned standardized testing. The non-Council group strongly identified colleagues who did not agree with the reform position and lack of planning time.

4. DISCUSSION

What can researchers and mathematics educators learn from the beliefs of these teachers? The data tell us that all the teachers believe collaboration and working with supportive AMP colleagues is critical to teacher change, which is consistent with the findings reported by Llinares (this volume) and Wilson and Cooney (this volume). The Council members believe it was important for them to have the ideas and strategies of reform modeled for them and that observing each other teach, debriefing together, and planning together is important. They believe that reflection nurtures the change process. The non-Council members point to the reform movement, in general; learning from their students; and becoming learners themselves (both within and outside of the project) as significant factors in their change. Given this, we need to consider two questions. First, what do these beliefs tell us about the teachers' experiences within reform in mathematics education? And, how does this information contribute to our understanding of the professional development of teachers? The beliefs these teachers hold offer a window of understanding into their professional development and change - not as much as a predictor of future behavior, but as an important insight into the teacher change process and to inform future professional development activities.

The importance of colleagues in changing beliefs and ultimately changing teaching behaviors is well documented. There is strong evidence that teachers develop and reconstruct their knowledge about learning and teaching through interaction and reflection with their peers (Hargreaves & Fullan, 1992; Schubert & Ayers, 1992). Aston and Hyle (1997) found that social networks impact the beliefs and practices of teachers and can, in fact, ameliorate the effects of traditional schooling experiences and cultural norms within a school setting that are counter to reform practices. This study confirms these findings through the teacher's voice. All the AMP teachers indicated on the survey and in the interview that they believe that collaboration in AMP and their colleagues in AMP were fundamental to their change. And, just as with the teacher who changed the way she taught fractions, we find confirming evidence of the belief by their behavior – by their continued participation in a project that is no longer funded and is maintained by the teachers for the sole purpose of collaboration and professional development.

More surprising is the belief AMP teachers shared about the impact of modeling in their change process. As was mentioned earlier in the discussion of modeling in the RTM, social learning theory tells us that much human behavior is learned through observation and modeling (Bower & Hilgard, 1981; Rotter, 1992).

Modeling is particularly effective in teaching abstract behaviors and standards of conduct [such as teaching within a constructivist framework of learning]. The more complex the skill, the greater the need to observe the behavior. Retention is increased through discussion of what was observed [reflection] and actual practice [teaching] (Bandura, 1977). I have added the emphases in brackets to make the connection to teacher development practices. Modeling also supports the notion of shared authority. When teachers see teacher educators in the act of teaching - exposing themselves to the pitfalls of the classroom - teachers acquire a new and more egalitarian relationship with the teacher educator. I would like to suggest that this perspective has not yet been explored in mathematics education teacher change research. Is modeling (couched with reflection) a way to enhance the formation of viable models of reform practices by novice and traditional teachers? Does it build those valuable relationships needed for change, i.e., researcher as critical friend (Clarke, 1997)? Through these experiences do teachers' beliefs about effective teaching begin to change?

The teachers also shared their belief that their day-to-day working conditions and the principal/school administration is only somewhat helpful or even hindered their change. This, too, comes as no surprise as Clarke suggested "to those experienced in the professional development of teachers" (Clarke, 1997). But together these beliefs provide important information in understanding the nature of teachers' professional development. It is this pursuit that drives research on teacher beliefs and teacher change. Ultimately we want to frame a professional development model that uses knowledge gained in research on teacher change. But this work must echo the voices of teachers' on their own change process. Their insights into how they change or have changed may be the key in preventing the traditional from being perpetuated. If teachers believe the greatest deterrent is within the school itself, this must be considered as we plan future teacher development projects.

5. A FINAL THOUGHT

One colleague suggested that by using a survey I had actually "given the teachers the answers". In some respect, he was right. Although teachers were given the opportunity to list other factors that they considered important and to expand their beliefs in the interview, the survey clearly was suggestive. However, as one of the developers of the RTM model, this allowed me to assess the aspects of the model that the teachers believe were most critical in their change. This information will be useful as my colleagues and I continue our research in teacher change and teachers' professional development.

6. REFERENCES

Annenberg Foundation. (1996). *The guide to math & science reform.* South Burlington, VT: The Annenberg/Corporation for Public Broadcasting-Elementary and High School Project for Mathematics and Science. http://www.learner.org/k12.

Aston, M., &. Hyle, A. E. (1997). *Social networks, teacher beliefs and educational change.* Paper presented at the annual meeting of University Council of Educational Administration, Orlando, Fl. Dialog, ERIC ED420621.

Bandura, A. (1977). *Social learning theory.* Englewood Cliffs, NJ: Prentice-Hall.

Bauersfeld, H. (1992). Classroom cultures from a social constructivist's point of view. *Educational Studies in Mathematics, 23,* 467-481.

Bower, G. H., & Hilgard, E. R. (1981). *Theories of learning.* 5th ed. Englewood Cliffs, NJ: Prentice-Hall.

Clarke, D. M. (1997). The changing role of the mathematics teacher. *Journal of Research in Mathematics Education, 28*(3), 278-308.

Cooney, T. J. (1999). Examining what we believe about beliefs. In E. Pehkonen and G. Törner (Eds.), *Proceedings of the workshop in Oberwolfach on mathematical beliefs and their impact on teaching and learning of mathematics* (pp. 18-23). Oberwolfach, Germany: Gerhard Mercator Universitat Duisburg.

Cooney, T. J. (1993). On the notion of authority applied to teacher education. In J. R. Becker and B. J. Pence (Eds.), *Proceedings of the 15th annual meeting of the North American Chapter of the Psychology of Mathematics Education* Vol. 1 (pp. 40-46). Asilomar, CA: University of California Press.

Cooney, T. J., & Shealy, B. (1997). On understanding the structure of teachers' beliefs and their relationship to change. In E. Fennema and B. S. Nelson (Eds.), *Mathematics teachers in transition* (pp. 87-110). Mahwah, NJ: Lawrence Erlbaum.

Ernest, P. (1992). The nature of mathematics: Towards a social constructivist account. *Science and Education, 1*(1), 89-100.

Ferrini-Mundy, J. (Ed.). (1997). *Making change in mathematics education: Learning from the field.* Reston, VA.: NCTM.

Flavell, J. H. (1979). Metacognition and cognitive monitoring." *American Psychologist, 34*(10), 906-911.

Furinghetti, F., & Pehkonen, E. (1999). A virtual panel evaluating the characteristics of beliefs. In E. Pehkonen and G Törner (Eds.), *Proceedings of the workshop in Oberwalfach on mathematical beliefs and their impact on teaching and learning of mathematics* (pp. 24-30). Oberwolfach, Germany, Gerhard Mercator Universitat Duisburg.

Goldsmith, L. T., & Schifter, D. (1997). Understanding teachers in transition: Characteristics of a model for the development of mathematics teaching. In E. Fennema and B. S. Nelson (Eds.), *Mathematics teachers in transition* (pp. 19-54). Mahwah, NJ: Lawrence Erlbaum Associates, Inc.

Grouws, D. A., & Schultz, K. (1996). Mathematics teacher education. In J. Sikula (Ed.), *Handbook of research on teacher education* (pp. 442-458). New York: Macmillan Library Reference.

Hargreaves, A. & Fullan, M. (1992). *Understanding teacher development.* New York: Teachers College Press.

Hart, L. C. (1996). The Decatur Elementary Math Project. In E. Jabuwoski (Ed.), *Proceedings of the 18th annual meeting of the North American Chapter of the Psychology of Mathematics Education* (pp. 491-496). Panama City, FL.: University of Florida.

Hart, L. C. (1993). Shared authority: A roadblock to teacher change? In J. R. Becker and B. J. Pence (Eds.), *Proceedings of the 15th annual meeting of the North American Chapter of the Psychology of Mathematics Education* Vol. 2 (pp. 189-197). Asilomar, CA: University of California Press.

Hart, L. C., & Najee-ullah, D. (1997). *The Reflective Teaching Model.* Atlanta, GA: Galileo, Inc. CDrom.

Hart, L. C., Schultz, K., Najee-ullah, D., & Nash, L. (1992). Implementing the Professional Standards for teaching mathematics: The role of reflection in teaching. *Arithmetic Teacher, 40*(1), 40-42.

Hermann, B. A. (1990) Teaching preservice teachers how to model thought processes. *Teacher Education and Special Education, 13*(2), 73-81.

National Council of Teachers of Mathematics. (1989). *Curriculum and evaluation standards for school mathematics.* Reston, VA.: Author.

National Council of Teachers of Mathematics. (1991). *Professional standards for teaching mathematics.* Reston, VA.: Author.

National Council of Teachers of Mathematics. (1995). *Assessment standards for school mathematics.* Reston, VA.: Author.

Pehkonen, E. (1999). Beliefs as obstacles for implementing an educational change in problem solving. In E. Pehkonen and G Törner (Eds.), *Proceedings of the workshop in Oberwalfach on mathematical beliefs and their impact on teaching and learning of mathematics* (pp. 106-114). Oberwolfach, Germany: Gerhard Mercator Universitat Duisburg.

Pehkonen, E., & Törner, G. (1999). Introduction to the abstract book for the Oberwalfach meeting on belief research. In *Proceedings of the workshop in Oberwalfach on mathematical beliefs and their impact on teaching and learning of mathematics* (pp. 3-10). Oberwolfach, Germany: Gerhard Mercator Universitat Duisburg.

Richardson, V. (1996). The role of attitudes and beliefs in learning to teach. In J. Sikula (Ed.), *Handbook of Research on Teacher Education* (pp. 102-119). New York: Simon & Schuster.

Rotter, J. B. (1992). *The development and applications of social learning theory: Selected papers*. New York: Praeger.

Schifter, D. (Ed.). (1996). *What's happening in math class? Reconstructing professional identities* Vol. 1. New York: Teachers College Press.

Schubert, W. H., & Ayers, W. (Eds) (1992). *Teacher lore: Learning from our own experiences.* New York: Longman.

Schultz, K., Hart, L. C., & Najee-ullah, D. (1995). *AMP Final Report.* Atlanta, GA: Georgia State University.

Simon, M. (1995). Reconstructing mathematics pedagogy from a constructivist perspective. *Journal of Research in Mathematics Education, 26*(2), 114-145.

Thompson, A. (1992). Teacher's beliefs and conceptions: A synthesis of the research. In D. A. Grouws (Ed.), *Handbook of research on mathematics teaching and learning* (pp. 127-146). New York: MacMillan.

von Glasersfeld, E. (1983). Learning as a constructive activity. In *Proceedings of the fifth annual conference of the International Group for the Psychology of Mathematics Education* Vol. 1. Montreal: U. of Montreal.

Wilson, M., & Lloyd, G. M. (2000). The challenge to share mathematical authority with students: High school teachers' experiences reforming classroom roles and activities through curriculum implementation. *Journal of Curriculum and Supervision, 15,* 146-169.

APPENDIX A

AMP Participant Survey

Check the response that most closely represents what you believe about your own change.

Item #	Description	Very Helpful	Somewhat Helpful	Somewhat Hindered	Absolutely Hindered
1	The reform movement in general				
2	The principal/school administration				
3	Colleagues in my schoo				
4	Support personnel within my school				
5	Colleagues in AMP				
6	Innovative curriculum materials				
7	External support personnel				

Item		Very	Somewhat	Somewhat	Absolutely
	(university)				
8	AMP in general				
9	The on-site support of AMP				
10	The long-term support of AMP				
11	The plan/teach/debrief cycle of AMP				
12	Modeling of strategies in AMP				
13	Observing other teachers in AMP				
14	Collaboration in AMP				
15	My day-to-day working conditions				
16	My math content knowledge				
17	Other				
18	Other				

CHAPTER 11

OLIVE CHAPMAN

BELIEF STRUCTURE AND INSERVICE HIGH SCHOOL MATHEMATICS TEACHER GROWTH

Abstract. The nature of belief structure and change are discussed in this chapter, as are reports on an interpretative study of the professional growth of four inservice high school mathematics teachers who made fundamental changes to their teaching over their teaching careers without the aid of a particular professional development program. However, because of space constraints detailed data are presented for only two of these teachers. The analysis of the study focused on the nature of the teachers' belief structures for mathematics and the relationships between these structures and changes in their teaching. The findings included that, for each teacher, the belief structure was built up over time, was centered around a pedagogical view of mathematics held as an experiential construct (e.g., play) and was characterized by attributes associated with this construct (e.g., fun). These attributes were important in allowing the teachers to recognize and resolve pedagogical tensions and in generating changes in their teaching. The way in which this occurred is discussed through narratives of two of the participants' experiences and schemas of their belief structures for mathematics. Implications of the resulting perspective of belief structure are discussed in relation to what should be attended to when trying to facilitate change in teachers' practice.

1. INTRODUCTION

It has become an accepted view that it is the teacher's subjective school related knowledge that determines for the most part what happens in the classroom. Consequently, any attempt to reform the teaching of mathematics will also require a corresponding "re-forming" of teachers' thinking as a necessary criterion for its success. This position is reflected in the literature in terms of a growing focus on understanding the mathematics teacher, with particular interest on how to effect change in his or her thinking and teaching to reflect the current reform perspective of mathematics, and its teaching and learning (e.g., Fennema & Nelson, 1997). However, how to effect change meaningfully continues to be a concern for the teacher and requires ongoing consideration, particularly in terms of understanding what might be unique in the case of mathematics teachers.

One aspect of teacher thinking that is viewed as being of importance in understanding and facilitating change/growth in teaching is belief. This chapter presents a study that examined beliefs in the context of inservice high school mathematics teachers who changed their practice on their own from a predominantly teacher-centered perspective to a more student-centered perspective. The focus here

G. C. Leder, E. Pehkonen, & G. Törner (Eds.), Beliefs: A Hidden Variable in Mathematics Education? 177-193.

is on the nature of the teachers' belief structure for mathematics and the relationship between this structure and changes in their teaching over their teaching career.

2. BELIEF STRUCTURE AND CHANGE

Achieving change in the teaching of mathematics is generally a challenging endeavor for teachers. Even teachers who are interested in change do not necessarily succeed at making substantive or fundamental shifts in their teaching. They tend to fall into three categories: those who change their teaching on their own, those who change their teaching with external support, and those who do not change their teaching in spite of involvement in professional development programs. The underlying reasons for this pattern of response to change are not clear-cut, but the nature of the relationship between thought and action seems to be a significant contributor to it. In particular, teachers' beliefs are being viewed as playing a role on if, when or how change occurs because of their apparent relationship to behavior. Ernest (1989), for example, suggested that beliefs are primary regulators for mathematics teachers' behaviors in the classroom. Although it is not clear that beliefs by themselves can account for teachers' classroom behaviors, the possible importance of them is recognized in studies of mathematics teachers (e.g., Cooney, Shealy, & Arvold, 1998; Chapman, 1997, 1999a; Lloyd & Wilson, 1998; Pehkonen, 1994; Raymond, 1997; Thompson, 1992). These studies generally focused, directly or indirectly, on understanding what teachers believe and the relationship to teaching or learning to teach mathematics.

This theme of teachers' beliefs continues in the studies presented in this section of this book with the relationship between belief and change being a significant focus. Each of these studies deals with this relationship in a unique way. For example, Lloyd points out that teachers' beliefs can change on the basis of experiences with innovative curriculum materials associated with a reform-oriented mathematics curriculum. This can happen through university courses or workshops in which teachers work collaboratively on the activities outlined in the materials or while they are implementing the curriculum materials in their classrooms. Hart, like Lloyd, also focuses on inservice teachers but in relation to their beliefs about factors influencing changes in their teaching. She found that teachers who participated in a teacher enhancement project with which she was involved believed that collaboration, colleagues in the project and modeling of the thinking and behaviors advocated were instrumental in their change process, while the principal/school administration, colleagues in their schools, and their day-to-day working conditions had very little effect or actually hindered their change. Philippou and Christou, on the other hand, argue that beliefs in one's ability to overcome obstacles and bring about a predetermined outcome function as a decisive motivation to undertake the endeavor and put forth time and resources. Finally, Llinares' study of preservice teachers draws attention to the importance of reification of beliefs as points of reflection or discussion to effect those beliefs. These studies also suggest or imply a variety of possible interventions to effect change or growth, e.g., use of reform-oriented curriculum materials (Lloyd), modeling reform-oriented teaching (Hart), use of content framed in the history of mathematics (Philippou & Christou) and use of cases of students' mathematical thinking in a situated-learning context (Llinares). The study in this chapter differs from these studies in that while it explicitly deals with change, it does not describe relationship of beliefs or change through external interventions to facilitate teacher growth/change and it focuses on belief structure.

Review of the literature on mathematics teachers' beliefs, as reflected in the preceding sample of studies, also reveals that less attention has been given to belief structure and, particularly, its possible relationship to change or growth of the teaching of experienced teachers. Cooney et al. (1998) incorporated belief structure in framing their study of preservice secondary mathematics teachers and indicated its importance to change. This is supported by theory on belief structure that suggests the importance of exploring its connection to teachers' change in order to understand the change process involved in teachers' transformation of their classroom behaviors. A discussion of this theory follows.

Belief structure in relation to belief system has been characterized in different ways depending on the perspective one assumes. From a cognitive perspective, factors that describe how the beliefs are configured or held tend to be similar to those that can be associated with change. The general assumption is that individuals develop a belief system that contains all of their beliefs in some organized psychological but not necessarily logical form (Rokeach, 1968). This system forms a basis for giving and receiving meaning to and from the way one experiences the world (Abelson 1979; Rokeach, 1968). It contains different types or structures of beliefs. For example, beliefs have been proposed to be primary or derivative (Green, 1971), evidential or non-evidential (Green, 1971), existential, shared and underived (Rokeach, 1986), and descriptive, inferential and informational (Fishbein & Ajzen, 1975). Primary and derivative beliefs refer to the quasi-logical relation between beliefs. Evidential beliefs are supported by evidence. Existential beliefs directly concern one's own existence and identity. Shared beliefs are beliefs about existence and self-identity that are shared with others. Underived or descriptive beliefs are learned by direct encounter/experiences. Inferential beliefs go beyond directly observable events and could be based on prior descriptive beliefs. Informational beliefs are established by some source. There are also classes of beliefs (Rokeach, 1986) or isolated clusters (Green, 1971), that is, beliefs are held in clusters, more or less in isolation from other clusters and protected from any relationship with other sets of beliefs.

According to Rokeach (1968) and Green (1971), not all beliefs are equally important to the individual. Beliefs are organized along a central-peripheral dimension that reflects psychological strength or degree of nearness to self. Inherent in this particular structure of belief is a considerable inertia against change that makes it important in understanding change. Rokeach (1986) and Green (1971) suggested that the more central a belief, the more it will resist change. Thus the relative importance or centrality of various types of beliefs within a given belief system is connected to change. Green (1971) considered the importance of a belief to the believer to be determined by whether it is psychologically central. Rokeach (1968) defined importance in terms of connectedness: the more a given belief is functionally connected or in communication with other beliefs, the more implication and consequences it has for other beliefs and, therefore, the more central the belief. Rokeach's criteria for connectedness are existential, shared and underived beliefs. However, whether a belief is primary or derivative does not necessarily determine how amenable it is to change (Green, 1971). For example, a teacher who holds a psychologically central belief that mathematics is a collection of facts may be less likely to change it regardless of whether it is primary or derivative or inferential. Change can also be restricted when isolation of belief clusters occur to facilitate contradictory beliefs or when beliefs are held from a non-evidential perspective, a perspective immune from rational criticism (Green, 1971).

The above description of belief structure focused on the relationship to change of belief. However, the view that one's beliefs strongly affect his or her behavior (Ernest, 1989; Fishbein & Ajzen, 1975; Rokeach, 1986) implies some possible connection between changing beliefs and changing teaching. In the context of teaching, the assumption is that beliefs reflect teachers' expectations and what they value. As a system, they form the bedrock of teachers' intentions, perceptions, and interpretations of a given classroom situation and the range of actions the teacher considers in responding to it (cf. Pajares, 1992). Thus, investigating fundamental changes in a mathematics teacher's practice in relation to belief structure/system is being considered in this study as a viable way of gaining insights of inservice teachers' growth.

3. RESEARCH METHOD

This study evolved from an ongoing project on mathematics teachers' thinking in teaching mathematical word problems. The first year of the project focused on four experienced (ranging from 16 to 33 years) high school mathematics teachers. They were from different schools. By the time of the study, they were involved in writing and/or reviewing mathematics textbooks, conducting inservice professional development sessions, and making presentations at teachers' conferences. The teachers were asked and willingly agreed to be involved in the study. They were very articulate and open about their thinking and experiences in teaching mathematics.

Data collection and analysis for the word-problem project followed a humanistic approach (Chapman, 1999b) framed in phenomenology (Creswell, 1998). Data collection involved interviews, role-play, and classroom observations. The interviews focused on paradigmatic and narrative accounts of the teachers' past, present, and possible future teaching behaviors and their thinking in relation to "word problems", problem solving, and mathematics in general. Paradigmatic accounts highlighted the teachers' theories about a situation (e.g., problem solving). These accounts were triggered in a variety of ways, using more than one context for the same situation at different times during the interviews. For example, for the situation mathematical word problems, the participants were asked to role-play giving a presentation on word problems at a teacher conference. They were asked to role-play a cooperating teacher having a conversation with a preservice teacher about word problems. They were asked to respond to what is most important and what is least important about word problems, framed as two separate questions. Narrative accounts highlighted the teachers' lived experiences and consisted of stories that described the experiences as they occurred. The participants were asked to tell stories for a variety of contexts and situations. For example, they were asked to tell stories of lessons they taught that (i) were memorable, (ii) they liked, and (iii) they did not like. Discussions of why they chose a particular story often triggered a description of other teaching scenarios. Classroom observations and follow-up discussions of the participants' teaching covered a complete unit that included word problems. The focus was on recording what the teachers said and did and to identify scenarios for the teacher to talk about or role-play.

The data, then, embodied the teachers' thinking and behaviors, directly and indirectly, in a variety of contexts and situations that were compared for consistency before considering them as valid evidence. The data analysis for the word-problem project started with a cursory reading of all of the data for a participant to get an overall sense of his or her thinking and perspective. At this point it became evident that underlying the teachers' stories about teaching word problems were stories of growth and change in their teaching and a decision was made to pursue the latter as a study by itself. The data were then scrutinized for situations indicating change in teaching and for beliefs related to teaching mathematics. Change was identified based on the participants' explicit statements about change, their explicit comparison of teaching behaviors (i.e., then and now situations), and their individual stories that implied change. This information was compared to classroom observations of current teaching. Beliefs were identified in terms of significant statements and actions that reflected judgments, intentions, expectations, and values of the participants in the context in which they were described. Belief about mathematics was made the focus of the analysis because of its explicit occurrence and apparent dominance in each of the teachers' story of change. A narrative format was used to describe the teachers' growth in teaching mathematics from the teachers' perspective in a holistic way. These narratives formed a basis for identifying relationships between belief structure and growth. The findings are thus described in terms of the stories of change and the relationship between change and belief structures.

4. STORIES OF CHANGE IN TEACHING HIGH SCHOOL MATHEMATICS

The story of change in teaching for each of the participants is very different, but there are many similarities in terms of the relationship between belief structure and change. Elise and Mark's (pseudonyms) cases are presented here to illustrate these differences and similarities. This section provides summaries of their stories of change.

4.1. Elise's Story

The dominant belief Elise held about mathematics was "mathematics is play/game". As a beginning teacher, Elise's expectation was that she would be able to teach to reflect mathematics as play even though she did not have a clear conception of what teaching looked like to realize this. But this expectation was quickly smothered when she started her practice. For the first two years of her practice, Elise was mentored by her experienced colleagues whose approach to planning for teaching, she explained, was:

> Today you'll teach graphing, tomorrow elimination, next day substitution ... and you follow the textbook. It was lock step.

She described their teaching as a "stand and deliver" approach. Elise was told by these colleagues that their approach was the only realistic way to deal with high school mathematics, there was nothing play-like about it and she should abandon any thoughts of wanting to make it that. This triggered a conflict for Elise between what she believed about mathematics and the way she was being steered to teach it. In order to deal with this conflict, she decided that teaching mathematics was different from doing mathematics. She explained:

> You know, that that's the time when I separated mathematics from teaching mathematics. That's when ... it became internalized to me that there must be a difference between teaching mathematics and doing mathematics. But those aren't the same thing and they can never be the same thing. And that to this day frustrates me because I don't want that to be the way it is.

With this ongoing desire to resolve the tension, Elise continued to think of how to make high school mathematics be play for her students. Two years later, when she no longer felt under the influence of her experienced colleagues, for the most part, she continued with their view of planning and teaching. However, since what she was doing lacked fun, she decided that if she added some fun activities, students might start to experience mathematics as play. These activities, she explained, were used "more on specific days, at the end of a unit or at the very beginning of a unit".

Initially, these activities seemed to reduce the tension between what Elise believed about mathematics and how she was teaching it. But after a few years, she realized that they did little to foster her beliefs about mathematics in her classroom. They were "fun" in an isolated way and did not give her a feeling of play or a sense of the students engaging in play in terms of the mathematics being taught/learnt or mathematics in general. They were too detached from the core content being taught and served more of a recreational purpose in the transition from one unit to the next. Elise also became aware that, for the most part, the students were simply mimicking her instead of engaging in play or being problem solvers. As she focused on how to help them to become problem solvers, she eventually made a connection between game and problem solving, in particular, viewing problems as games and emphasizing the importance of strategies. She explained,

> I thought, if I'm going to be a good problem solver, I have ...to think about what strategies to try. And I really firmly believe as a learner, what I need to do is look at them [problems] as a game. When I play monopoly, I know the rules but it's dynamic, it changes. When I solve a problem, I have my strategies that color the rules, but it's a dynamic situation, and so sometimes I use this strategy, sometimes I use that strategy, but I'm more relaxed because it's a game.

Elise's interpretation of strategy included a way of thinking, seeing patterns, making connections, and reasoning and was seen as relevant to all areas of mathematics. For her, strategies were not just techniques to solve problems, but a way of viewing and learning mathematics. They were also "something you must see for yourself". This perspective of strategies provided a way for Elise to think of high school mathematics and her teaching of it differently. For her, in addition to fun and fun activities, high school mathematics and doing mathematics became being and focusing on strategies, respectively, and consistent with her belief of mathematics as play/game.

With these changes, Elise's teaching shifted from being "stand and deliver" to being more student-centered, but teacher guided. She tried to guide students to seeing strategies, for example, looking for patterns in developing a procedure for themselves. She started to use more questioning and less telling. She followed the textbook less and selected or developed activities in which students could discover strategies through discussion. Students worked in small-groups to figure out strategies and shared them in whole-class discussions. However, Elise often intervened in the groups with questions to guide them to a strategy or led interactive, whole-class discussion to do so. This strategy-based approach created for Elise an acceptable level of harmony between her belief about mathematics as play and her teaching. Thus, there was no further conscious effort to change her teaching until she encountered a new conflict.

This new conflict arose when Elise's view, that her approach of focusing on mathematics in terms of strategy was helping students to think for themselves, was challenged by her students' performance on the Grade 12 provincial diploma examination (required for graduation in mathematics). In the mid-90s, this exam was revised to include genuine problem-solving items and Elise was surprised when even her best students did not perform at the level she expected. In trying to resolve this new tension, Elise eventually concluded that guiding students to see strategies was not enough for them to be good problem solvers. They must be able to think for themselves to be successful. Elise decided to facilitate this by getting students to think about their thinking through writing and self-questioning. She noted,

> I've focused much more over the past five years on reflective thought for each person, each individual, and trying to not only encourage but in many ways force kids to do it. ... Reflecting on what it is that you know and what does it mean to understand the remainder theorem [for example] is very important. ... If I get them to reflect as learners then they're going to be doing better thinking, because they'll be asking themselves the questions. ... [That] is what we used to do in proof, which was, you know, if you make a statement, can you hold it to be true in all cases, and if you can't, if I can find one flaw, then you've got to refine that a bit.

The reflective process Elise added to her teaching was, for her, both a strategy and a way of making sense of strategies. With this change, her teaching started to become more student-centered and students were allowed to "play" more. This increased the level of harmony she was developing between her belief about mathematics as play/game and her teaching.

At the time of the study, Elise's teaching had evolved from a "straight stand and deliver" at the beginning of her practice to a focus on teacher-facilitated situations in which she explained,

> They [students] get to discover for themselves or try things or take risks. ... I like the kids to do more play than me giving them the process.

She, herself, also took more risks in handling the content in the classroom and was transcending her concerns about covering the curriculum by being too open-ended in her teaching. She pointed out:

> I'm much more spontaneous in class. I handle the "what ifs" much better. ...I don't mind if the kids get off topic because now I have the confidence that no matter what, ... I could get through the content if I had to. So I'm much more relaxed now about it. I still have that I have to get through the content thing but I'm much more flexible ...and all of those aspects in mathematics that I really love come into the classroom more often.

4.2. Mark's Story

The dominant belief Mark held about mathematics that was related to changes to his teaching was "mathematics is experience". As a beginning teacher, Mark accepted, what he called, the traditional routine of teaching high school mathematics, a routine that focused on direct instruction. He explained,

> I knew the routine to follow. I knew what was expected, the expected traditions. ...When I got my own classes to teach, that's what I started doing and continued to do for many, many years.

"What mathematics is" was not important to Mark. His focus was on satisfying "the expected traditions" treatment of mathematics. This was:

> We presented the skills and processes to students. ... We created for them a structure, they'd follow the structure and then we knew that they knew mathematics, because they could model the structure.

Mark, however, did not consciously consider this treatment of mathematics to be a personal belief, but the way it was for high school. He considered his personal belief to form several years later when he started teaching grades four and six in addition to high school grades. His experience with these elementary grades led to the construction of his belief that mathematics is experience. However, while this belief made sense for elementary school mathematics, when Mark decided to apply it to teaching high school mathematics, he was not sure how to do this. His interpretation of experience in relation to mathematics, at this point, was in terms of hands-on activities and these seemed impossible with high school mathematics. This triggered a conflict for Mark between what he now believed about mathematics and the way he had been teaching it. To deal with this conflict, initially, he decided that teaching elementary school mathematics was different from teaching high school mathematics. He explained:

> At the time, I saw them as two distinct levels of students with very different content that required different approaches to teaching it.

Mark soon realized that this conclusion did not resolve the conflict for him and continued to think of how to make high school mathematics "be experience" for his students. He eventually shifted his focus from the content to the students' behavior, which he found to be problematic because of the lack of active participation. He was able to associate this with experience in that it should involve sharing/communication, which was not happening in his class. This allowed him to make his first significant change to his classes. He explained:

> The first thing I did when I started to decide that mathematics had to be more of an experience and a community experience was to take my desks out of rows, and put them in clumps. ... In my high school classes, whether they're in grade 10 or they're in calculus, they do not sit in isolated rows, they sit in groups of three and four desks.... It created the opportunity for them to communicate.

Initially, this change seemed to reduce the tension between what Mark believed about mathematics and what was happening in his classroom. But after some time, he realized that while allowing students to communicate made some difference to his classroom, by itself, it was not resolving the conflict. The communication was occurring in an isolated way, limited to students' intra-group talk about getting the answer to practice exercises. This did not give him a sense of the students engaging in experience in terms of the mathematics concepts being taught or mathematics in general.

Mark now felt that the ongoing tension was because of how he was dealing with the content. He was still doing a lot of telling and manipulating symbols on the chalkboard, which he realized limited students' active/experiential involvement with the mathematics. With this awareness, he felt that high school students should work with "non-symbolic [mathematical] situations and approaches". Non-symbolic was his way of relating high school mathematics to hands-on or concrete situations. But to implement it, he eventually associated it with another quality he attributed to experience – connections. He explained:

> Experience is how people make connections and see and react to a situation ... [and] connect to the world that they see around them.

In relation to mathematics, Mark initially considered connection as context, in particular, personal real world experiences. Thus,

> You try to bring your experiences of the world to connect the math concept or problem back to the world. Kids need to connect the [math] problems to the world, otherwise it's just hanging in a space, an empty space.

With this view, Mark started incorporating real world scenarios into his presentation of the mathematics concepts. He did this by telling students about the connections, i.e., giving them real-world examples. When after a couple of years Mark realized that this did not resolve the conflict between mathematics as experience and his teaching satisfactorily for him, he started focusing on it again. He eventually realized that while the way he implemented connection and communication in his teaching was helping him to live mathematics as experience, his students were not living it in the same way. He was still doing much more talking than the students, telling them about the mathematics concepts and connections based on his experience or what he thought were their real-world experiences. This became problematic for him and allowed him to revisit his initial way of implementing communication. He now felt that it should be more than having students talk with a neighbor in class about how to get the answer and expanded communication to a way of viewing and doing mathematics, as he would explain to his students:

> I also said to them in my classroom, that mathematics to me was a communication skill … I keep saying to my high school kids, my request from you is that you try and be a contributor and not just an observer to what goes on in this classroom. It's [math] something we share, not something I share.

Mark's teaching thus shifted from lecturing to leading discussions with the class where students got to share their thinking.

Accompanying this expanded view of communication was a related expansion of Mark's view of connection, which he now considered as "how students should see and react to the concepts of the mathematics" for it to be experience. Specifically, connection now became personal experience, real world situations, a way to see the value of mathematics, and students' active involvement in developing the mathematics concepts. Thus, there were times when, as Mark explained, he got his students to solve mathematical problems, even in his calculus class, completely without paper and pencil working with real or imagined physical objects.

In the midst of expanding his understanding of communication and connection to better reflect mathematics as experience in his teaching, Mark associated problems with experience. Problems were "concerns" one encountered and tried to resolve. Such concerns emerged from physical/real situations. Mark also associated this view of problems with the creation of mathematics.

> It was the problem, every time in mathematics that's what I often say [to students], somebody came across a concern, there was a snag in their need, they had a need … and they didn't know what to do with it, so how did they solve that problem, it was a problem, it was a physical problem or a world problem, … they created new kinds of mathematics to solve that.

Thus, in relation to experience, Mark extended problems in mathematics to situations involving concerns or "snags" for students. This became a focus on questioning instead of telling in his teaching. Mark started to place significant emphasis on questioning as a teaching strategy. He pointed out:

> It's how I ask them [students] the questions. … It's how you ask the questions that stops them. It's when you make them focus on something that they had just sort of brushed over, and they have to back up, and they have to talk to you about it. It's the questions that I think make a difference. It's how you ask what questions as the teacher that also make math experience for them.

Similarly, problem solving also became more than solving written problems and was perceived in an experiential way, as Mark explained:

> Problem solving is an attitude. It's how you, how they think about what you're talking about. So if I'm introducing a topic, I'm asking questions all the way through and those questions for those kids are problem solving.

With this perspective of problems, Mark made both problem posing and problem solving an important part of his teaching. They became a way of facilitating communication and connection, and a way of investigating and coming to know mathematics.

With these changes, Mark's teaching reflected mathematics as experience in terms of communication, connection and problem solving/posing and allowed him to achieve an acceptable level of harmony between his belief about mathematics and his teaching. At the time of the study, his teaching had evolved from a lecture/direct instruction perspective at the beginning of his practice to a student-centered, inquiry perspective. He now saw his high school teaching as being compatible with his experience with the elementary grades.

5. REFLECTION ON BELIEF STRUCTURE FOR MATHEMATICS AND CHANGE

Figures 1 and 2 are schematic representations of Mark and Elise's beliefs about mathematics as reflected in their stories. These schemas embody those aspects or relationships of the belief structure that seemed to be important in the evolution of their teaching of high school mathematics.

Fun Fun-Activities
↑ ↗
Mathematics → Play/Game → Strategy
↘
Reflection
[Self-Monitoring]

Primary Belief:	Mathematics is play/game
Primary Attribute:	Play/game
Inferential Beliefs:	Math is fun, fun activities, strategy, reflection
Inferential Attributes:	Fun, fun activities, strategy, reflection

Figure 1. Elise's beliefs

Mark and Elise's beliefs of mathematics as experience and play, respectively, emerged from and were supported by their personal experiences. For Elise, the belief emerged from her experience as a student of mathematics. She explained,

> As a student in the classroom, I just thought it [math] was a blast... something that you absolutely love to do for no other reason. ... I mean, I just loved the thrill of the chase. I loved proof. You know, it's the, it's a game, like it's just play, and it has a set of rules but it doesn't really have a set of rules.

Hands-on Communication
↑ ↗
Mathematics → Experience → Connection
↘
Problem Solving

Primary Belief:	Mathematics is experience
Primary Attribute:	Experience
Inferential Beliefs:	Math is Hands-on, communication, connection, problem solving
Inferential Attributes:	Hands-on, communication, connection, problem solving

Figure 2. Mark's beliefs

Thus her lived experience with doing mathematics, independent of how it was taught, provided the context for her belief. For Mark, the belief emerged from his experience as a teacher, specifically, teaching elementary school mathematics. As described in Mark's story, several years into his career, he started teaching elementary school mathematics along with teaching high school mathematics at the same school. The elementary teaching was supposed to be tentative, but after two years, Mark developed very strong positive emotions for it and chose to continue with grade six for a few more years. He felt that for the first time he was able to help students to enjoy and understand mathematics and saw his use of hands-on learning activities as playing a significant role. This context eventually framed the view of mathematics Mark called his own, i.e., mathematics is experience.

In Fishbein and Ajzen's (1975) language, Elise and Mark held their beliefs of mathematics as a *relationship* with mathematics being the *object* and play and experience, respectively, the *attribute* of the belief. This relationship is also in the form of a metaphor with mathematics being the first domain and play/experience the second, with the former being understood in terms of the latter. The beliefs were also primary (Green, 1971) because they were developed directly from Mark and Elise's lived experiences. More importantly, the beliefs seemed to be central, i.e., psychologically strong (Green 1971) and functionally connected (Rokeach, 1968). The beliefs seemed to satisfy Rokeach's criteria for connectedness in that they could be considered as *existential*, *shared* and *underived* (Rokeach, 1968). They were existential in that Mark and Elise perceived them as their own and their way of being and valued them as self-enhancing. Thus, they were related to Elise and Mark's identities. The beliefs were also shared with others, in particular, Mark and Elise's students. They both explained that they consciously tried to make their students aware of their views of mathematics. Finally the beliefs were developed from direct encounter with, or active participation in, teaching and learning situations as previously described.

The psychological strength of the belief was also evident in the passion and conviction Elise and Mark displayed for their belief. Each of them was very critical of any thinking or actions of other teachers that was not consistent with her/his belief. They were very judgmental of their own teaching when it was perceived to be in conflict with their belief and not vice versa. Finally, there were *inferential* (Fishbein & Ajzen, 1975) or *peripheral* (Green, 1971) beliefs associated with mathematics based on the beliefs/attributes about play/experience as represented in Figures 1 and 2. The attribute of the primary belief was also the object of another belief with attributes that made sense to deal with the classroom phenomenon. These attributes generated inferential beliefs for mathematics, for example, strategy and reflection for Elise and communication and connection for Mark. This structure of Elise and Mark's beliefs of mathematics seemed to play an important role in shaping their teaching. In addition, the humanistic perspective of mathematics embodied in the beliefs was also important to facilitate the shift to a student-centered classroom.

Pedagogical conflict played a significant role in when and how changes occurred in Mark and Elise's practice. *Pedagogical conflict* is being used in reference to perturbations grounded in classroom experiences. In Mark and Elise's cases, this relates to conflicts between teaching act (e.g., clustering students' desk), teacher's expectation/intention of teaching act (e.g., students' collaboration) and outcome of teaching act reflected by students' performance, behavior or attitude (e.g., no meaningful discourse). Substantive changes in Mark and Elise's teaching were preceded by pedagogical conflicts and a desire to resolve them. This is consistent with the position that emotional and cognitive response when confronted by strong conflict is a condition of change (e.g., Rokeach, 1968). However, both the conflicts and resolutions in Elise and Mark's cases seemed to be influenced by the content and structure of the beliefs about mathematics. Fundamental shifts in their teaching were generally directed to eliminate a conflict between their teaching and their beliefs about mathematics that created a state of disequilibrium for them. The belief schemas in Figures 1 and 2 provide a way of understanding these tensions and their resolution.

The psychological strength with which Mark and Elise held their primary beliefs (Figures 1 and 2) about mathematics likely influenced their strong desire and persistence in trying to have their teaching reflect these beliefs. However, while this attitude was necessary to bring about change, it was not sufficient to allow Mark and Elise to implement it. Similarly, while the primary belief dictated their goals for their teaching, by itself, it did not enable the initial implementation and subsequent changes to their teaching. It was not applied directly to their teaching. Instead, the primary belief seemed to be held as a theoretical construct that required bridging to action. The inferential beliefs (Figures 1 and 2) were the necessary components of the belief structure required to bridge the primary belief with practice.

The primary belief seemed to form an overarching theoretical perspective connected to the context from which it was derived (the primary context), i.e., doing mathematics for Elise and teaching elementary mathematics for Mark. This connection was initially articulated as an inferential attribute (Figures 1 and 2) – fun, for Elise and hands-on for Mark, which made sense in relation to the primary context. The connection, however, seemed to create the initial barrier to implementing the primary belief to a new context – teaching high school mathematics. "Fun" and "hands-on" did not allow Elise and Mark, respectively, to connect the primary belief to teaching high school mathematics and initiated a pedagogical conflict. The conflict, however, did not lead them to question the primary belief but the appropriateness of the new context that initially seemed incompatible with the primary context. For Elise, doing mathematics (the primary context) was apparently different from teaching high school mathematics and for Mark, elementary school mathematics and students (the primary context) were different from that for high school. But Mark and Elise's persistence to resolve their pedagogical conflicts allowed them not to make this initial perception of the contexts a barrier. Instead, they seemed to become more reflective about the nature of the primary context, personal experience related to the primary attribute (Figures1 and 2) and the high school classroom (the new context), and eventually noticing and identifying what was present in the first two and missing from the latter to establish the connections. The connection was articulated as an inferential belief (Figures 1 and 2). Once the connection was made, implementation of the primary belief became possible for them. Change in teaching occurred in response to implementing what was identified as missing.

In Elise's case, the initial pedagogical conflict led to her associating game as fun activities, noticing the lack of them in her class and subsequently adding them to her teaching. In Mark's case, he associated experience with communication, noticed the lack of students' sharing in his class and subsequently added it to his teaching. With the conflict not fully resolved, or the emergence of a new conflict, the teachers developed or became aware of other connections/inferential beliefs, noticed the corresponding differences in their classrooms, and then added them to their teaching - Elise added strategies and reflection and Mark added connection and problem solving. Each addition to their teaching likely increased the psychological strength of the primary belief and their commitment to it as a basis of their teaching.

The process of noticing and associating the inferential attributes to mathematics seems to be similar to the generative process of metaphors, i.e., the primary belief was behaving as a generative metaphor. Generative metaphor (Lakoff & Johnson, 1980; Schön, 1979) facilitates learning through a process that involves generating or structuring one concept in terms of another. As Schön (1979) noted, "Metaphor refers to a certain kind of product … and to a certain kind of process, a process by which new perspectives on the world come into existence (p. 254)". The latter situation refers to the generative quality of metaphor in terms of helping individuals to generate new perceptions, explanations and inventions to understand and deal with their world. Since Mark and Elise's primary beliefs were held as metaphors, this generative process makes sense in terms of the generation of attributes of one domain (play and experience) to broaden perspective and understanding of another domain (mathematics) in a way that made it applicable to a perceived different context (high school mathematics teaching). As with generative metaphors, articulation of the inferential attributes, i.e., the new perceptions or explanations, was necessary to understand and deal with mathematics as play and experience in the context of teaching high school mathematics.

The study did not consider the relationship between the inferential beliefs. However, these beliefs did not necessarily exist independent of each other. For example, in Elise's case, fun can be associated with identifying and applying strategy and self-monitoring/reflection can be associated with understanding strategy and enhancing ability to select or develop and use them. The importance of such connections was more evident in Mark's case in that they directly influenced the changes in his teaching. For example, his interpretation of connection influenced his interpretation of communication and the two played a role in his associating experience with problem posing and solving. This suggests the need for harmony among the inferential beliefs for the changes to teaching to be in harmony with the primary belief. The study also looked at only one part of Mark and Elise's belief systems that was related to their teaching, i.e., beliefs about mathematics. As Ernest (1989) pointed out, there are other beliefs, for example, about teaching, learning and assessment, which play an important role in determining teaching.

Given the assumed relationship between belief and action, e.g., focusing on the teachers' actions is indirectly dealing with the beliefs they embody, changes in teaching could likely be a reflection of changes in beliefs about teaching and learning. On this basis, one could argue that an alternative way of understanding the pedagogical conflicts in Mark and Elise's cases is in terms of possible incompatibility between their beliefs about mathematics and their beliefs about teaching and learning. This, however, will require further evidence than is available in Mark and Elise's cases to validate such a position. Thus, unlike this study that evolved unintentionally from data for another project, a study that focuses explicitly on change over a teacher's career should consider these other beliefs to provide a fuller picture of the situation. The other two participants of this study had stories of change with an underlying belief structure similar to that described for Mark and Elise. But ongoing work with a range of teachers, perhaps at different stages of their careers, is necessary to gain further understanding of the belief structure-change relationship. Finally, while the teachers' beliefs about mathematics in a pedagogical context were explicitly dealt with, their mathematical beliefs were not explored. As Wilson and Cooney (this book) point out, examining mathematics teachers' pedagogical beliefs and their mathematical beliefs separate from each other could be limiting our understanding of their beliefs and practice. However, this study can provide a basis to build on in terms of some important factors to consider in order to examine teachers' growth by considering the interplay between the teachers' pedagogical and mathematical beliefs.

6. SOME IMPLICATIONS

Elise and Mark's cases reflect growth in their practice in that the same belief and corresponding perspective of teaching were being built up, but change is embodied in this growth. Data presented in the study suggest that belief about mathematics plays an important role in determining or defining teaching, particularly in terms of influencing how students are engaged in the content. In addition, the belief structure for mathematics can influence the teaching of mathematics and vice versa, to keep them in equilibrium. Thus, the findings support the position that in order to help teachers to change their teaching, it is important to understand not only what they believe but also how the beliefs are held – the belief structure. The cases of Mark and Elise provide one way of understanding this structure and the implication for change. They also suggest that a primary, central belief may be a necessary focus of the belief structure in relation to change in teaching mathematics.

Thus, in considering change in teaching mathematics, e.g., to make it compatible with a reform perspective, two likely situations to be dealt with are: (i) deconstructing an existing set of primary and inferential beliefs that are incompatible with a reform perspective and constructing a new set, and (ii) constructing new or additional inferential beliefs for an existing primary belief that is compatible with a reform perspective. In both situations, it would be important to understand the set of primary and related inferential beliefs underlying the aspect of teaching to be changed and the functional connection (Rokeach, 1968) and centrality of these beliefs. It would also be necessary to understand from the teacher's perspective the context from which the existing primary belief was derived or how an alternative one could be derived and the relationship between this context and the context to which it will be implemented. It may also be important to understand a possible generative process that connects these contexts since having a certain belief may not necessarily mean that one can convert it into action in the classroom without such a process.

In determining an approach to influence change, then, it will be of importance to take into account the relation between the aspect of teaching to be changed and the set of primary and inferential beliefs that are affected most immediately by the approach. Such an approach should involve experiential contexts from which to derive the primary and inferential beliefs, consistent with Rokeach's (1968) criteria for connectedness. One dilemma, however, is knowing what alternative experiences one should be exposed to in order to influence all of the appropriate beliefs. As the findings of this study suggest, belief structure and teaching are unique and personal to the individual. Thus, the same learning activities or situations may not allow different teachers to deal effectively with their personal circumstances even if a teacher holds beliefs consistent with a reform perspective. In addition, such teachers would likely require different experiences from those with beliefs incompatible with reform perspectives. For example, they may need experiences to allow them to notice and form new or additional inferential beliefs for themselves. Being exposed to cases like those of Mark and Elise could help to make them more sensitive about what to notice in their individual situations.

In general, however, the study suggests that belief structure can be complex and simply exposing mathematics teachers to alternative beliefs or contexts may not be sufficient to alter their teaching accordingly. Another factor that could contribute to this is that the belief structure could include connections to aspects of the teachers' lives other than teaching. Thus undoing it in the context of teaching could also mean undoing a teacher's identity in other fundamental ways. Depending on what mathematics, for example, is being related to and how important or functionally connected that is to other centrally held beliefs that are significant to the person's identity could be an important determinant of change of beliefs about mathematics and teaching. Thus while a teacher may be willing to change her/his teaching, she/he may not be willing, consciously or subconsciously, to change other aspects of her/his identity. This could help to explain why some teachers do not change their teaching fundamentally in spite of apparently meaningful experiences and support. It also implies the need for ongoing consideration and investigation of meaningful ways of facilitating change within a humanistic perspective of conceptualizing the teacher.

7. NOTES

This chapter is based on a project funded by a grant from the Social Sciences and Humanities Research Council of Canada.

8. REFERENCES

Chapman, O. (1999a). Inservice teachers' development in mathematical problem solving. *Journal of mathematics teacher education, 2*, 121-142.

Chapman, O. (1999b). Researching mathematics teachers' thinking. In O. Zaslavsky (Ed.), *Proceedings of the 23rd conference of the International Group for the Psychology of Mathematics Education (PME)* Vol. 2 (pp. 185-192). *Haifa, Israel: The Technion - Israel Institute of Technology.*

Chapman, O. (1997). Metaphors in the teaching of mathematical problem solving. *Educational Studies in Mathematics, 32*, 201-228.

Cooney, T. J., Shealy, B. A., & Arvold, B. (1998). Conceptualizing belief structures of preservice secondary mathematics teachers. *Journal for Research in Mathematics Education, 29*(3), 306-333.

Creswell, J.W. (1998). *Qualitative inquiry and research design.* London: Sage Publications.

Ernest, P. (1989). The knowledge, beliefs and attitudes of the mathematics teacher: A model. *Journal of Education for Teaching, 15*, 13-33.

Fennema, E., & Nelson, B. S. (Eds.). (1997). *Mathematics teacher in transition.* Mahwah, NJ: Lawrence Erlbaum.

Fishbein, M., & Ajzen, I. (1975). *Belief, attitude, invention, and behavior: An introduction to theory and research.* Menlo Park, CA: Addison-Wesley Publishing Company.

Green, T. (1971). *The activities of teaching.* New York: McGraw-Hill.

Lakoff, G., & Johnson, M. (1980). *Metaphors we live by.* Chicago: University of Chicago Press.

Lloyd, G. M., & Wilson, M. (1998). Supporting innovation: The impact of a teacher's conceptions of functions on his implementation of a reform curriculum. *Journal for Research in Mathematics Education, 29*(3), 248-274.

Pehkonen, E. (1994). On teachers' beliefs and changing mathematics teaching. *Journal für Mathematik-Didaktik, 15*(3/4), 177-209.

Pajares, M. (1992). Teachers' beliefs and educational research: Cleaning up a messy construct. *Review of Educational Research, 62*(3), 307-332.

Raymond, A. M. (1997). Inconsistency between a beginning elementary school teacher's mathematics beliefs and teaching practice. *Journal for Research in Mathematics Education, 28*(5), 550-576.

Rokeach, M. (1968). *Beliefs attitudes and values.* San Francisco: Jossey-Bass Inc., Publishers.

Schön, D. (1979). Generative metaphor: A perspective on problem solving in social policy. In A. Ortony (Ed.), *Metaphor and thought* (pp. 255-283). New York: Cambridge University Press.

Thompson, A. (1992). Teachers' beliefs and conceptions: A synthesis of the research. In D. A.Grouws (Ed.), *Handbook of research on mathematics teaching and learning* (pp. 127-146). New York: Macmillan.

CHAPTER 12

SALVADOR LLINARES

PARTICIPATION AND REIFICATION IN LEARNING TO TEACH: THE ROLE OF KNOWLEDGE AND BELIEFS

Abstract. The role of student teachers' beliefs and knowledge in learning to teach is studied from a situated learning perspective, considering participation and reification as processes involved in the negotiation of meanings. From the situated learning perspective, what is learnt depends on what is problematized and reified in learning environments. Hence, knowledge and beliefs, as both reference and target (focus) in the reification process, are discussed as factors influencing how and what may be learnt in mathematics teacher education programs. Finally, issues on becoming a teacher and learning to teach at university are analyzed based on learning in practice.

1. TEACHER LEARNING AND BELIEFS

Becoming a mathematics teacher is a process that involves developing knowledge, beliefs, knowledge-based skills, and awareness of oneself as being a teacher able to engage in teaching mathematics. In this process, the beliefs and knowledge that student teachers bring to the teacher education program are references in their learning to teach, since they affect what and how they learn. The general assumption is that, as in any active process, when learning to teach, student teachers interpret experiences through their existing conceptual structures and according to their ways of participation (Brown & Borko, 1992; Calderhead, 1996; Pajares, 1992). From this point of view, what student teachers consider to be problematic in their experience in the teacher education program turns into a learning space for them. How and on what student teachers focus their attention will determine what is learnt and how it is learnt. Although student teachers make sense of their learning experience through the lens of what they already know and believe, at the same time student teachers' knowledge and beliefs must become a target (focus) of their learning (Borko & Putman, 1996; Richardson, 1996). This twofold aspect of the role of knowledge and beliefs in student teacher learning poses issues for our understanding of the process of learning to teach. If knowledge and beliefs can be both the "focus" and "lens" of learning to teach, it is necessary to determine how these two roles interplay, in order to understand better how student teachers learn and why some activities in mathematics teacher education are more effective than others.

G. C. Leder, E. Pehkonen, & G. Törner, (Eds.), Beliefs: A Hidden Variable in Mathematics Education? 195-209.

New ideas about the nature of knowledge, thinking and learning, from situated learning perspectives, afford new analytical and conceptual constructs to study the roles of knowledge and beliefs in teacher learning. There are two assumptions in situated learning perspectives that have important implications for teacher learning. First it is assumed that the physical and social contexts in which the activity takes place are an integral part of the activity, and that the activity is an integral part of the learning that takes place within it. Secondly, there is a focus on interactive systems that include individuals as participants interacting with each other as well as with materials and representational systems (Putman & Borko, 1997, 2000). A situated view provides a way to integrate context, the learning activity and development of the identity of being a teacher, with a view to understanding the relationship between student teachers' beliefs and knowledge and learning to teach. An understanding of how student teachers become attuned to the constraints and possibilities of mathematics teaching, and become more centrally involved in the practices of the community of primary teachers, can be developed by considering how knowledge and beliefs affect the student teachers' practice in their learning environments.

In the next section, I describe the notions of community of practice and negotiation of meanings in the teaching of mathematics, and introduce the reification process, as an analytical tool in the study of how student teachers' knowledge and beliefs become the targets of their learning. Next I sketch an example, to illustrate how student teachers' beliefs and knowledge become points of focus around which they organize the negotiation of meaning of elements of mathematics teaching practice. This characterization, involving an analysis of the reification process and the forms of participation considered as constituting the process for learning to teach, provides room to reflect on the role of beliefs and knowledge in learning in practice. Finally, the process of becoming a primary teacher is discussed in relation to the different forms of participation in the practice of mathematics teaching in which they can take part at university and at school.

2. A SITUATED VIEW

2.1. Community of Practice

From situated learning perspectives, learning to teach can be hypothesized as becoming attuned to the constraints and possibilities of mathematics teaching and becoming more centrally involved in the practices of the primary teachers' community (Lave & Wenger, 1991). Lave and Wenger argue that:

> ...learners inevitably participate in communities of practitioners and ... the mastery of knowledge and skill requires newcomers to move toward full participation in the social and cultural practices of a community ... A person's intentions to learn are engaged and the meaning of learning is configured through the process of becoming a full participant in a socio-cultural practice. This social process includes, indeed it subsumes, the learning of knowledgeable skills. (p. 29)

Hence, the elementary student teacher's learning may be seen as a shift from peripheral to fuller forms of participation in different aspects of mathematics teaching, and as the acquiring of knowledge-based skills as the student teacher comes to participate in the decision-making in communities of mathematics teachers. Viewing learning to teach as a shift from peripheral to fuller forms of participation in mathematics teaching practice refers both to the development of knowledgeable, skilled identities in practice and the reproduction and transformation of communities of practice. Therefore a student teacher should be involved in performing new tasks and functions and mastering new understanding as a mathematics teacher in the teacher education program. In this sense, activities, tasks, knowledge, beliefs, and social relationships do not exist in isolation, but rather they are part of broader systems of relationships in which they have meaning. Hence, it is natural to think that the meaning of the mathematics teaching practice that the student teacher generates in learning to teach depends partly on how knowledge and beliefs become the focus of attention in the negotiation of meanings for elements in mathematics teaching practice.

Viewing teacher learning from the perspective of negotiation of new meanings for teaching mathematics involves understanding the teaching of mathematics and discussing different interpretations of what mathematics teaching is about, so as to produce and adopt tools, artifacts and representations to engage in the practice of mathematics teaching. It is precisely in the development of these processes that the dual role of student teachers' prior knowledge and beliefs, as focus and lens, might indicate the nature of what is learnt.

Furthermore, the notion of community of practice emphasizes the social process for adopting shared goals in the practice of mathematics teaching. Since recent reforms in mathematics teaching challenge the beliefs that support existing school practice, reflecting on student teachers' beliefs in mathematics teacher education programs is a key issue. For example, the new frameworks for mathematics teaching stress that all students should be engaged in mathematical thinking and reasoning, challenging the prevailing norms for mathematics instruction and imposing new demands on the teacher. These new suggestions often run counter to the experiences that student teachers themselves had at school. The challenge arises when student teachers come into a teacher education program as learners with the very beliefs that the new reform is trying to overcome.

2.2. Negotiation of Meanings

Studying the process for learning to teach mathematics involves understanding the overlapping roles of beliefs and knowledge in relation to how student teachers participate and on what they focus the negotiation of meaning. Wenger (1998) argues that through the negotiation of meanings people gain experience of the world and their engagement in it as being meaningful. A sense of continuous interaction, gradual achievement and give-and-take is conveyed. Wenger (1998) situates the meaning in a process, which he calls negotiation of meaning, that involves the interaction of two main processes: participation and reification. By participation, he refers to a process of taking part and also to the relationships with others. In relation to the concept of reification, Wenger refers to

> the process of giving form to our experience by producing objects that congeal this experience into 'thingness'. In doing so, we create points of focus around which the negotiation of meaning becomes organized ... A certain understanding is given form. This form then becomes a focus for the negotiation of meaning. (pp. 58-59)

For Wenger,

> Any community of practice produces abstractions, tools, symbols, stories, terms and concepts that reify something of that practice in a congealed form ... with the term reification I mean to cover a wide range of processes that include making, designing, representing, naming, encoding, and describing, as well as perceiving, interpreting, using, reusing, decoding and recasting. (p. 59)

For example, reifying the notion of fractions as a teaching-learning object, or beliefs about mathematics learning as an object of reflection, may change the student teachers' experience of their world by focusing their attention in a particular way and affording new kinds of understanding. So learning to be a teacher might be described through the processes of participation and reification of meanings about the teacher's role, the nature of school mathematics, the teaching/learning process and the school subject-matter. The reification process means that student teachers convert some aspects of the practice of mathematics teaching into a problem. For a belief or some aspect of knowledge to become a focus for the negotiation of meaning it is necessary to consider it as being problematic in the context of practice (Wilson & Cooney, this volume).

3. LEARNING IN PRACTICE

3.1. An Example

The situated perspectives approach to teacher learning suggests the importance of activities in teacher education programs that foster ways of thinking and practising as an expert. The following analysis of an episode from a course with activities requiring these characteristics is used to illustrate the role of knowledge and beliefs in the reification processes during the negotiation of meaning. This episode involves a small group of student teachers, from a cohort of 45, in the second year of the primary teacher education program (grades 1-6, with 6-12 year-old pupils).

The activities in the course were designed to illustrate the notion of partial participation in segments of mathematics teaching. The general assumption was that the learning environments and the ways that the student teachers participated would influence their beliefs and knowledge in various domains. One of the goals of the course was for the student teachers to analyze pupils' ways of thinking, drawing on their mathematical knowledge, specific mathematical pedagogical content knowledge, and pedagogical reasoning. The learning environments were organized around multifaceted problems in which the attempts to solve them showed the need for acquiring new "conceptual tools" and the development of a set of beliefs in order to teach mathematics in line with the reform, viewed as a socio-cultural practice. A prior analysis of the activities presented to the student teachers was undertaken to see whether the topics that we wanted them to discuss would emerge from their work (Llinares, 1999). One of the activities within the course was analysis of cases –

practical teaching problems - describing the difficulties that Primary School pupils have with mathematics. The purpose of these activities was to increase the student teachers' knowledge about the pupils' ways of thinking, thus constructing a space where they would be able to discuss their beliefs about school mathematics, teaching and learning, and develop a discourse on practice. The cases described real world events and were designed to appeal to student teachers. These activities were situated in the context of the overall social practice of teaching mathematics. Hence they may be deemed to be authentic activities. Student teachers discussed the cases in small groups and then their solutions with the whole class. One of the cases presented to student teachers consisted of two cartoon strips and a videotape showing the difficulties of two pupils from the 5th grade with the concept of fractions and the modes of representation. The first cartoon showed one pupil's response to the task "represent 5/4 of a rectangle". The pupil drew a rectangle and divided it into five parts as if he were making divisions in a circle (from a point in the center trying to represent circular sections), shading them in to show that it had 5 quarters. The task in the second cartoon strip asked: "How many counters were two thirds of 6 counters?" It showed one pupil's attempt to solve this by drawing a circle and then dividing it into three parts by placing one counter on each part and giving two counters as the answer. The student teachers also watched a videotape that captured the interactions between the teacher and the pupils, and then discussed this in small groups.

In the next section I use the notions of dimensions of practice, to organize the analysis of the role played by student teachers' beliefs and knowledge in creating points of focus around which the negotiation of meaning is organized in the discussion in a small group.

3.2. Dimensions of Practice

The interactions between student teachers with the purpose of endowing meaning to a case address the relationships between knowledge of subject matter, specific mathematical pedagogical content knowledge and their beliefs. Analysis of these interactions revealed different ways of participating in the practice and differences in how student teachers interact with and engage with others, understand and tune goals for mathematics teaching and develop and negotiate the meaning of elements of practice in mathematics teaching. The content of the interactions showed the role played by the student teachers' beliefs and knowledge in the process for creating target points on

- task definition (e.g., the analysis of pupils' ways of thinking)
- renegotiation of the meaning of the different elements (e.g., the different interpretations of the fractions)
- representation modes in mathematics learning, and
- beliefs (e.g., what it means to do mathematics at school, the teacher's role in teaching).

The relations between the case and the activity generated (whether cognitive, affective or social) determined a learning environment that revealed characteristic conflicts, common meanings and intersecting interpretations.

In this sense, mastering new knowledge and developing beliefs in relation to mathematics teaching may be seen as aspects of learning that involve the construction of an identity as a primary mathematics teacher. It is in this process of the negotiation of meaning during learning in practice that the student teachers' beliefs and knowledge may play different roles. This section describes characteristics of learning in practice using Wenger's (1998) dimensions of practice as analytical tools to reveal different "slices" of activity, and the role played by beliefs and knowledge in the generation of this practice.

Wenger argues that in order to become a peripheral member of a community of practice, one must do some learning along three dimensions that involve the following processes:

i) evolving forms of mutual engagement, such as the ability to engage with other members and respond in kind to their interactions, and discover how to engage in practice;
ii) understanding and tuning an undertaking of the community, struggling to define the undertaking and reconciling conflicting interpretations of what the undertaking is about, and
iii) developing the repertoire, styles and discourse of the community, renegotiating the meaning of different elements, producing or adopting tools, artifacts, and representations.

3.3. Mutual Engagement.

The group included four student teachers, Carmen, Rocio, Beatriz and Raúl. When they analyzed the case, the engagement in the task was different for each and showed the different responsibilities that each student teacher takes on to explore new domains. On the one hand, Carmen raised issues about mathematical meaning, whilst Raúl provided explanations from his mathematical understanding. Rocio attempted to follow Raúl's explanations, while Beatriz derived indications on how to teach. Raúl came to play the role of the "expert" in the group on particular aspects of the fraction as a teaching-learning object. But mutual engagement in practice should be connected to other dimensions to understand learning in practice. The evolving forms of mutual engagement do not provide enough information. What is really important is identifying how, through mutual engagement, these student teachers defined the task and developed meanings for the different elements of mathematics teaching.

3.4. Understanding and Tuning the Case

There were several points of attention in tuning the task. The initial focus was to understand the pupils' way of thinking. However, there was a continuous shift between this focus and the meaning of the mathematical content. When the focus was the pupils' way of thinking, the interactions concentrated on making the pupils' actions meaningful.

Rocio: The thing is that he (the boy) really did not understand what a fraction meant... a fraction is a division, and it is a way of sharing things out .... What he really says is "you have to divide it into four and take five" ...

Carmen: What the boy really thinks is ... How am I going to divide it into four and pick up five?

However, at other times the focus shifted to the mathematical meaning of the concept and the procedures. So, for example, after first trying to make the pupils' results meaningful, Carmen questioned the meaning of the mathematical content in both tasks "Draw 5/4 of a rectangle", and "what are 2/3 of 6 counters?"

Carmen: Just a minute, what is 5/4 of that?
Rocio: Dividing it into four parts and picking up five.
Carmen: Then how do you put in another part?
Rocio: Don't you remember that we did it (the task) with the math teacher last year. We spent a while trying it, but in the end I did not follow it ...

And then at another time

Carmen: Because, for example, you tell the pupil that 2/3 of 6 counters is the same as 6 times 2, divided by 3, and the child tells you, 'Well, what has that got to do with anything?' Oh, it doesn't? Because I don't even know what it has to do with anything.
Rocio: Let's go back to the first one (to the first activity). Until I understand exactly what that first child did, I can't (carry on).

The process through which student teachers understand and tune their enterprise illustrates their attempts to negotiate the meanings of pupils' mathematical activity and the mathematical content involved. However, the beliefs of the student teachers only appeared on a discursive level in this process. The student teachers did not create points of focus around which to negotiate the meanings of their beliefs as they did with the meanings of the fraction concept and about different roles for representation modes in learning. The difference that arose in the discursive use of beliefs, without beliefs becoming a target for the negotiation of meanings, will be analyzed in the next section.

3.5. Developing their Repertoire, Styles and Discourse

The struggle to endow meaning made the student teachers try to clarify their knowledge of fractions (fraction as an operator and as a part-whole), renegotiating the meanings for the concept of fractions from the teaching perspective. But student teachers also derived "what one must teach". These interactions may be understood as attempts to gain some control over meaning for a specific mathematical notion.

Some of the interactions between the student teachers, mentioned in the preceding sections as examples of the process for defining the task, may also be seen as renegotiating the meaning of different elements of the subject matter. Other interactions show how the student teachers try to develop a shared repertoire for the practice of mathematics teaching. For example, after a heated discussion about the first task in the case, the student teachers return to Carmen's initial question.

Rocio: ...then you can draw a picture, and divide it into 3 parts and you take away two

Carmen: : ...and why?

Raúl: Because, what the child has been told is that the lower part (number) is what divides and the upper part (number) is what you take away.

Carmen: So, why do we say that when we find this out, that 6 times 2 is 12 and it is divided by 3?

Raúl: Because if 6 is the unit..., wait a minute (counting 6 chips)...Six is the unit, and you take away, you divide it into 3 parts...one part, two parts and three parts (making groups with the chips). Now, you take away two...one and two. That's 4 (chips). It's not that, the trick is to multiply 6 by 2, and then you divide it by 3, that's it isn't it? It is that, what you do is also what the fraction means. Divide it into 3 parts and take 2. What happens is that the unit isn't 1, but 6 ...

Beatriz: So, one must teach it that way to the children ... With things like...

Raúl: ... this would be the second step in teaching children fractions, do you understand? What I don't understand is this part up here (he refers to the first task in the case).

After leaving the question raised by Carmen unsolved, they focus their attention on the first task in the case.

Carmen: How is he going to say "add one"?

Raúl: OK, I know what it is. It is what I was talking about before in the unit, but in a different way. Here what one must teach is the fraction concept linked to the unit ...

In this interaction, the meaning of unit and producing units, and interpreting fractions as measures and operator, become a focus for the negotiation. The student teachers talk about the meaning of fraction as an operator, as an element of teaching practice. In response to Carmen's question, Raúl struggles to make the interpretation of a fraction as an operator meaningful. Together with the meanings of fraction as an operator and a measure, student teachers derive "routines" for their future practice (how one should teach this to a pupil). Furthermore, beliefs about teaching and learning, such as "the teacher should" provide the information, and "the pupil should" repeat what he has been told, are present in these interactions - when, for example, Raúl says that the teacher should tell the pupils what they must do. These beliefs about teaching-learning and the teacher's role appear in different interactions. However, what is really important to point out here is not the content of the student teachers' beliefs, but rather how they intervene in the learning process. While the meanings of fractions are the target of the interactions - fraction as a measure and an operator - beliefs only appear on a discursive level and they are not discussed. These student teachers seem to match their beliefs to new experiences. For example, in the following interaction:

Raúl: 2/3 of 6 counters ... then ... the child must know how to do it. But this (he is referring to the first task) Do you know what it is? It means taking 5/4 of a unit. Instead of saying 6 (counters), it is this piece (pointing to the picture of the rectangle). Then, you have to take ...

Rocio: That is because ...

Raúl: .... draw this rectangle five times, and now divide these five rectangles by 4 ...
Carmen: But... they've never done it before

and then at another time,

Carmen: What I don't understand is why they (the teachers) give them these exercises if they don't know how to do them.

Carmen's statement is supported by her belief that the purpose of any activity at school is to apply the information provided beforehand by the teacher. Nobody from the group questions this belief and there is no discussion about it. For these student teachers school-level mathematical activities form part of their shared reality, hence they do not need to question it. They all believe that the teacher should first provide the information needed for performing the tasks. One consequence of this belief is that the activities that are presented to the school pupils are seen as applications.

As beliefs appear only on a discursive level in the student teachers' practice, we may think they are transparent and do not intervene in the learning. In this sense, this activity for these student teachers does not make them aware of their beliefs, so they do not feel the need to discuss their beliefs. The student teachers' beliefs do not create points for them to focus on for negotiating meaning. Although the cases in this mathematics methods course were designed with the possibility of creating space to discuss beliefs, it is possible that other kinds of support activities might be required. This episode demonstrates the difficulty of translating beliefs as references for making the situation meaningful to beliefs as a focus for negotiating meaning.

What is worth pointing out from this situation are the different roles that beliefs play in learning in practice and what implications may be drawn from the various episodes. If learning in practice means considering what is reified, talking about beliefs without the link to practical situations, or problem situations, might possibly pose one use of beliefs, on a discursive level only and acting only as a reference for the situation but not as a target for learning. This question will be analyzed in further detail in the next section.

We can see here several aspects from the three dimensions of what produces practice as an emergent structure (see above; Wenger, 1998). First, the ability to engage with other members and respond in kind to their reactions in order to display the roles of target and reference in student teachers' knowledge and beliefs. Beliefs about mathematical activity and the teacher's role appear here as a reference endowing meaning to the situation but without being reified. Secondly, different aspects of subject matter knowledge and pedagogical content knowledge were reified and a new meaning was generated. Finally, the negotiability of the repertoire enables participants to make use of the repertoire of the practice and also to engage in it. For example, the conflicts between a specific practice and school culture (in line with the pupils' practice) show the way in which some beliefs were supported. Furthermore, student teachers' engagement may be viewed as a peripheral form of participation in teaching mathematics that may be considered legitimate for becoming a teacher without fulfilling all the conditions for full membership in a community of practice.

4. BELIEFS AND KNOWLEDGE IN LEARNING IN PRACTICE

4.1. Reification

In learning to teach, student teachers bring beliefs and knowledge that intervene in endowing meaning to the situation, but at the same time, their knowledge and beliefs should also be the target of learning. In this sense, as student teachers work toward shared goals (i.e., in the learning environment described, they endow meaning to the pupils' different ways of thinking about the fraction concept and their role as teachers) together they create new meaning and understanding. Hence the object of the discussion will determine what may be learnt. It is precisely this fact that distinguishes the roles of knowledge and beliefs in learning in practice.

When the focus of interactions between student teachers was the subject matter knowledge, this illustrated the discrepancies between the knowledge about the fraction concept as an operator and the use of the counters as a mode for representation. During the interactions, the student teachers' "struggle" to make sense of the pupils' answers brought into play their knowledge and their understanding of school mathematics. The interactions amongst student teachers were focused primarily on mathematical knowledge, and much less on their beliefs about the teacher's role, learning and teaching. In this situation, the points of focus around which student teachers negotiated what matters to them were related to knowledge of subject matter (different interpretations of fractions) and pedagogical content knowledge (the link of different modes of representation to activities with fractions) but beliefs were not the focus of their interactions and were not reified.

When the student teachers attempted to understand the pupils' difficulties, it seems that their understanding of the different interpretations of the fraction concept strengthened certain beliefs about learning ("but ... they [the pupils] have never done it before") or about teaching and the teacher's role ("why do they [teachers] give these exercises to the pupil if the kids don't know how to solve them"). These beliefs were derived from their own life experience. The way in which beliefs are held (derived from other beliefs or from evidence) opens up a space in the way in which they may be reified. One feature in the reification process is to create a focus around which to negotiate the meaning. When there are no discrepancies about the meanings of the "things", no negotiation is involved, and it is simply accepted that things are that way. The student teachers "matched" their beliefs to the practice, but they did not reify them as points for discussion. The student teachers' discourse in practice shows their beliefs (by making a discursive use of them), but from a situated perspective of learning we cannot consider that they really intervened in the learning. From this perspective, the shortfalls and the difficulties identified by the student teachers turn out to be the key for generating the reification process. Hence, only when beliefs are questioned in the process for solving the activities proposed in the training program is there a possibility that they might be reified (Simon & Schifter, 1991).

When it is assumed that the reification process is an essential part of learning in practice, it highlights the difference between the role of beliefs as a reference for endowing the situation with meaning and beliefs as the "target" for the learning. Cooney, Shealy, and Arvold (1998) argue that through the encouragement of reflective practices in teacher training programs, it is possible for the student teacher to develop both as autonomous and as a participant in the mathematics education program. The duality of becoming a mathematics teacher and a member of a broader community is recognized as an aspect of developing the identity as a teacher. Reflexive activity may help to reify certain meanings about beliefs by providing the student teachers with the opportunity to discuss them. In this context, the interactions between the student teachers should be seen as a way of participating in the reification of beliefs. From this perspective, the interactions between student teachers should be developed when solving practical tasks, not merely by reflection on a discursive level. Furthermore, the way in which the student teachers believe and how they believe determines their capacity to reify the beliefs, insofar as the way in which one believes determines the extent to which one is open to evidence that potentially conflicts with what one believes (Cooney et al., 1998).

4.2. Participation

The development of identity as primary teachers shows the complexity of the relationship between learning knowledge, knowledge-based skills and the reification of beliefs as elements of mathematics teaching. The settings for learning, the different ways of participating in them and the focal points around which student teachers negotiate the meaning - what to reify - provide us with another perspective for examining beliefs in the process of learning to teach mathematics. The shift from peripheral to a fuller form of participation in mathematics teaching practice involves the possibility for student teachers to develop different aspects of practice in different learning environments. In the above example, the task (case analysis) provides opportunities for student teachers to think about the meanings of some elements of their subject matter knowledge and pedagogical content knowledge. Other aspects of teaching practice may be developed by concentrating on the learning environments, more specifically on the mathematical content (Wilcox, Schram, Lappan, & Lanier, 1991), on their beliefs (Simon & Schifter, 1991), on the link between the use of "manipulative" and general beliefs regarding how mathematics is learnt, what kind of mathematics is of the most value (Llinares, 1994), and student teaching (Borko et al., 1992). However, these different studies have shown that results are complex with several factors intervening to influence beliefs on learning to teach.

Change in the student teachers' beliefs in the course of the training programs has been shown to be a complex process in which different factors would seem to intervene (Llinares, García, Sánchez, & Escudero, 2000). Nesbitt and Bright (1999) mention the possibility that intensity of experience and a focus on children's thinking in the mathematics methods course may be the keys for helping preservice teachers to change their beliefs. However, these authors were not sure about the effect that different factors might have on student teachers' changing beliefs. The use of participation and reification constructs provides new elements for studying the effects that the different factors in the training program may have on the change in beliefs. The fact that the student teachers make use of their beliefs in order to endow an experience with meaning is not sufficient to understand learning in practice. As long as student teachers do not reify their beliefs for solving the situations in which they find themselves, it might just be that there is only a change in their discursive use and that beliefs are not included in their decision-making in practice. The question that needs to be addressed in teacher programs is how different forms of student teacher participation might be designed so that students' own beliefs might be recognized as problematic. It seems that it is not sufficient to talk about beliefs. Instead we should talk about approaching beliefs as focal points around which to negotiate meanings. These must be converted into problems in the process of solving different activities. Possibly the different roles of beliefs in learning in practice may explain the fact that student teachers use certain beliefs in their discourse that are not reflected later in their teaching (Nesbitt & Bright, 1999). Information from research into students' learning, and a wide range of activities, should be included in the training program, with explicit emphasis on negotiation and reification of the beliefs.

5. ISSUES IN TEACHER EDUCATION

5.1. Becoming a Primary Teacher

The above observations illustrate the notion of partial participation in segments of mathematics teaching practice and different aspects of the process for learning to teach, highlighting the different roles of student teachers' beliefs and knowledge. In the various learning environments experienced throughout the teacher education program, the student teachers are encountering new work and unfamiliar territory, reasoning and solving problems in a new domain through social relationships generated around a meaningful activity. We argue that, to some extent, they are attempting to understand the pupils' ways of thinking and are engaged in aspects of pedagogical reasoning as part of becoming a mathematics teacher. Knowledge and beliefs, as they affect participation and reification, can be seen to be essential aspects in the development of an identity as an elementary teacher. This is underlined by the shift from understanding the pupils' ways of thinking to developing meaning for mathematical topics and to identifying aspects of the teacher's role.

In the preceding section, we looked at aspects of learning in the community of practice: mutual engagement, negotiating with one another about what they are doing there, and the meanings of the artifacts used (mathematical notions, modes of representation, different instructional modes, and so on). However, the different forms of participation and identity of the persons who engage in sustained participation in a community of practice, as shown here, should be used to complement the analysis of the changing forms of participation in different phases of teacher education programs. This gives rise to new issues about the relationships between learning in practice at university, in mathematics methods courses, and student teaching in schools.

5.2. Learning to Teach at University

These notions about forms of participation and identity development may be applied to the study of student teacher learning as they engage in the variety of practices related to mathematics teaching. However, it is necessary to make a distinction between the community of practice for student teachers in the mathematics teacher education program and communities of practice for mathematics teachers, e.g., teachers engaged in a reform project such as that described by Stein, Silver, and Smith (1998). The "practice" and learning activities in the two communities are different. The difference lies in the "responsibility for learning" defined by the goals of the activity. In the case of student teachers, the fundamental goal is to become a teacher. However, in the teacher's case, the responsibility of learning is to teach. This distinction poses the issue of belonging, or not, to a community, as pointed out by Lampert (1998) who says:

> At the same time that students are arguing about what knowledge is true or useful in relation to a problem at hand, they need to be acquiring a repertoire of the tools that professional knowledge makers have made available ... we ask learners - both in the school and teacher education - not only to know the practice, but to be able to represent what they know and connect their representations with those created by other communities of discourse. (p. 61-62)

The practice for a student teacher in a community in the mathematics teacher education program is different from the practice in a community of practice for mathematics teachers (community of mathematics teaching). If the meaning of practice is what defines the community, then the meaning of practice is different when the communities of practice are also different. As Greeno (1998) points out:

> It is useful to think of the activities as being organized so that different functional purposes are central in different communities. These differences in functional organization lead to differences in the kinds of problems that are addressed in the practices of communities. (p. 80)

This situation raises many questions on teaching and learning in mathematics teacher education programs. One difference between learning to teach at the university level and learning in practice is that in the former there is professional knowledge useful for the practice of teaching - worth knowing – "situated" in a different context (university) from where it is used (school) to solve problems and create new knowledge. Undoubtedly, schools and universities are different as situations for learning. Nevertheless, in some aspects mathematics teacher training programs may be articulated so as to involve student teachers in evolving forms of mutual engagement, negotiating teaching goals and developing a shared repertoire.

In this chapter, I have attempted to illustrate that some aspects of teacher learning at the university level can be analyzed from a situated perspective of learning. When we present meaningful activities to our student teachers, certain aspects of learning can be analyzed using notions such as participation and reification as two constituent processes of the negotiation of meaning. The negotiation of meaning generated in our classrooms shows this duality of learning, demonstrating the different roles played by the knowledge and beliefs that student teachers bring to the teacher education program. From this perspective, it is possible that we may better understand the process for participation and reification in learning environments if we take into account the different roles that knowledge and beliefs can play. However, incorporating knowledge and beliefs as a means for understanding participation and reification does not remove the fact that many questions remain unanswered. In particular, the design of learning environments, so that student teachers can legitimately and peripherally participate in authentic social practice in rich and productive ways, still remains a challenge (Brown & Duguid, 1996).

Other questions raised from this point of view are: Can we carry on thinking about this learning as learning in practice? What is the meaning of "ways of knowing linked to practice" derived from a social practice framework? What is the relation between knowledge "in" the university and the knowledge "in" the practice of teaching? Is it possible to have "compatibility" between learning in the university and learning through the practice of teaching in a school? What is the meaning of "knowledge useful for teaching" in this situation? How can we best help student teachers to reify their beliefs in the practice developed in different types of courses in a teacher training program? How are beliefs reified in learning at the university linked to practice at school? Other issues raised over the last few years within the perspectives of situated learning include: the debates between instructional design and situated learning versus the experiences of intentionally designed and situated learning (Winn, 1996); the issue of learning inside or outside a community of practice – for example, creating knowledge and solving problems in schools (Lampert, 1998); the relationship between the individual's understanding and the community's knowledge. The paradoxes raised by these questions are challenges to us as mathematics teacher educators, since they provide new ways for addressing old issues.

6. REFERENCES

Borko, H., & Putman, R. (1996). Learning to teach. In D. Berliner & R. Calfee (Eds.), *Handbook of educational psychology* (pp. 673-708). New York: Macmillan.

Borko, H., Eisenhart, M., Brown, C., Underhill, R., Jones, D., & Agrad, P. (1992). Learning to teach hard mathematics: Do novice teachers and their instructors give up too easily? *Journal for Research in Mathematics Education, 23*(3), 194-222.

Brown, C. A., & Borko, H. (1992). Becoming a mathematics teacher. In D. A. Grouws (Ed.), *Handbook of research on mathematics teaching and learning* (pp. 209-239). New York: Macmillan.

Brown, J. S., & Duguid, P. (1996). Stolen knowledge. In H. Mclellan (Ed.), *Situated learning perspectives* (pp. 47-56). Englewood Cliffs, NJ: Educational Technology Publications, Inc.

Calderhead, J. (1996). Teachers: Beliefs and knowledge. In D. Berliner & R. Calfee (Eds.), *Handbook of educational psychology* (pp. 709-725). New York: Macmillan.

Cooney, T., Shealy, B., & Arvold, B. (1998). Conceptualizing Belief structures of Preservice Secondary Mathematics Teachers. *Journal for Research in Mathematics Education, 29*(3), 306-333.

Greeno, J. (1998). Trajectories of participation and practice: Some dynamic aspects of the thinking practices of teaching, educational design and research. In Greeno, J. & Goldman, S. (Eds.), *Thinking practice in mathematics and science learning* (pp. 79-88). Mahwah, NJ: Lawrence Erlbaum Ass. Inc.

Lampert, M. (1998). Studying teaching as thinking practice. In Greeno, J. & Goldman, S. (Eds.), (1998) *Thinking practice in mathematics and science learning* (pp. 53-78). Mahwah, NJ: Lawrence Erlbaum Ass. Inc.

Lave, J., & Wenger, E. (1991). *Situated learning. Legitimate peripherical participation.* Cambridge: Cambridge University Press.

Llinares, S. (1994). The development of prospective elementary teachers□ pedagogical knowledge and reasoning: The school mathematical culture as reference. In N. Malara & L. Rico (Eds.), *First Italian-Spanish research symposium in mathematics education* (pp. 165-172). Dipartimento di matemática, Università di Modena, Italy.

Llinares, S. (1999). Preservice elementary teachers and learning to teach mathematics. Relationships among context, task and cognitive activity. In N. Ellerton (Ed.), *Mathematics teacher Development: International perspectives* (pp. 107-119). West Perth: Australia: Meridian Press.

Llinares, S., García, M., Sánchez, V., & Escudero, I. (2000). Changes in preservice primary school teachers□ beliefs about teaching and learning mathematics. In S. Götz & G. Törner (Eds.), *Research on Mathematics Beliefs. Proceedings of the MAVI-9 European Workshop* (pp. 89-110). University of Vienna, Austria.

Nesbitt, N., & Bright, G. W. (1999). Elementary preservice teachers□ changing beliefs and instructional use of children□s mathematical thinking. *Journal for Research in Mathematics Education, 30*(1), 89-110.

Pajares, M. F. (1992). Teachers' beliefs and educational research: Cleaning up a messy construct. *Review of Educational Research, 62*, 307-332.

Putman, R.T., & Borko, H. (1997). Teacher learning: Implications of new views of cognition. In B.J. Biddle et al. (Eds.), *International handbook of teachers and teaching* (pp. 1223-1296). Dordrecht: Kluwer Academic Publishers.

Putman, R. T., & Borko, H. (2000). What do new views of knowledge and thinking have to say about research on teacher learning. *Educational Researcher, 29*(1), 4-15.

Richardson, V. (1996). The role of attitudes and beliefs in learning to teach. In J. Sikula, T. Buttery, & E. Guyton (Eds.), *Handbook of research on teacher education* (pp. 102-119). New York: Macmillan.

Simon, M., & Schifter, D. (1991). Towards a constructivist perspective: An intervention study of mathematics teacher development. *Educational Studies in Mathematics, 22*, 309-331.

Stein, M. K., Silver, E,A., & Schwan, M. S. (1998). Mathematics reform and teacher development: A community of practice perspective. In J. Greeno & Sh.V. Goldman (Eds.), *Thinking Practices in mathematics and science learning* (pp. 17-52). Mahwah, NJ: Lawrence Erlbaum.

Wenger, E. (1998). *Communities of practice: Learning, meaning, and identity.* New York: Cambridge University Press.

Wilcox, S., Schram, P., Lappan, G., & Lanier, P. (1991). The role of a learning community in changing preservice teachers□ knowledge and beliefs about mathematics education. *For the Learning of Mathematics, 11*(3), 31-39

Winn, W. (1996). Instructional design and situated learning: paradox or partnership? In H. McLellan (Ed.), *Situated learning perspectives* (pp. 57-66). Englewood Cliffs, NJ: Educational Technology Publications, Inc.

CHAPTER 13

GEORGE PHILIPPOU AND CONSTANTINOS CHRISTOU

A STUDY OF THE MATHEMATICS TEACHING EFFICACY BELIEFS OF PRIMARY TEACHERS

Abstract. The affective domain has in recent years attracted much attention from the mathematics research community; empirical data seem to increasingly support expert opinion that affect plays a decisive role in the process of cognitive development. One of the less researched dimensions of the affective domain is teachers' beliefs about the efficacy of their mathematics teaching. Though there are studies examining efficacy-beliefs with respect to mathematics learning, we have not been able to locate any study related to efficacy-beliefs with respect to teaching mathematics. Belief in one's ability to overcome obstacles and bring about a predetermined outcome may be decisive in motivation to undertake the endeavour, and to put in time and resources. In this chapter, based on a sample in Cyprus, we examine primary school teachers' efficacy-beliefs with respect mathematics teaching. From an analysis of self-reported questionnaires and interview data, it was found that teachers feel quite competent to teach mathematics, and that the level of efficacy improves, after diminishing during the initial career period. The preservice program seemed to make a difference to beliefs about the efficacy of their mathematics teaching; however, in general the teachers seemed to be critical about the preservice program they passed through. The findings of this study might have important implications for teacher preparation and teacher development in general.

1. INTRODUCTION

The chapter is organized in two parts. The first part consists of three sections, each reviewing respective research on: a) the affective system of teachers, including perceived self-beliefs, b) recent trends in preservice primary education and the need to account for the affective representation system, and c) the construct of "efficacy beliefs" with respect to teaching mathematics. The second part of this chapter briefly presents the rationale and the findings of an empirical study, in which we examined teachers' efficacy beliefs about teaching mathematics.

2. AFFECT AND TEACHER EDUCATION

The first part of this chapter considers the most recent conceptualizations and developments concerning three related topics: the affective domain with respect to mathematics, including efficacy beliefs; preservice primary mathematics teacher education; and efficacy beliefs in the area of mathematics education. This provides the framework and context for the empirical research that is reported in the second part.

G. C. Leder, E. Pehkonen, & G. Törner (Eds.), Beliefs: A Hidden Variable in Mathematics Education? 211-231.

2.1. The Affective System

The affective system encompasses emotional and cognitive components. The term is used to include emotions and feelings as well as attitudes, beliefs and conceptions. Beliefs, in particular, can be conceived as the personal appraisals, judgments, and views that constitute one's subjective knowledge about self and the environment. Beliefs are clearly distinct from what we include in the term "knowledge" - whether its meaning is true knowledge, viable or socially accepted knowledge - in that beliefs are subjective, do not need formal justification, and, apart from the cognitive component, they also involve an affective component. Wilson and Cooney (Chapter 8, this volume) make the point that beyond verbal proclamations and dispositions to act, beliefs could also be seen as one's conceptualizations of a certain situation. Beliefs, like attitudes, are organized around an object or a specific situation, predisposing one to respond in a favorable or unfavorable way; they are contextual, experientially formed, and emerge during action.

Richardson (1995) identified teachers' beliefs as their own theories, consisting of sets of interrelated conceptual frameworks, and as connecting self-knowledge and the act of teaching i.e., a kind of "knowledge-in-action". Llinares (Chapter 12, this volume) views teachers' learning from a situated perspective of negotiating new mathematics meanings; in this sense he argues that the student teacher's prior knowledge and beliefs should be both a focus and a target. The importance of teachers' beliefs derives from expert consensus that a teacher's knowledge is translated into practice through the filter of his/her own related belief system (Swafford, 1995). In general, beliefs are supposed to drive action, while experience and reflection on action can modify beliefs. In other words, there is a two-way interaction and mutual influence among one's belief system and one's behavior in the respective field.

Several dimensions of the construct "mathematical beliefs" have recently been extensively studied. From the philosophical point of view, beliefs about the nature of mathematical knowledge and the development and learning of mathematics are of primary pedagogical importance. Beliefs and conceptions of what mathematics is really all about, and what it means to know and learn mathematics, is a determinant of the way one views involvement with the subject, that is, the process of developing understanding and competency in doing mathematics. The students' mathematical belief systems were repeatedly found to make a difference in determining the level of their motivation and persistence in the face of difficulties. Likewise, the teachers' mathematical beliefs and views influence their general pedagogical outlook, the learning climate they will contribute to, and specifically their choices of teaching strategies and learning activities. For instance, if a teacher subscribes to the platonic philosophical view that mathematical knowledge exists "a priori", is immutable and eternally truth, he/she will quite naturally try to convey this eternal truth to students as a firm and finished product in a more or less didactic-receptive way. If, on the other hand, a teacher perceives mathematical knowledge as a fallible human creation along the quasi-empiricism view, he/she will rather be inclined to invest more resources and time in an effort to facilitate the students to make conjectures, test hypotheses, provide reasons and, in general, construct mathematical knowledge on their own through individual involvement.

In recent years, research on affect has concentrated on issues that relate to the conceptualization and analysis of the beliefs system and its connections to the learning process, including the genesis and change of mathematical beliefs. Several review papers in the area (McLeod, 1994; Pajares, 1992; Thompson, 1992) have supported the conclusion that affective responses depend on one's experiences, which include factual knowledge, feelings, and emotions that are tightly connected with the specifics of the situation.

Once attitudes and beliefs about an issue are developed, they have a degree of stability and intensity. Emotions, attitudes and beliefs form a hierarchical scale, or a linear continuum, which is characterized by an increasing level of the cognitive component and by stability, and by a decreasing trend of the affective component. That is, emotions are highly affective and of short duration, while beliefs are based on cognitive experiences, normally less affective and quite stable. The degree of intensity varies irrespective of the construct. That is, an emotion, an attitude or a belief can be strong, intense, or weak and loose, depending on the provoking situation and the degree of personal involvement and interest. Change in one's beliefs is feasible but not easy; it can occur only under certain conditions in which the individual is faced with new information and experiences that come in conflict with established beliefs. In particular, change in the beliefs of mathematics teachers could occur as a consequence of pedagogical conflict resolution and is influenced by the content and structure of his/her beliefs about mathematics per se (Chapman, this volume).

Self-beliefs and self-awareness about cognitive strategies and one's ability to achieve a specific goal, such as to solve a particular problem, have also been studied in connection to motivation, self-regulation, orienting and self-judging. Boekaerts (1999) draws a distinction between feelings that relate to the learning process, such as feelings of knowing on the one hand, and the "self-referenced cognitions and affects that bring to bear one's task value, interest, goal orientation and self-efficacy on the task" on the other (p. 573); that is between feelings about some knowledge and value judgments with reference to one's interests, satisfaction, etc. In a study focusing on two characteristics of competent problem solvers considered as teachable learning tools, that is, orienting and self-judging, Masui and De Corte (1999) draw attention to two equally important factors, the control of cognition, and the emotional motivational control. Emotional motivational control involves generating, fostering or repairing positive feelings and reacting in constructive ways to negative feelings.

The importance of affective variables in mathematics learning has been stressed in various contexts. Goldin (1998) included affect as one of the five possible representation systems that he considered as sufficient components for a unified psychological model for mathematical learning and problem solving. The other four components of Goldin's model were: the imagistic system (including visual/spatial, auditory/rhythmic, and tactile/kinesthetic), the formal notational system, and the system of planning, monitoring, and executive control. These systems develop over time and interact as humans engage in mathematical activities. Within affective representation Goldin included two types of attitudinal constructs, the relatively stable type called "global affect", and the changing states of feelings and emotions, which he termed the "local affect". In an evaluative statement Goldin (1998) regarded "the affective system of representation as the most fundamental to understand the structure of mathematical ability in students and adults" (p. 155).

One of the benefits of accepting affect as a representation system is that it "entails the recognition that affect is not simply an emotional state or an attitude, it is configured" (p. 154), which means that it could serve as either a signifier or a signified. This dual role could be traced in problem solving situations, where, for some students, mere confrontation with a mathematical problem might signify frustration, while a moderate negative feeling might result in recalling information of recent unsuccessful experiences, signify lack of progress, and lead to a new strategy.

Another consequence of the representational nature of affect is that it leads to a further level of seeking symbolic relationships among affective configurations and between affective and cognitive ones. Affect is not considered as an objective parallel to cognitive goals, but rather as standing on its own. Individuals have feelings about their feelings and beliefs about their own beliefs. Therefore, when we discuss metacognitive processes, we also need to discuss meta-affect processes in relation to both affect and cognition. If we accept that learning can be operationally defined as the "acquisition of competencies", and if we include in competency structures all types of representation systems, we believe that two questions of major practical importance arise: (a) How could one go ahead teaching these competencies? (b) What kind of competencies do teachers themselves need to develop, in order to be able to facilitate students' construction of meaning and their acquisition of competencies including the affective ones?

2.2. *Beliefs and Pre-service Primary Mathematics Education*

Even though the affective domain has been described fairly well since the 1950s and its significance has been increasingly recognized in educational practice, it has been viewed, typically, as a by-product of the learning that occurs in the cognitive domain. Developing positive attitudes, for instance, has been associated with the learning climate, the wider social environment, and the opportunities for successful experiences that all students should be provided with. The development of affective competencies through teaching has been emphasized only recently and we are still far from being able to propose a fairly clear practical way in which to do this teaching, if such a way really exists.

Over the past decades the teaching profession has managed to attract relatively few prospective teachers with positive attitudes toward mathematics (Philippou & Christou, 1998) and even fewer of them were able to teach affective competencies on graduation, simply because there was hardly any provision for learning to do this in their training program. A question, therefore, that could be addressed to teacher educators, who are responsible for designing and implementing preservice programs, is the following: what kind of activities and experiences should preservice programs provide, in order to enhance prospective teachers' belief systems and help them develop those competencies that will enable them to teach affective competencies to, or facilitate the development of, affective competencies in their own students?

A major concern of teacher educators springs from the fact that schoolteachers often lack the requisite conceptual understanding, defined in terms of the nature of mathematics and the teachers' mental organization of mathematical knowledge. Modern mathematics curricula have increased demands on teachers, who are expected to select worthwhile mathematical tasks, to orchestrate classroom discourse, to seek tasks for strengthening students' mathematical understandings, to help students use technology, and to assess progress (Swafford, 1995). The teaching of affective competencies is now entering the field as a new and strange element added to an already heavily loaded program. Furthermore, this new element has the additional characteristic of being unknown to teachers and learners, vague and rather unspecified.

In general terms, the debate on preservice teacher mathematical education continues, though some of the earlier established concepts and perspectives have been widely recognized as well grounded. For instance, Shulmans' (1986) seven domains of cognitive schemata: subject matter, specific and general pedagogical content, other related matter, the curriculum, the learners, and educational aims, were positively received and analyzed. Lappan and Theule-Lubienski (1994) focused on a shorter list of three domains that should be considered in the teacher programs, namely, knowledge of content, knowledge of pedagogy, and knowledge of students. The same authors considered the integration of knowledge in these three domains with beliefs as critical in the education of professional teachers. Cooney (1994) referred to similar requirements for a teacher to be an "adaptive agent" in the classroom; he further stressed the need for a certain orientation toward knowledge and change, since one's actions and behavior are directly influenced by the lenses through which the world is viewed. In addition, Cooney argued that content knowledge, pedagogical content knowledge, and one's orientation toward these types of knowledge are different entities. Orientation towards the content and its teaching is connected to one's outlook and evaluations. An implication for teacher education is that understanding and teaching any subject are fundamentally connected to the meaning that the teacher assigns to the nature of the subject and its learning and teaching, that is, to the teacher's related epistemological views and beliefs.

Preservice mathematical education includes goals such as enhancing teachers' mathematical and pedagogical knowledge, offering teachers opportunities to experience alternative ways of learning challenging mathematics, fostering sensitivity toward students' feelings about learning mathematics, and promoting teachers' ability to reflect on their learning and teaching experiences (Zaslavsky & Leikin, 1999). Krainer (1999) proposed a model for teacher professional practice in which the affective factor is a main component; the model focused on four factors "action, reflection, autonomy, and networking". The important point for our discussion is that Krainer views all four factors in connection to attitudes; beliefs and competence are jointly sought along each of the four factors identified. Specifically, he stressed that action refers to “attitudes towards and competence in” experimental, constructive, and goal directed work. Reflection refers to “attitudes and competence” in self-critical work that reflects systematically on one's own actions. Autonomy refers to “attitudes and competence” in self-initiated, self-organized and self-determined work. Networking refers to attitudes and competence in communicative and co-operative work with public relevance. Each of the two pairs "action-reflection" and "autonomy-networking" expresses both contrast and unity, and can be seen as complementary dimensions which have to be kept in a certain balance. In general, more reflection contributes to a higher quality of actions and a higher quality of actions and autonomy promotes the quality of reflection and networking. Experience shows that teachers' practice is usually characterized by a lot of action and autonomy and less reflection and networking (Krainer, 1999).

In the above analysis we have tried to make the point that a preservice program should consider the structure of beliefs the students bring to teacher education and provide experiences that help students overcome common myths and misconceptions about mathematics, its teaching and learning. Students should be led to transform unexamined beliefs in relation to classroom actions into objective and reasonable beliefs. Belief systems are change resistant; change can occur only when students engage in personal explorations and are involved in powerful experiences in mathematical thinking and conceptual understanding that motivate a new perspective on students' views towards learning. This subsequently leads to modified classroom practices, though a change in beliefs does not necessarily translate into changes in practice. Apart from positive and well-grounded beliefs about the discipline and its teaching, teachers need also a sense of competence, or self-efficacy, that is, a sense that they are capable of succeeding in specific mathematics teaching tasks.

2.3. *Efficacy Beliefs and Mathematics Teaching*

Self-beliefs, such as self-esteem, self-concept and self-efficacy, comprise components of the general beliefs system. Numerous studies have supported the theoretical claim that confidence in one's ability to undertake a certain action is the best predictor of behavior for accomplishing the task (Bandura, 1986, 1997; Guskey & Passaro, 1994; Pajares, 1996). In this context, the teachers' sense of efficacy in teaching is becoming a focus of research in teacher education.

The concept "teacher efficacy" has grown out of two psychological strands; the first is based on Rotter's expectancy theory for internal versus external control of reinforcement, and the second on Bandura's social cognitive theory (Tschannen-Moran, Hoy, & Hoy, 1998). Bandura (1997) defined perceived self-efficacy as "beliefs in one's capabilities to organize and execute the courses of action required to produce given attainments" (p. 3). In the same sense, teaching efficacy can be defined as teachers' beliefs in their capabilities to design and apply effective teaching activities. In other words, efficacy with respect to teaching any subject can be defined as one's confidence in one’s capabilities to organize and orchestrate effective learning environments. Self-efficacy is generally viewed as a future oriented belief, a belief about the level of expected success in executing a specific task. A teacher's sense of efficacy constitutes a motivational factor influencing the amount of effort the teacher will expend and the persistence shown in the face of obstacles. Teachers' activities and actions are mostly dependent on their beliefs rather than on what they know or are really competent to achieve.

The concept of self-efficacy is not synonymous with self-concept, even though some researchers use the terms interchangeably. Bong and Clark (1999) concluded that academic self-concept incorporates both cognitive and affective responses "heavily influenced by social comparison, while self-efficacy concerned primarily cognitive judgments based on mastery criteria" (p. 139). However, the distinction made by most researchers between the two constructs concerns the degree of specificity rather than the balance between cognitive and affective components. Self-concept judgments are more global and less context specific than self-efficacy judgments; they are general descriptions of one's self-evaluations rather than task specific competencies. Self-concept can be described as a generalized form of self-efficacy (Pajares, 1996), though self-efficacy need not necessarily affect one's self-concept. For example, one may feel quite efficacious to solve quadratic equations, without this making any difference to one's academic self-concept, nor even to one's mathematics self-concept. Efficacy beliefs might also be multifaceted, but the distinctive characteristic refers to the specific ability of the individual to accomplish the task in question. This is in line with research findings related to efficacy and learning, which have indicated that general measures of self-efficacy are weak predictors of academic performance, while specific measures of self-efficacy have been found to be linked to a wide range of related teaching variables (Pajares, 1996; Tschannen-Moran, et al., 1998).

Feelings of competence depend on one's experiences in connection to related actions. Efficacy beliefs about the teaching of mathematics are mostly, but not entirely, shaped by one's experience and knowledge of mathematics and its pedagogy. Concerning the genesis and growth of efficacy, Bandura (1997) postulated the following four sources of self-efficacy information: "mastery experiences", "vicarious experiences", "social persuasion", and "physiological and emotional arousal". These sources contribute to both the analysis of the teaching task and the self-perception of teaching competence.

In brief, the effect of mastery experiences refers to the fact that outcomes that are interpreted by individuals as successful raise their self-efficacy beliefs, while those that are evaluated as failures lower these beliefs. This source of efficacy information includes all related experiences, such as those provided in the preservice program and fieldwork. Vicarious experiences refer to the effects produced by the success level of the efforts of significant others. When people have no personal experience and direct involvement with a certain task or when they feel uncertain about their own capabilities, then they tend to rely on the outcomes produced by model individuals. This construct is directly related to social comparison and peer modeling, which influence one's self-perceptions, particularly if competencies are compared with the model. Social persuasion has to do with positive or negative appraisals by peers and other significant others. Appraisal of an individual's behavior and achievement is a source of efficacy information that results in reinforcing self-efficacy, while negative judgments have an adverse consequence. Physiological and emotional states provide information related to efficacy, in the sense that anxiety, mood, and fatigue provide information related to efficacy in the task under consideration. These sources of information are not directly translated into self-competence evaluations, but they form the basis of the events the individuals interpret, and provide information directly relating to efficacy on which they draw (Pajares, 1999).

Early research in the field identified two dimensions of teacher efficacy beliefs: personal teaching efficacy (PTE) and general teacher efficacy (GTE). The PTE measures teachers' own conviction in their power to control students' motivation and achievement, while the GTE refers to the possibility of teachers, in general, to affect students' learning. The former is confidence in one's own capabilities, whereas the latter refers to the perceived effectiveness of a teacher population in a certain environment. Another dimension found in the literature refers to internal versus external control of reinforcement (Tschannen-Moran et al., 1998). The former concerns beliefs in the possibility of teachers and the school in general to control students' learning, irrespective of the cultural environment (Internal Control), while the latter accepts the predominance of environmental and heredity factors in learning (External Control).

The most widely used scale for measuring teacher efficacy (TES) was the one developed by Gibson and Dembo (1984) and consisted of 30 Likert-type items. Using this scale the two dimensions of teacher efficacy mentioned above, the PTE and the GTE, were repeatedly identified. Empirical results seemed to indicate that individuals' judgments concerning their ability to promote learning outcomes differ from their evaluations of the capabilities of other teachers in general to do as well as they do on the same task. Subsequently, Soodak and Podell (1993) developed an improved version of the TES with 16 items, which had the additional characteristic that each item loaded on either the PTE or the GTE factor. These first scales for measuring efficacy were not balanced in terms of items of the PTE and GTE dimensions and the internal-external interpretation of learning. The PTE items were biased along the internal interpretation and the GTE items along the external interpretation of learning. To eliminate this bias, Guskey and Passaro (1994) proposed that the internal-external dichotomy should be crossed with the PTE and GTE dimensions to produce four categories of items: Personal-Internal (P-I), General-Internal (G-I), Personal-External (P-E), and General-External (G-E) (for indicative items, see Table 1).

In a more recent study Deemer and Minger (1999) tested a version of a revised TES in which the positive-negative dimension was available for each item. Even though their analysis indicated that the TES is a uni-dimensional construct, when confounding wording was eliminated, they felt uncertain about their findings and concluded that the construct might be multidimensional. Despite the long cherished belief about the usefulness of the GTE dimension, one could be skeptical about the relevance of such a factor, particularly when the point of issue concerns self-beliefs. A consistent use of the term self-efficacy seems to exclude one's evaluations about the capabilities of others to pursue a certain task, irrespective of the importance of such evaluations for the teaching learning process.

Educational outcomes depend on multiple factors and teacher involvement goes far beyond instruction. In this context, Bandura (1997) proposed a teaching efficacy scale with 30 Likert-type items of nine-points that comprises the following seven sub-scales: Efficacy to handle or influence the "decision making process", the "school resources", the "instruction process", the "discipline of students", the "parental involvement", and the "school climate". Efficacy beliefs can be characterized in terms of level, generality and strength. Various research studies (Pajares, 1996; Tschannen-Moran et al., 1998) confirmed the relationship between efficacy beliefs and significant educational factors, such as professional commitment, enthusiasm, instructional experimentation, implementation of progressive and innovative methods, level of organization, and certainly, students' affective growth and achievement.

Teacher efficacy is conceived to be subject specific as well as task specific. For instance, a teacher may feel more efficacious in teaching mathematics than geography, and may feel quite comfortable and capable in teaching arithmetic operations, but not equally capable of teaching division of fractions. Self-efficacy with respect to learning and teaching mathematics is a little-researched area. Some researchers examined students' efficacy in learning mathematics (Soodak & Podell, 1996; Hackett & Betz, 1989; Randhawa, Beamer, & Lundberg, 1993). Hackett and Betz (1989) found that the self-efficacy beliefs of psychology students were moderately correlated to mathematics performance and positively correlated to attitudes and to gender. Randhawa et al. (1993), testing a structural model involving attitude, self-efficacy, and mathematics achievement, found that generalized mathematics attitudes had a strong direct effect on self-efficacy and that attitudes had an equally strong and direct effect on performance as self-efficacy beliefs. In a path-analysis model that controlled for the effects of anxiety, cognitive ability, mathematics grades, self-efficacy for self-regulatory learning, and gender, Pajares and Miller (1994) found that self-efficacy made an independent contribution to the problem solving performance of students. In conclusion, there is empirical evidence, which relates efficacy to mathematics learning, but there is very little known about mathematics teaching efficacy. The question of teacher efficacy with respect to teaching mathematics remains open.

Ongoing discussion about preservice mathematics programs of primary teachers includes integration of cognition and affect. Despite the general interest and the intense research publications in the field of beliefs, a decade ago Schoenfeld (1992) argued that "(w)e are still a long way from a unified perspective that allows for the meaningful integration of cognition and affect or, if such integration is not possible, from understanding why it is not" (p. 364). Since then, empirical research has provided the educational community with more insight and more understanding of the domain. We have probably come closer to, but we are still far away from, a satisfactory level of understanding of the many implications of affect in educational practice and in particular in mathematics teacher education. The second part of this chapter describes a study, which aimed to shed some light on one dimension of affect and cognition, with respect to teaching mathematics, contributing to efforts toward the description of a unified perspective, if such a perspective is really feasible.

3. A STUDY OF EFFICACY BELIEFS IN TEACHING MATHEMATICS

In the above analysis we took the stance that affective competencies are learnable and consequently teachable. Teacher educators have to design preservice programs that will help future teachers feel comfortable and confident that they are capable of overcoming difficulties and succeed in bringing about the targeted outcomes. The study described in this chapter aimed to examine the construct of teacher self-efficacy in teaching mathematics and relate it to preservice mathematics education programs. The objectives of the study were to examine teachers' self-beliefs and evaluations; specifically, the personal and general teaching efficacy beliefs, the perceived competence to influence students' mathematical learning, the ability to control the school climate, enjoyment or anxiety associated with teaching, the development of efficacy beliefs over time, and the beliefs about the preservice mathematics program the teachers passed through.

The cultural context: The study took place in Cyprus at a time when the teacher population was going through a transition period, since a university degree became mandatory for primary teaching in 1992. The university accepted its first students in the same year and the Department of Education was assigned the responsibility for educating primary teachers. The preservice mathematics program[1] that was implemented, made extensive use of the history of mathematics. After the first graduates of this program were employed in schools, one question of interest concerned the effectiveness of this new preservice education program. In this respect, beyond the more general questions, this study could also satisfy a concern of local interest, namely evaluating the specific mathematics program against the programs of other institutions. The evaluation involved a comparative element, in the sense that it searched for differences among groups of the teacher population in terms of preservice education.

3.1. Methodology

3.1.1. The Instrument

We used a five-point Likert-type scale with 28 items based on the scale developed by Gibson and Dembo (1984) and modified by subsequent researchers. Motivated by Bandura's ideas, we included within the PTE dimension items reflecting the subjects' efficacy beliefs about their preservice program, the school climate, and anxiety and enjoyment beliefs arising from mathematics teaching. Specifically, the factor personal teacher efficacy (PTE) was measured through five dimensions: the internal and the external interpretation of learning control, the mathematics teaching anxiety, the mathematics teaching enjoyment, the school climate, and the efficacy beliefs of the preservice mathematics program. Four indicators, all reflecting the external interpretation of learning control, measured the general teaching efficacy dimension (GTE) of the subjects. The highest efficacy level was coded as 5 and the lowest as 1, meaning that number 3 signified the neutral level.

3.1.2. The Subjects

The primary teacher population of Cyprus consists of graduates from the Pedagogical Academy (PA) and the University of Cyprus (UC), and of graduates from several Greek universities (GU). A proportion of the Pedagogical Academy graduates have received a degree from a foreign university (PAG). The questionnaire was mailed to selected schools on the basis of the composition of the teaching staff. Specifically, we sent questionnaires to schools whose teaching staff included at least 40% of teachers with less than 15 years of teaching experience. The school principal was requested to ask only those teachers who had less than 15 years of service to complete the questionnaire. We received 157 completed questionnaires, a percentage of about 65% of the total number of questionnaires sent out. One hundred and six of them (58.6%) were PA graduates including 15 (9.7%) degree holders (PAG), 21 (13.4%) GU, and 28 (17.8%) UC graduates. This distribution was quite representative of the teachers within this "age group".

3.1.3. The Interviews

About two months later, we interviewed 18 of the participants. Ten of them were UC graduates, six were PA graduates, and two were GU graduates. The bias toward UC graduates was due to our effort to look particularly for the effect of the preservice program we implemented. The interviews were semi-structured, focusing on letting teachers talk about their mathematics teaching experiences, their concerns, and their evaluations of the preservice program they attended. The data collection process was completed by the end of the academic year 1996-1997, just before the first UC graduates completed their first year at school.

4. RESULTS

4.1. Analysis of the Questionnaire Data

The results showed a high level of teacher self-confidence in teaching mathematics, even though most teachers did not feel able to control pupils' learning. Specifically, more than 80% of the subjects felt confident in helping students "make progress" and "think mathematically", felt comfortable to "consult experienced colleagues", capable "to answer pupils' questions" and "to correct pupils' assignments", and they would not like to "give away mathematics" if they could. On the other hand, the majority of the subjects did not endorse the items indicating efficacy "to cover the subject matter", "to help the weak students", "to discipline students who are not used to this from home", and they did not believe in "the possibility of weak pupils to get help".

Table 1 shows indicative items from each dimension of the scale, the mean value for the whole sample, and the percentages of positive endorsement (above the neutral point 3) by UC graduates, PA (including PAG), and GU graduates, respectively. At first sight it appears that there is a considerable variability of item endorsement among the subject groups as well as in the overall mean efficacy level.

Table 1. Proportions of Endorsement to Indicative Items by Institution and Overall Mean

	Personal Teaching Efficacy (PTE)
Internal Interpretation of control (P-I)	1. I can teach successfully and achieve good results, even in mathematical topics considered difficult (54%, 62%, 67%)*, $\overline{X}$ = **3.62**
	3. I have no doubt that I can help children improve their mathematical thinking (89%, 91%, 90%), $\overline{X}$ = **4.14**
External Interpretation of control (P-E)	11. When a child becomes better in mathematics, I believe that it was due to the variety of different ways I found to help him/her (82%, 76%, 67%), $\overline{X}$ = **3.79**
	2. Some children face so many difficulties in mathematics, that I feel unable to help them (68%, 63%, 54%), $\overline{X}$ = **3.33**
	10. Quite often I have the feeling that it is difficult to help some of my students to become better in mathematics (64%, 77%, 90%), $\overline{X}$ = **3.66**
Teaching Anxiety	17. I cannot help a weak student without neglecting the other students (25%, 33%, 43%), $\overline{X}$ = **2.74**
	21. Sometimes I feel anxious because I am afraid that a student might ask me a question that I do not know how to answer or I cannot explain (53%, 84%, 81%), $\overline{X}$ = **4.12**
Teaching Enjoyment	25. If I could permanently give away one subject, that would have been mathematics (74%, 68%, 71%), $\overline{X}$ = **3.67**
	27. Mathematics is among the subject I love teaching (74%, 68%, 71%), $\overline{X}$ = **3.77**
School Climate	28. If I were to choose one subject in a colleague's class, I would have opted for mathematics (57%, 30%, 29%), $\overline{X}$ = **3.16**
	16. I feel that I can communicate productively with the other teachers in my school (52%, 67%, 55%), $\overline{X}$ = **2.73**
	18. I feel uncomfortable when the school principal or the inspector observe my mathematics class (64%, 28%, 62%), $\overline{X}$ = **2.83**
Preservice Program	9. My preservice mathematics program, offered me the basics to become an efficient mathematics teacher (50%, 25%, 28%), $\overline{X}$ = **2.55**

External interpretation of control (G-E)	*General Teaching Efficacy (GTE)*
	4. The influence of the students' environment is so decisive that the teacher of mathematics can very little to help (96%, 77%, 57%), $\bar{X}=\mathbf{3.79}$
	As things now really stand, the weak students cannot get the necessary help in mathematics (50%, 44%, 35%), $\bar{X}=\mathbf{2.83}$
	13 Taking into account all factors influencing mathematics learning, then the possibilities of the teacher are very limited (78%, 59%, 67%), $\bar{X}=\mathbf{3.41}$

The figures in brackets refer to CU, PA and GU graduates, respectively

In general, teachers expressed a high level of self-confidence in teaching mathematics, while they did not feel as capable with regard to controlling pupils' learning; the mean score on P-I items was higher than the mean on the P-E items. On the teaching anxiety and teaching enjoyment dimensions, it was found that the general feelings of the subjects were quite high. The highest efficacy levels were observed in items 3 and 21 ($\bar{X}=4.141$ and $\bar{X}=4.195$, respectively) indicating that the subjects feel capable of helping students learn and feel comfortable with students' questions. The mean efficacy level was found below the neutral point on four items. One of them reflected students' beliefs that weak students are helpless (item 6, $\bar{X}=2.832$), two of them reflected the school climate (items 16 and 20, $\bar{X}=2.731$ and $\bar{X}=2.537$, respectively), and finally item 9 reflected beliefs about the "preservice mathematics program" ($\bar{X}=2.548$).

The level of teacher efficacy beliefs varied, both between the personal and the general efficacy beliefs, as well as among the dimensions within the PTE factor. The teachers favored the internal against the external interpretation of learning control ($\bar{X}_{P\text{-}I}=3.893$, SD = .494, and $\bar{X}_{P\text{-}E}=3.436$, SD = .637, respectively), felt no anxiety (ANX) and enjoyed (ENJ) teaching mathematics ($\bar{X}_{ANX}=3.957$, SD = .6839, and $\bar{X}_{ENJ}=3.684$, SD = .824, respectively). The mean efficacy in managing the school climate (SCL) was found to be the lowest among the PTE dimensions ($\bar{X}_{SCL}=3.355$, SD = .604). The mean value of the GTE factor was found to be lower than any of the dimensions of the PTE factor ($\bar{X}_{GTE}=3.265$, SD = .688), indicating that teachers as individuals feel more efficacious than they consider the rest of the teacher population to be. As mentioned above, the lowest efficacy level was observed for item 9 ($\bar{X}_{PSP}=2.548$, SD = 1.112), meaning that the teachers expressed dissatisfaction with the preservice mathematics program (PSP) they attended.

We used Analysis of Variance to test for differences in the efficacy level among the four sample groups (UC, PA, UPA, and GU graduates) and among "age groups". We found significant differences among the sample groups on the GTE dimension of the scale and on the item reflecting the evaluation of their preservice mathematics program. Specifically, UC graduates, more than the rest of participant groups, held the belief that students are teachable i.e., they support the internal interpretation of learning control ($\underline{F}_{(2, 156)} = 3.150$, $\underline{p} = .027$). UC graduates showed a higher level of efficacy beliefs on all four items of the GTE dimension of the scale, meaning they believed that students are teachable. Finally, the UC graduates expressed significantly higher efficacy beliefs about their preservice mathematics program than did the rest of the sample ($\underline{F}_{(2, 156)} = 8.992$, $\underline{p} = .000$). PA graduates expressed the least appreciation of the preservice mathematics program they had attended (the mean values were $\underline{\bar{X}}_{UC} = 3.29$, $\underline{\bar{X}}_{PA} = 2.25$, $\underline{\bar{X}}_{GU} = 2.90$). Much of the difference found above could be attributed to the factor of mastery experience, which was supported by the analysis of interviews that follow.

Significant differences were also found on the total efficacy scale in terms of the teachers' years of teaching experience. The analysis of variance of the overall efficacy by three groups of teachers showed a difference ($\underline{F}_{(2, 156)} = 3.257$, $\underline{p} = .042$). The mean efficacy values by group indicated that the teachers' efficacy beliefs tend to get worse during the first period of professional life and improve subsequently, during their career, through experience ($\underline{\bar{X}}_1 = 3.59 \Downarrow \underline{\bar{X}}_2 = 3.37 \Uparrow \underline{\bar{X}}_3 = 3.65$). Hoy and Woolfolk (1993) also found improvement of efficacy through experience, but the initial decline is, to the best of our knowledge, a new result, which needs further consideration.

4.2. Analysis of the Interviews

During interviews, the subjects were encouraged to talk about their self-beliefs, feelings and evaluations in relation to mathematics teaching. In particular, the aim was to clarify matters that seemed unclear from the analysis of the questionnaires. In the following paragraphs we analyze the interviews, focusing on four dimensions: the general mathematics teaching efficacy along the personal dimension (PTE), perceived efficacy to influence non-motivated pupils (NMS), the evaluation of the preservice program (PSP), and the school climate (SCL). In general, the interviews support the findings from the quantitative analysis and provide a clearer picture of the interviewees' beliefs on each of the above four dimensions. The teachers' answers concerning efficacy judgments were classified as positive, neutral, or negative, and the participants were listed in one of these three categories. The subjects' beliefs according to the above classification are summarized in Table 2. To provide a clear understanding of the views expressed and the meaning of each category, indicative excerpts of the interviews follow:

4.2.1. Mathematics Teaching Efficacy Beliefs (PTE)

Q1- PTE: How confident do you feel when you teach mathematics? Are there any topics that you don't feel comfortable to teach?

Positive view: No, I feel quite confident that I can teach any topic of the curriculum.

Neutral view: Em ..., well, I can tell that I am quite comfortable to teach lower grades mathematics, ... am not so sure for the upper grades. One teacher mentioned that she used to be afraid of mathematics at school, another worries about the teaching of some topics in upper grades, e.g., fractions.

Negative view: I do have difficulties (I had at the beginning), and I am anxious whether the students will be able to really understand what I am talking about. Yet, surprisingly enough, the same teacher stated that she felt quite capable to get through.

4.2.2. Efficacy to Help even the Non-Motivated students (NMS)

Q1-NMS: Now, suppose that you have some non-motivated students in your class, how confident do you feel to help them?

Positive view: I am sure that I can help all students to make progress. One has to make special arrangements for the 2-3 non-motivated students, normally found in every class, by simplifying activities, allowing for more time, seeing the parents, etc.

Neutral view: I can help slow learners, but it is not possible for all students to reach the same level.
There are a few special cases for which it is very hard to do anything.

Q2-NMS: If a child makes progress, to whom do you think this should be credited?

Positive view: See, a student's learning is affected by many factors; the final outcome is a combination of joint efforts. In my view, however, the teacher is the first to be credited.

Neutral view: I think that a student's progress is equally due to the teacher, the parents and the student himself/herself.

Negative view: Teachers' influence on students' learning is limited; they are at school for only 4-5 hours a day.
Everything depends on the child.
There are some cases in which the teacher cannot do much.

4.2.3. Efficacy of the Preservice Mathematics Program (PSP)

Q1-PSP: How do you judge your preservice program?

Positive view: I believe it offered me the background to teach mathematics effectively. Sometimes, when I was a student, I wondered whether several of the issues and ideas discussed were practically applicable, now I am convinced they are useful.

Neutral view: The most useful course was the methods course. The history of mathematics course helped me to appreciate the developmental nature of mathematical ideas, but I think that the tutorials should have been more profitable.

Negative view: Indeed, we only did one course on teaching methods, which was just an introduction, rather irrelevant to teaching.

Q2-PSP: A frequently raised point concerning preservice mathematics programs refers to the balance between theory and practice; how do you judge your program in this respect?

Positive view: I think that the teacher needs a strong theoretical basis. We covered a wide spectrum of topics and this provided me with the basis, which is necessary for choosing and creating didactical situations on my own.

Neutral view: I think it could have offered us more. There were quite a few useless elements for the primary teacher in the content courses, which caused us additional anxiety. I think we should stick to methods of teaching specific topics.

Negative view: It paid too much attention to mathematics instead of the methods of teaching mathematics, it was very poor...Instead of sets and probability, we could have done more didactics of mathematics.
We did a lot of mathematics at the high school, we should stick with teaching methods.

4.2.4. Efficacy to Manage the School Climate (SCL)

Q1- SCL: How do you feel about the subject matter coverage?

Positive view:
(a) I don't worry about that, though I am quite behind schedule... I do not like to rush and lose some pupils. I will finally be able to get through the subject matter.
(b) In my view it is not necessary to cover all the proposed content. The issue is to let pupils understand what we do, so I don't worry.
(c) ... The inspector asserted that I was behind schedule, but I cannot push the children... They need time to understand.

Neutral view:
(a) Yes, because there is so much in the textbooks and I realize that the level of my pupils is not high.
(b) One has to insist on the basics, not to rush, and that worries me a lot.
(c) We rush to cover the prescribed matter without going in depth.

Q2- SCL: How do you feel when the principal or the inspector attends your class?

Positive view: Well, it's natural not to feel as easy as when you are on your own; it may cause me some tension, but not really anxiety. I have nothing to hide; I want them to get the real picture of the class, to bring possible problems on the surface.
It is a matter of self-confidence.

Neutral view: There is a certain degree of anxiety. After all, you are under scrutiny, they are examining your results.

Negative view: I feel anxious and uneasy...I think it is in my character.

The percentages of teachers with positive, neutral, and negative reactions on each of the four scale dimensions examined for each of the three sample groups, are shown in Table 2. It is evident that the whole picture presented tends to affirm the results found from the questionnaire data, though the sample size does not allow for generalizations. The overall efficacy level was higher on the PTE dimension than on the NMS dimension which is a reflection of the GTE factor; efficacy on the PSP was again found to be the lowest of the four. Concerning the differences between groups, it can be observed that UC graduates were found to have a superior efficacy level on the PTE, the PSP and the NMS dimensions, and an inferior level on the SCL dimension. It seems that UC graduates were relatively more positive about their preservice program, they felt capable to help even the unmotivated pupils, and believed that pupils in general are teachable. Their efficacy level about the school climate could be a result of the fact that they were the first of a new group of UC graduates entering the profession. In the third column we have included fractions, instead of the percentages, showing a more or less balanced distribution between the positive, the neutral and the negative endorsements by the subjects who graduated from GU. PA graduates appeared to have the least positive efficacy beliefs in general and they were the most critical about their preservice program.

Table 2. A Summary of Responses in the Interviews by Scale Dimension and Preservice Education

		Univers.of Cyprus	Pedagogical Academy	Greek Universities
Personal efficacy beliefs (PTE)	Positive	80%	50%	3/6[1]
	Neutral	15%	33%	3/6
	Negative	5%	17%	2/6
Ability to help Non-Motivated (NMS)	Positive	85%	33%	2/6
	Neutral	15%	50%	4/6
	Negative	0%	17%	2/6
Manage the School Climate (SCL)	Positive	54%	78%	4/6
	Neutral	32%	17%	2/6
	Negative	14%	5%	2/6
Appreciate the Preservice Program (PSP)	Positive	50%	0%	2/6
	Neutral	50%	50%	4/6
	Negative	0%	50%	2/6

[1] *In the case of Greek Universities, we use fractions instead of percentages because of the sample size*

5. CONCLUSIONS

Affective variables are nowadays recognized as playing an increasing role in education. In this chapter we have reviewed earlier research and presented empirical evidence that endorsed the idea that affective competencies develop much like cognitive competencies. It seems, therefore, crucial to educate prospective teachers to teach affective competencies. We examined the mathematics teaching efficacy beliefs of young teachers and searched for differences among groups in terms of preservice education and length of service. The results seemed to indicate that in Cyprus young teachers felt quite efficacious in teaching mathematics, though there were some components of the scale in which the level of their efficacy was not satisfactory. The overall mean efficacy as measured by the questionnaires was found to be above the neutral level and the interviews indicated quite positive teacher efficacy beliefs, even though not completely in line with the quantitative data. Both analyses indicated three dimensions of teacher efficacy that seem to be of primary importance and warrant further consideration: first, efficacy about the preservice mathematics program, second, managing the school climate, and third, the so called "initial confidence shock", which caused a decline in self-efficacy in teaching mathematics.

The differences among groups by years of service indicate that the newly appointed teachers have a high initial level of mastery experiences and vicarious experiences that make them feel over-optimistic. This judgment becomes more moderate and realistic when they first come across the complexities and practical difficulties of teaching mathematics, and improves with experience. Hoy and Woolfolk (1993) found similar improvement in efficacy beliefs as the teachers continue in their career. This could be explained as part of a teacher's development, a sign of reflection and adaptation to the needs of the students. Whether the decline during the initiation period springs from limited teaching practice, or from unrealistic conceptions acquired during preservice education, needs further examination. It seems that it takes young teachers some trial period before they develop the necessary cognitive, behavioral, and self-regulatory strategies for creating and executing appropriate courses of action. Another explanation might be due to the uneven distribution of the UC and the UG graduates in the sample in terms of length of service. Since all UC graduates were in the first age group, an effect of preservice education on the difference in efficacy by age group could not be ruled out.

UC graduates expressed higher efficacy beliefs than PA and GU graduates in their ability to teach mathematics effectively, in the children's possibility of learning mathematics, and in the capacity of teachers to influence students' progress. The higher level of efficacy beliefs of the UC graduates might reflect confidence in their preservice program and its implementation. The preservice program at the UC involved more mathematics than the programs of PA and GU. The distinctive feature, however, of this program was its design, which followed historical lines in an effort to improve students' beliefs about mathematics. The present results seem to reaffirm earlier studies indicating that the specific program was effective in developing positive attitudes toward mathematics (Philippou & Christou, 1998). A systematic follow up of our graduates through observation of their teaching practices will provide clearer evidence about their efficacy beliefs and their general attitudes toward mathematics and the specifics of the program.

The low efficacy level of UC graduates on the SCL dimension, which was observed particularly in the interviews, should be given serious consideration, and considered in terms of the respective items of the scale. This dimension of the scale referred to efficacy beliefs with respect to collegiate relations, the relations with the principal and other superiors, and the expectations of the system relative to the subject matter. Why the teachers in general, and particularly UC graduates, have moderate to low efficacy beliefs in the former two domains is hard to explain. One possible source could be the highly centralized educational system, which leaves little room for individual initiatives. This is even clearer in the case of the subject matter coverage. A plausible explanation regarding the weakness of the UC graduates to manage the school climate might be a feeling of being alien to the system, representatives of a new group of teachers with different characteristics.

In the interviews it became clear that the teachers felt quite competent to facilitate students' mathematical learning (not so much the NMS), they tended to be critical about their preservice program, and they did not seem to approve of the school climate. The differences among UC graduates and the graduates from other institutions were affirmed and even highlighted with respect to the preservice program.

In brief, the results seem to support the hypothesis that a preservice program such as the one implemented at the UC might be successful in improving affective goals. In absolute terms, however, the mean endorsement of the respective items as well the analysis of the interviews indicate that there is much room for improvement of the preservice program. One might also argue that these findings, preliminary as they are, could be limited to the specific cultural environment. It is true that we used history of mathematics, mainly of ancient Greek mathematics, to motivate Greek students to get involved in mathematics. We can only speculate whether a similar curriculum would have equally positive effects in other cultural environments. We consider replication studies in different cultures to be useful.

The main issue is how teacher educators might facilitate improving the efficacy beliefs of prospective teachers in regard to teaching mathematics. What kind of experiences do prospective teachers need to get involved in and what kind of activities ought the student teachers to be able to organize in order to improve their own students' affective structures? As one possibility, we tried the history of mathematics in a constructivist environment and it seems that we had some success.

5. NOTES

[1] The program consists of three mandatory courses, two content and one method course. Its difference from other preservice programs is that it is based on the history of mathematics. The course design provides for the examination of mathematical ideas, concepts and methods, starting from the time and the cultural setting of their genesis and proceeding to later developments. The students work on problems and activities taken from well-known mathematicians, while a constructivist approach to learning is followed throughout. A longitudinal study indicated that during the exposure to the program, there occurred significant improvement in students' attitudes toward mathematics. For a comprehensive description of the program and a detailed analysis of the study in attitude change, see Philippou and Christou (1998).

6. REFERENCES

Bandura, A. (1986). *Social foundations of thought and actions: A social cognitive theory*. Englwood Cliffs, NJ: Prentice-Hall.

Bandura, A. (1997). *Self-efficacy: The exercise of control*: New York: Freeman.

Boekaerts, M. (1999). Metacognitive experiences and motivational state as aspects of self-awareness: Review and discussion. *European Journal of Psychology of Education, 14*(4), 571-584.

Bong, M., & Clark, R.E. (1999). Comparison between self-concept and self-efficacy in academic motivation research. *Educational Psychologist*, 34(3), 139-153.

Cooney, T.J. (1994). Teacher education as an exercise in adaptation. In D. B. Aichele and A. Coxford (Eds.), *Professional development for teachers of mathematics*. Yearbook of the National Council of Teachers of Mathematics (pp. 9- 22). Reston, VA: NCTM.

Deemer, S. A., & Minke, K. M. (1999). An investigation of the factor structure of the teacher efficacy scale. *Educational Research, 93*(1), 3-11.

Goldin, G. A. (1998). Representational systems, learning, and problem solving in mathematics. *Journal of Mathematical Behavior, 17*(2), 137-165.

Guskey, T. R., & Passaro, P. D. (1999). Teacher efficacy: A study of construct dimensions. *American Educational Research Journal, 31*(3), 627-643.

Gibson, S., & Dembo, M. H.(1984) Teacher efficacy: A construct validation. *Journal of Educational Psychology, 76*, 569-582.

Hackett, G., & Betz, N. E. (1989). An exploration of the mathematics self-efficacy /mathematics performance correspondence. *Journal for Research in Mathematics Education, 20*(3), 261-273.

Hoy, W. K, & Woolfolk, A. E. (1993).Teachers' sense of efficacy and the organizational health of schools. *The Elementary School Journal, 93*(4), 355-372.

Krainer, C. (1999). Promoting reflection and networking as an intervention strategy in professional development programs for mathematics teachers and mathematics teachers' educators. In O. Zaslavsky (Ed.), *Proceedings of the Conference PME23* Vol. 1. (pp. 59-168). Haifa: Technion.

Lappan, G., & Theule-Lubienski, S. (1994). Training teachers or educating professionals? What are the issues and how are they being resolved? In D. Robitaille, D.H. Wheeler & C. Kieran (Eds.), Selected lectures from the 7th International Congress on Mathematical Education (pp. 249-261). Quebec: Les Presses de L' Universite Laval.

Masui, C., & De Corte, E. (1999). Enhancing learning and problem solving skills: Orienting and self-judging, two powerful and trainable learning tools. *Learning and Instruction, 9* , 517- 542.

McLeod, D. B. (1994). Research on affect and mathematics learning in the JRME: 1970 to the present. *Journal for Research in Mathematics Education, 25*.(6), 637-647.

Pajares, M. F. (1992).Teachers' beliefs and educational research: Cleaning up a messy construct. *Review of Educational Research, 62*(.3), 307-332.

Pajares, F.(1996). Self-Efficacy Beliefs in Academic Settings. *Review of Educational Research, 66*(4), 543-578.

Pajares, F. (1999). Current directions in self-efficacy research. In M. Maehr & P. R. Pintrich (Eds.), *Advances in motivation and achievement*. Vol. 10 (pp. 1-49). Greenwich, CT: JAI Press.

Pajares, F., & Miller, D. 1994). Role of self-efficacy and self-concept beliefs in mathematical problem solving: A path analysis. *Journal of Educational Psychology, 86*.(2), 193-203.

Philippou, G. N., & Christou, C. (1998). The Effects of a preparatory mathematics program in changing prospective teachers' attitudes toward mathematics. *Educational Studies in Mathematics, 35*, 189-206.

Randhawa, B. S., Beamer, J. E., & Lundberg, I. (1993). Role of mathematics self-efficacy in the structural model of mathematics achievement. *Journal of Educational Psychology, 85*(1), 41-48.

Richardson, V. (1996). The role of attitudes and beliefs in learning to teach. In J. Sicula (Ed.), *Handbook of Research on Teacher Education* (pp. 102-119). London: Prentice Hall International.

Schoenfeld, A. (1992). Learning to think mathematically: Problem solving, metacognition, and sense making in mathematics. In D. A. Grouws (Ed.), *Handbook of Research on Mathematics Teaching and Learning* (pp. 334-370.) New York: Macmillan.

Shulman, L. S. (1986). Those who understand: Knowledge growth in teaching. *Educational Researcher, 15*(1), 4-14.

Soodak, L., C., & Podell, D. M. (1996). Teacher Efficacy: Toward the understanding of a multi-faceted construct. *Teaching and Teacher Education, 12* , 401-411.

Swafford, J. O. Teacher preparation. (1995). In I. M. Carl (Ed.), *Prospects for School Mathematics* (pp. 157-174). Reston VA: National Council of Teachers of Mathematics.

Thompson, A. (1992). Teachers' beliefs and conceptions: A synthesis of the research. In D. A. Grouws (Ed.), *Handbook of Research on Mathematics Teaching and Learning* (pp. 127-145). New York: Macmillan Publishing Company.

Tschannen-Moran, M., Hoy, A. W., & Hoy, W. K. (1998). Teacher Efficacy: Its meaning and measure. *Review of Educational Research*, *68* , 202-248.

Zaslavsky O., & Leikin R. (1999). Interweaving the training of mathematics teacher educators and professional development of mathematics teachers. In O. Zaslavsky (Ed.), *Proceedings of the Conference PME23* Vol. 1 (pp. 143-158). Haifa: Technion.

CHAPTER 14

STEPHEN LERMAN

SITUATING RESEARCH ON MATHEMATICS TEACHERS' BELIEFS AND ON CHANGE

Abstract. The six chapters in this section focus on studying teachers' beliefs with a particular concern for how those beliefs might change in order to improve the teaching of mathematics in our schools. Five of the chapters draw on empirical studies whilst the sixth gives an overview of recent research to investigate how researchers interpret teachers' beliefs and what evidence they use for their claims. There are important theoretical analyses offered and fascinating stories told of teachers and by teachers concerning their beliefs and their practices. In the first part of this review I will draw out some of the key themes in the six chapters and indicate commonalities and differences. My contribution in the second section will be to try to situate these studies within research in the mathematics education community in order to point out some alternative orientations for the study of teachers and teaching-and-learning mathematics.

1. RESEARCHING BELIEFS

The authors all draw a strong link between beliefs and practices and suggest that changing teachers' practices will depend on changing their beliefs, and they indicate that changing what teachers do, through inservice courses or other projects, can lead to changes in their beliefs. A number of the authors indicate that beliefs may change without practices changing, however. They all argue that this is an important but complex field and the work they have reported on here is at the cutting edge. Their work is, in general, methodologically strong in that their empirical studies inform their developing theories as their theories inform the design and nature of their empirical studies.

Wilson and Cooney argue for a conceptual link between what teachers believe and what they do, drawing on Scheffler's claim that beliefs are 'dispositions to act in certain ways'. This is quite a challenge to many researchers in the field. Too much research has treated beliefs revealed in interviews and questionnaires (espoused beliefs) and beliefs as interpreted by observing teachers' actions (enacted beliefs) with the expectation that where teachers are consistent these will match. Researchers then try and explain why they are frequently very different. Wilson and Cooney look to what evidence researchers produce for their claims about teachers' beliefs and, in particular, for changes in their beliefs. They rightly highlight the methodological weakness demonstrated by the assumption that using interviews and questionnaires reveals the presence of an identifiable object called a belief, or a system of beliefs, that is the main determinant of a teacher's actions in the classroom.

G. C. Leder, E. Pehkonen, & G. Törner (Eds.), Beliefs: A Hidden Variable in Mathematics Education? 233-243.

For all the authors beliefs are mental objects, an implicit realism, albeit concerning an unobservable element. Chapman investigates the idea of a belief *system* which contains belief structures or types and she offers an analysis of the literature of belief structures describing connections between these beliefs, centrality and peripherality and so on. This adds a welcome complexity to research on beliefs. In her chapter she uses these notions to analyze the change in two teachers' practices as described by the teachers themselves as they look retrospectively at their teaching histories. She outlines the primary beliefs and inferential beliefs for each of the two teachers and their stories describe their struggle to bring their teaching into coherence with their beliefs. Philippou and Christou place beliefs within an overall affective analysis, arguing that affect is no longer seen in research as the poor relative of cognition but is now a key field of study in its own right. This leads them to examine beliefs in the form of teachers' feelings about the *efficacy* of their mathematical knowledge and their pedagogical knowledge. Hart sees beliefs as part of one's subjective knowledge but, like Philippou and Christou, she also recognizes a strong affective component. Her research draws on teachers' own stories of what affected their ideas and their teaching that led to change. Hart indicates that we know so little about reform-oriented change that teachers' stories are a key source for research.

Change is a key focus in all the chapters and a number of factors are identified as precipitating change. For Hart and Llinares one of the key factors is the engagement with colleagues and Wilson and Cooney argue that engagement in reform-oriented instructional settings at the very least enhances teachers' beliefs about reform, and presumably therefore also their practice. For Chapman and for Philippou and Christou conflicts between personal goals and the perceived situation in the classroom can bring about change in beliefs. This constructivist orientation is also picked up by Wilson and Cooney and Hart who claim that beliefs are justified knowledge for an individual and that is the nearest any of us come to knowledge, the fit with an individual's experience. Hart takes the constructivist position further in arguing that teacher educators can only observe behavior and conjecture what beliefs are held, a second-order construction (Steffe & Thompson, 2000). Thus beliefs are an indicator of prior experiences, facilitating practical decisions about future development and change for that teacher. Hart refers to the role of University tutors modeling the good practice they advocate for schools, a theme picked up by Wilson and Cooney also. Lloyd points to teachers working with innovative curriculum materials, both as learners and as teachers, as productive of change and whilst Wilson and Cooney concur they further suggest that support in the classroom is needed. Reflection on one's actions is another factor that can bring about change, as Hart and Wilson and Cooney explicitly indicate and other authors imply.

Too many researchers in the field of research on teachers fail to address theory in terms of the issue of the nature of teachers' learning (Lerman, 2001a). This is not the case here. In five of the chapters the authors take either an explicitly constructivist position or an implied one when theorizing the nature of the learning process involved in teacher change. It is in the recognition of conflict between what one wishes to do, or believes oneself to be doing, and the perceived reality of one's teaching that can bring about change. Reflection is also assumed to play a key role in change of practice, or further learning about practice. One author, Llinares, takes a different view, originating in the work of Lave and Wenger, whereby objects that are reified in a practice become available for negotiation. He points to different aspects of subject matter knowledge and pedagogic content knowledge, the repertoire of the practice; and a participant's, in this case a teacher's, beliefs. He conjectures that change in practices comes about only when these objects are reified, including the beliefs. Beliefs remain, then, psychological objects even when all other features of a person's engagement in a practice are about developing identity in activity.

All the authors see a cyclical relationship between changing beliefs and changing practices; wherever one starts they affect each other. In terms of an emphasis on where the process starts my reading of these chapters suggests that Chapman, Llinares, and Philippou and Christou come down on the side of changing beliefs leading to change in practices, Lloyd and Hart on the side of changing practices leading to changing beliefs, and the review chapter by Wilson and Cooney remains with the dialectic. For Chapman the goal is to describe and map the links but she subscribes to the view that "beliefs strongly affect his or her behavior". In calling for the reification of beliefs Llinares appears to point to the need for a change in beliefs if teaching practices are to change. Philippou and Christou, in studying the essential role of teachers' beliefs about their efficacy, the role of efficacy beliefs in reform-oriented teaching, and the potential for a preservice course to change efficacy beliefs also seem to point in the same direction. Hart and Lloyd emphasize factors of practice: collaboration, colleagues and modeling of thinking and behaviors in the research of Hart; working with innovative curriculum materials in the research by Lloyd. Perhaps readers will feel that Wilson and Cooney's position on this is the most fruitful!

As indicated above, Chapman and Hart draw on teachers' own stories for their data and give a clear rationale for that choice. The task of the researchers is one of interpretation in order to illuminate and support the foci of their research. Llinares analyses transcripts of student teachers with the aim of looking in particular for evidence of reifying beliefs in order to renegotiate those beliefs. Lloyd uses the texts of a series of interviews with a teacher carried out over a long period and demonstrates the link between changing beliefs and innovative instructional materials. Philippou and Christou use statistical analysis of questionnaires backed up by a series of interviews. Their study has a number of elements, including examining the effects of a course on the history of mathematics on changing beliefs about teaching. The main point of their study, researching efficacy beliefs, is revealed in the quantitative data analysis and its triangulation in the interviews.

In summary, these chapters provide rich evidence of changing practices in teaching mathematics as seen through a lens of a discourse of beliefs as mental objects that are both cognitive and affective. They are constituted in teachers' prior experiences and they need to become the subject of reflection and analysis. A whole range of activities can bring about change towards reform and that change will come about as beliefs change. The picture is rightly a complex one to which this final summary does not do justice but I hope the previous paragraphs do.

2. SITUATING THE RESEARCH

The literature of research on mathematics teachers' beliefs and their teaching practices is extensive and well developed and I have played my own small part in this. Beliefs and belief systems (also conceptions, cognitions and other terms) are psychological constructs and it is not surprising that they have played such a significant role in research on teaching since psychology has been the major source of theory and research practice within our community (Kilpatrick, 1992). Wilson and Cooney have questioned the assumption that what is revealed in responses to questionnaires and interviews somehow captures the object being searched for, the belief system, on both methodological grounds and on conceptual grounds. We could argue that those research techniques are *productive* of the findings, given the interpretation that is constituted within the 'beliefs' orientation. Further, as a reviewer of this book proposal pointed out, there is a circularity about the assumption of beliefs driving actions. Since beliefs are private and therefore hidden from the gaze of the researcher one can only infer a teacher's beliefs from her or his actions, including answering questionnaires or responding in interviews. One then claims that the actions are determined by the beliefs. Wilson and Cooney, through Scheffler, have moved us closer to a unified notion with a greater focus on social practice and practice forms a major element of my alternative account.

In this section I will offer another way of looking at the issue of changing practices in mathematics teaching and learning, applying a *sociological* gaze rather than a psychological one. In recent years sociological and sociocultural theories have entered the field of mathematics education research as alternative intellectual resources (Lerman, 2000a; see also Atweh, Forgasz, & Nebres, 2001) to those offered by psychology.

All the authors of these chapters adopt a reform-oriented perspective; that is, students will learn mathematics better and will enjoy their mathematics better when the teaching and learning are oriented to inquiry-based approaches. This argument is expressed in different ways by the authors but the orientation is the same. I do not want to question the assumption here[1]. Instead I want to ask how this orientation has come about and what effects it has had and is having on the mathematics education research community. I want to examine how this orientation is supported and justified theoretically. It seems that there is a range of positions available to teachers, from those who do not engage with the reform in their classrooms at all, through those whose rhetoric is in line but whose practices are not, and finally to most of those described in these chapters, whose practice is coming into line as well, as reported in their own stories or by observers in their classrooms. I want to ask how we can explain these positions. Positions can be seen as discursively constituted: that is to say that it is in language games that meaning is carried and ways of being are framed. 'Positioning' in discursive practices and the study of identity constituted in social practices provide a social form of beliefs (Evans, 2000). Studies of beliefs rarely engage with the origins of beliefs but without a theory for how they come about one can hardly account for change.

It must be admitted here that my account is unlikely to replace the psychological perspective. The field of production of mathematics education research is what

Bernstein (1999) has called a horizontal knowledge structure. Unlike science, an example of a vertical knowledge structure in which each new theory can subsume its predecessors and can account for the same phenomena and more (indeed such a criterion is necessary for the adoption of a new theory), in mathematics education a new theory sits alongside its predecessors. Each discourse has its own language and loyalties, making refutations of one by another almost impossible (see also introductory remarks in Lerman, 2000b). For this reason I refer to the ideas presented in this section as alternative.

A reform along the lines of that in the USA could not take place in the UK because of the current dominance of the official pedagogical field (Bernstein, 2000) as realized through Government control of what takes place within each classroom, particularly in mathematics classrooms[2]. There is almost no space for resistance or initiative by teachers in the UK whereas the reform in the USA was initiated and driven by the National Council of Teachers of Mathematics. In earlier decades the situation in the UK was different and one can see, for example in the introduction of the New Mathematics in the early 1960s or of problem solving and investigations in the 1970s, the initiatives of mathematics teachers and researchers. The political structure of education in the USA is quite different to the UK, being decentralized to individual states and beyond, to school districts; hence, despite the ‘math wars’, efforts by the mathematics education community to propagate a particular approach to teaching and learning mathematics continue and indeed develop. A further element in the difference between the UK and the USA is the status of academics active in educational research. Whilst there are critics of that research in the USA, in the UK there are moves from within Government to take control of the funding for educational research and concentrate its use into relevant research. ‘Relevance’ as a criterion is presented as common-sense and populist, whereas it in fact represents an opportunity for the implementation of the Government’s program. There are a number of other factors, but this is not the place to develop these further. Suffice to say that the concatenation of a range of social and political circumstances have resulted in these differences, and this can be seen in the unquestioned place of the reform in the four papers by the North American authors and the lack of concern with social and political issues - a fortunate situation in comparison with the UK. The other two chapters are written by researchers from Spain and Cyprus and whilst the political situations in those countries do not resemble either the USA or the UK they are both in positions of independence between the two fields of the official and the unofficial.

Most classrooms in the US are still dominated by a traditional performance model (Bernstein, 2000) whereby what is emphasized in classrooms is the production of appropriate mathematical texts through explicit (visible) pedagogical means. Assessment looks for what is missing in the student’s production (for a deep analysis of assessment using Bernstein’s theories see Morgan, Tsatsaroni, & Lerman, in press). This mode is supported by behaviorist psychology. There are clear demarcations between mathematics and other subjects on the school curriculum (what Bernstein calls strong classification) and the classroom is strongly controlled by the teacher (strong framing) in terms of what is legitimate as mathematics. The teacher is positioned within a hierarchical discourse where she or he represents the view that the mathematics curriculum is designed such that students learn mathematics that they will need for further mathematics when they are older. The teacher knows best; in fact the teacher is not as free as this sounds and is herself or himself controlled by the hierarchical discourse. This discourse positions the teacher as the traditional teacher with a dualist view of mathematics and of pedagogy (Wilson and Cooney), of working with school mathematics versus inquiry mathematics (Lloyd), as teacher-centered rather than student-centered (Chapman), or using conventional lecture-orientated teaching (Hart). The authors of the chapters in this section characterize this position as the starting point for teachers, from which to change.

The 'reform' is a quite different model, a competence model, where one looks for the student demonstrating mathematical competences. Competences are much harder to identify than in a student's performance since partial understandings, what a student says rather than writes, and so on, might still demonstrate a degree of competence. Hence teachers require much greater training for this approach, which is of course much more expensive. One might conjecture that educational systems that move back from competence models to performance models may be driven at least partly by cost[3]. The particular version of competence model that is represented by the reform is a liberal/progressive mode (Bernstein, 2000). One can identify other modes within the competence model: a populist mode, such as an ethnomathematical approach, and an emancipatory mode such as criticalmathematics (Lerman & Tsatsaroni, 1998). Other modes are certainly possible. The liberal/progressive mode emphasizes what is present in the student's text and is supported by a Piagetian, constructivist psychology. Whilst there is still a strong distinction between mathematics and other subjects in the curriculum, strong classification, the framing is much weaker. The control of the production of what are thought of as appropriate mathematical texts appears to be much more open to negotiation and interpretation, especially from an everyday mathematics perspective although it is not so open in actual fact (invisible pedagogy). Students can approve other students' productions and, within limits, what is legitimate mathematically can be classroom specific. The mathematical authority is shared within the classroom, again within limits.

We could continue this analysis of the classrooms in the two modes, comparing forms of assessment, the pacing of the teaching, what is included in the mathematics curriculum, the relations between everyday mathematics and school mathematics, further elaborations of the pedagogical approaches, and so on. We could also look at the consequences for students within these modes. For example, the study by Cooper and Dunne (1999) of the use of everyday contexts in mathematics questions, typical of the liberal/progressive mode, can disadvantage children from working class backgrounds. What concerns me here, though, is the position of the teacher in this mode. I have described the 'traditional' teacher as positioned within the traditional performance model. Here, the teacher can be positioned differently, again as is clear from the examples and stories in the chapters in this section. The teacher conforming to the reform is a facilitator of students' learning of mathematics, the guide[4], and the designer of pedagogic materials suitable for students to work with in developing their understanding. There are personalized forms of control, dependent on the social relations developed by the teacher and the students, rather than positional as in the traditional performance mode. But this is not the only position available to the teacher. The following diagram presents the relationships between the various fields acting in the arena of mathematics education:

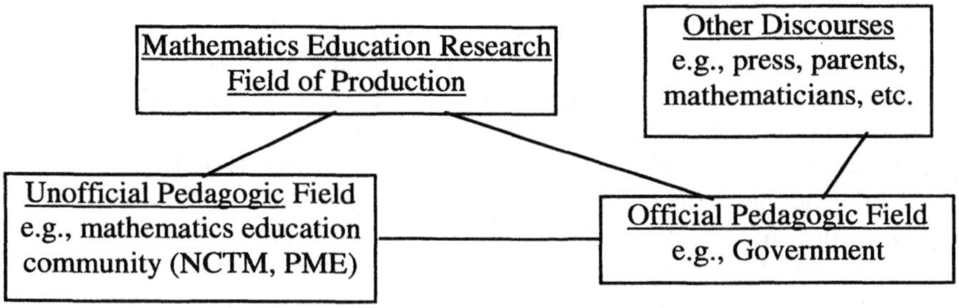

Figure 1: Fields and sub-fields in the production of positions of teachers[5]

In the US, as I mentioned above, the official pedagogic field in terms of Government is not strong. However in states such as California it is realized in some leading mathematicians together with the conservative press supporting their views. Teachers who identify with that official pedagogic field are positioned within a traditional performance mode. Teachers who resist can be positioned within one of the modes that appear within the mathematics education community. These, too, are socially constituted, at the very least through reading the Standards.

Broadly, then, we can characterize the positions available to teachers as in Table 1, although the first form within the competence model, the liberal/progressive, is the one that has been discussed in this chapter.

Table 1: Teachers' positions within different pedagogic modes

Model	Mode	Supporting theory	Teaching
Performance model	Traditional	behaviorist	Visible pedagogy, strong framing, teacher as transmitter
Competence model	Liberal/progressive e.g., reform mathematics	constructivist	Invisible pedagogy, weak framing, teacher as facilitator
	Populist e.g., ethnomathematics	sociology	Invisible pedagogy, weak framing, teacher as co-ordinator
	Emancipatory e.g., criticalmathematics	critical theory	Invisible pedagogy, weak framing, teacher as co-ordinator

3. CHANGING PRACTICES

In terms of changing classroom practices along the lines described by the authors of these six chapters the question we must address now is what are the conditions that might lead to a shift in positioning to that appropriate to the competence model and, in particular, to the liberal/progressive mode? Bernstein draws links between dominant groups in society and the dominance of particular modes in schools. Here, where the discussion is about the orientations of teachers, we need also a focus on the more micro-level picture, that of the dominant social groups in which the teachers work or with which they identify. As Daniels (2000) points out, what Bernstein's sociology does not offer is the "interpersonal" (p. 21), the possibility of analyzing the micro-level of the classroom and the process of knowledge acquisition. He also argues that whilst socio-cultural theories offer that micro-level they do not give accounts of the forms of culture that are appropriated in learning, the "institutional" (p. 21), for which he proposes a turn to Bernstein's work. Thus we have an account of the origins and influences of the range of classroom environments that emerge from a sociological analysis of the available discourses. What we need is an analysis of how teachers' actions, including their responses in interviews and to questionnaires, might change from one formation to another. A cultural analysis will need also to refer to the individual teacher, since some of these chapters (Chapman's and Lloyd's in particular) focus on individuals who change the way they teach.

I have already referred to the social constitution of these teaching practices. What strikes the reader very strongly in the chapters in this section is the range of activities that have been observed to support and encourage change in teachers' practices. I mentioned above the engagement with colleagues; conflicts between personal goals and the perceived situation; the role of University tutors modeling the good practice they advocate for schools, teachers working with innovative curriculum materials; and support in the classroom. Elsewhere (Lerman, 2001a) I have referred to teachers participating in inservice courses, taking higher degrees, or being involved in research projects as settings that research indicates can lead to changes in teachers' practices. All these settings seem to me to point to participation in social practices and developing identities as the common theme, although only Llinares draws on these ideas in this particular collection of chapters (see also Stein and Brown, 1997).

Lave and Wenger (Lave, 1988; Lave & Wenger, 1991; Wenger 1998) have shifted the discussion on learning in the education community and beyond from issues of cognition to those of participation and identity. Skills, knowledge and role become unified in the situating of knowledge in practices. To learn is to participate and to become an-other. They argue that the notion of a decontextualized individual carrying a set of beliefs, meanings or identity from one practice to another is too limited a notion for the social world. They suggest that we draw on a unit of analysis that places people in activity, person-in-practice. Wertsch (1991, p. 12) uses the same idea when he writes:

> It is more appropriate, when referring to the agent involved, to speak of "individual(s)-acting with mediational means" than to speak simply of "individual(s)". (p. 12)

One can extend the unit of analysis, however. When a person steps into a practice, there is a sense in which she or he has already changed. The person has an orientation towards the practice, or has goals that have led the person to the practice, even if she or he leaves the practice after a short time. One can express that change by noting that the practice has become in the person. We can therefore extend the unit of analysis to person-in-practice-in-person (Lerman, 2001b). In the different forms of participation, peripheral, central, and in the overlapping of that practice with other practices, people's identities develop in different ways. The trajectory into and through the practice are modified by virtue of prior experiences, personal goals and the form of participation.

As an example of research that monitors changes in beliefs during a program of intervention, Schifter (1998) explores two "avenues for promoting teachers' mathematical investigations. The first avenue is *exploration of disciplinary content* ... The second avenue is *examination of student thinking*" (p. 57, italics in original). A total of 36 elementary school teachers and six staff members from three institutions are engaged in a four-year teacher-enhancement project. During sessions in college, teachers engage in mathematical investigations at their own level, and experience models of different mathematical learning experiences, including problems being set without prior instruction, group problem solving, and so forth. Through reflection on the nature of their own learning experiences in mathematics some, at least, of the teachers recognize possibilities for transforming their own classrooms. At the same time, the journals that they keep include records of their students' mathematical activities and analysis of their thinking. Schifter also refers to assigned readings as an element that contributes to teachers' changing practices. Schifter demands yet more, though, the development by teachers of "a new ear, one that is attuned to the mathematical ideas of one's own students" (p. 79). Narratives of incidents in their classes, as well as narratives from the staff members of the project team, offer opportunities for the teachers to discuss their students' learning with other teachers in the project. The final element is that of staff visits to their classrooms, in which the staff help teachers to work on aspects of their lessons which the teachers identified at the beginning of the year.

The reports of teachers' development are impressive, as they are in other of Schifter's writings (for example, Schifter, 1995). The range of experiences in which teachers engage, during the project, clearly has a significant effect on their perceptions of their own teaching. From the accounts in these papers it seems clear that many of the teachers have learned a great deal about mathematics and about listening to and analyzing their students' mathematical activity, and their practice has been transformed. I want to ask what is the effect of the writing, of the work on their mathematics, of the length of the project, or of the group discussion between the teachers? In short, what leads to teachers' learning (or not)? One can imagine that new language and specific practices will have developed, new friendships, new orientations, and merging goals will have emerged. Studies of individual teachers would reveal stories of developing identities in new forms of practice.

This is not to say that practices are monolithic; individuals' paths through the activities will result in different forms of identities close to or further away from that modeled[6] by the research team. Interviews and questionnaires could be analyzed in terms of beliefs but I suggest that the sociological analysis of the origins of forms of practice and of the shifting identities of participants is at least an alternative account to the psychological one, and one that is situated in cultural theories of recent times (Schutz, 2000).

4. NOTES

[1] For instance, one can argue that from the perspective of the postmodern, reform discourses are productive of particular subjectivities in a manner that can be seen to be just as oppressive of teachers and students as any other regulating discourse (Popkewitz, 1996).

[2] Since September 1999 a National Numeracy Strategy has led to the implementation of a daily mathematics lesson in every primary (elementary) classroom with content and teaching style prescribed. Whilst it is not legally required, all schools are inspected for achievement of the aims of the Strategy and it is a brave and rare school that will continue to use its own curriculum and teaching styles. An extension of the Strategy is also being implemented in the first 3 years of secondary schools.

[3] Bernstein (2000) described the current directions in the UK as a new form of performance model, quite separate from the traditional and supported by Vygotskian psychological theories. See also Merrtens and Wood (2000).

[4] The change from traditional to liberal/progressive has been described as "From the sage on the stage to the guide on the side".

[5] Adapted from Morgan, Tsatsaroni and Lerman (in press).

[6] Ensor's (1999) study of teacher education demonstrates the need for a complex analysis of the modelling notion focussed on the distinctions between the different sites of activity and the recontextualisation process that takes place as people move between them.

5. REFERENCES

Atweh, W., Forgasz, H., & Nebres, B. (Eds.). (2001). *Socio-cultural aspects in mathematics education: An international perspective.* Mahwah, NJ: Lawrence Erlbaum & Associates.

Bernstein, B. (1999). Vertical and horizontal discourse: an essay, *British Journal of Sociology of Education, 20*(2), 157-173.

Cooper, B., & Dunne, M. (1999). *Assessing children's mathematical knowledge.* Buckingham: Open University Press.

Daniels, H. (2000). Bernstein and activity theory. Paper presented at symposium *Towards a sociology of pedagogy: The contribution of Basil Bernstein to research,* June, book of papers published by Department of Education and Centre for Educational Research, School Science, University of Lisbon.

Ensor, P. (1999). *A study of the recontextualising of pedagogic practices from a South African university preservice mathematics teacher education course by seven beginning secondary mathematics teachers.* Unpublished PhD dissertation, University of London.

Evans, J. T. (2000). *Adults' mathematical thinking and emotions: a study of numerate practices.* London: Falmer.

Kilpatrick, J. (1992). A history of research in mathematics education. In D. A. Grouws (Ed.), *Handbook of research on mathematics teaching and learning* (pp. 3-38). New York: MacMillan.

Lave, J. (1988). *Cognition in practice: Mind, mathematics and culture in everyday life.* Cambridge: Cambridge University Press.

Lave, J., & Wenger, E. (1991). *Situated learning: Legitimate peripheral participation.* New York: Cambridge University Press.

Lerman, S. (2000a). The social turn in mathematics education research. In J. Boaler (Ed.), *Multiple perspectives on mathematics teaching and learning.* (pp. 19-44). Westport, CT: Ablex.

Lerman, S. (2000b). A case of interpretations of social: A response to Steffe and Thompson. *Journal for Research in Mathematics Education, 31*(2), 210-227.

Lerman, S. (2001a). A review of research perspectives on mathematics teacher education. In F-L Lin & T. Cooney (Eds.), *Making sense of mathematics teacher education* (pp. 33-52). Dordrecht: Kluwer.

Lerman, S. (2001b). Getting used to mathematics: alternative ways of speaking about becoming mathematical. *Ways of Knowing, 1*(1), 47-52.

Lerman, S., & Tsatsaroni, A. (1998). Why children fail and what mathematics education studies can do about it: The role of sociology. In P. Gates (Ed.), *Proceedings of the First International Conference on Mathematics Education and Society* (pp. 26-33). Centre for the Study of Mathematics Education, University of Nottingham.

Merrtens, R., & Wood, D. (2000). Sea changes in mathematics education. *Mathematics Teaching*, No. 172, 13-17.

Morgan, C., Tsatsaroni, A., & Lerman, S. (in press) Mathematics teachers' positions and practices in discourses of assessment. *British Journal of Sociology of Education.*

Popkewitz, T. S. (1996) Rethinking decentralization and state/civil society distinctions: The state as a problematic of governing. *Journal of Education Policy, 11*(1), 27-51.

Schifter, D. (1998). Learning mathematics for teaching: From a teachers' seminar to the classroom. *Journal of Mathematics Teacher Education, 1*(1), 55-87.

Schifter, D. (Ed.). (1995). *What's happening in math class? Reconstructing professional identities* Vol. 2. New York: Teacher's College Press.

Schutz, A. (2000). Teaching freedom? Postmodern perspectives. *Review of Educational Research, 70*(2), 215-251.

Steffe, L. P., & Thompson, P. W. (2000). Interaction or intersubjectivity?: A reply to Lerman. *Journal for Research in Mathematics Education, 31*(2), 191-209.

Stein, M. K., & Brown, C. A. (1997) Teacher learning in a social context: Social interaction as a source of significant teacher change in mathematics. In E. Fennema, & B. S. Nelson (Eds.), *Mathematics Teachers in Transition* (pp. 155-191). Mahwah, NJ: Lawrence Erlbaum Associates.

Wenger, E. (1998). *Communities of practice: Learning, meaning and identity.* Cambridge: Cambridge University Press.

Wertsch, J. V. (1991). *Voices of the mind: A sociocultural approach to mediated action.* Cambridge, MA: Harvard University Press.

PART 3

STUDENTS' BELIEFS

CHAPTER 15

PETER KLOOSTERMAN

BELIEFS ABOUT MATHEMATICS AND MATHEMATICS LEARNING IN THE SECONDARY SCHOOL: MEASUREMENT AND IMPLICATIONS FOR MOTIVATION

Abstract. Students' beliefs about mathematics and mathematics learning can have a substantial impact on their interest in mathematics, their enjoyment of mathematics, and their motivation in mathematics classes. This chapter has a dual focus with respect to such beliefs. First, an interview instrument to measure personal and environmental beliefs that influence student motivation in mathematics is discussed. Drawing from the mathematics education as well as psychological literatures, the instrument's questions focus on topics including feelings about school in general, non-school influences on motivation, self-confidence, perceptions of ability, goal orientation, study habits, mathematics content, assessment practices, and expectations of teachers. Second, findings from the instrument that are specific to beliefs about the nature of mathematics are described in the chapter. Findings include the fact that the nature of mathematics is not something secondary students think about. When pressed, however, most of their comments deal with the procedural rather than conceptual aspects of the subject. Students also tend to feel that memorization is an important part of mathematics even though they feel individuals who do not memorize well can still succeed. Taken as a whole, the chapter documents the importance of considering a wide variety of beliefs when trying to understand individual student interest and motivation in mathematics.

Interest in beliefs about mathematics and mathematics learning has increased substantially in the last ten years. As can be seen from the contents of this volume, some of that interest focuses on the concept of beliefs and some focuses on the beliefs themselves. In this chapter, I focus on the beliefs of secondary school students under the assumptions that beliefs are an important influence on motivation and that motivating students is a major goal of instruction. Specifically, I will describe an interview instrument for assessing students' motivation-related beliefs and then summarize data collected with the instrument that answer the research question "What do students think mathematics is and how does one learn mathematics?"

1. BELIEFS AND MOTIVATION IN MATHEMATICS

Early researchers on motivation treated it like an inner drive (Graham & Weiner, 1996) but, starting with Atkinson's (1957) discussion of motivation as the product of

G. C. Leder, E. Pehkonen, & G. Torner (Eds.), Beliefs: A Hidden Variable in Mathematics Education? 247-269.

expectation of succeeding on a task and perceived value of accomplishing that task, research on motivation has been increasingly treated as a function of cognitive decision making. Attribution theory (e.g., Graham & Weiner, 1996; Weiner, 1984), self-worth theory (e.g., Covington, 1992; Graham & Wiener, 1996), self-efficacy theory (Bandura, 1997), and goal-orientation theory (e.g., Ames, 1992; Blumenfeld, 1992; Elliott & Dweck, 1988; Graham & Weiner, 1996) all assume that individuals only put forth effort when they perceive that effort will result in fulfilment of their personal goals. Despite the need for better understanding of motivation in secondary school mathematics, these theories have rarely been applied to this subject area.

In terms of mathematics instruction, there is a need for better understanding of how psychological theories of motivation apply to the mathematics classroom. Of particular interest is the situation where mathematics instruction is changing to include increased focus on reasoning, problem solving, and conceptual understanding (National Council of Teachers of Mathematics [NCTM], 2000). In such reform-oriented classrooms, cognitive theories of motivation suggest that students may quit trying to learn when the instructor stresses comprehension of concepts over memorization of steps to get an answer (Kloosterman, 1996).

In this chapter, the assumption is made that motivation is a cognitive activity and that students make specific choices about how and when they apply effort to learn (Kloosterman, 1996; Stipek, 1996). These choices are based on students' beliefs, defined as personal conceptions that mediate action. That is, a student's belief is something the student knows or feels that affects effort – in this case effort to learn mathematics. This definition is similar to some of the psychological work on beliefs (e.g., Bartsch & Wellman, 1989; Pintrich & Schrauben, 1992) but differs from definitions elsewhere in this volume. Moreover, the term belief is sometimes absent from current motivational work (e.g., attribution theory, self-worth theory, goal-orientation theory) but it is often a synonym for terms such as attitudes and dispositions that are used. Finally, the distinction between knowledge and beliefs is important in some contexts (see Golden, this volume) but that distinction is not important for this chapter because the emphasis is on how knowledge and beliefs influence action.

It is also important to note that this chapter will focus on how motivation to learn mathematics can result from beliefs using several cognitive motivation theories. Given the subtle differences in the cognitive motivation theories (attribution theory, self-worth theory, self-efficacy theory, goal-orientation theory, etc.), one could argue that mixing them in a single study is inappropriate. However, teachers and researchers need to know the extent to which one of these theories is more appropriate than others with respect to motivation to learn mathematics. None of the theories has been adequately tested in the realm of mathematics learning and thus, when looking at motivational beliefs, it is appropriate to work from a variety of perspectives.

2. ASSESSING MOTIVATIONAL BELIEFS IN MATHEMATICS

At the present time, there are quantitative attitude scales (e.g., *Fennema-Sherman Mathematics Attitude Scales*, Fennema & Sherman, 1976) that can be used to assess some factors related to motivation in mathematics. Such scales can give researchers information about the beliefs that students hold but they are severely limited in their ability to explain how such beliefs formed or how beliefs are likely to influence action. Thus, an interview instrument was developed to assess key motivational factors outlined in earlier theoretical work (Kloosterman, 1996). In brief, my work follows some of the general psychological work (e.g., Pintrich & Schrauben, 1992) in that I have postulated that mathematics learning is affected by motivation and that motivation is the result of beliefs about (a) mathematics as a discipline, (b) the self as a mathematics learner, (c) the role of the mathematics teacher, and (d) other beliefs about mathematics learning. The instrument follows this conceptual framework.

The first stage in the development of the interview instrument was a search of the literature on academic motivation. For beliefs that were either known to relate to student effort in mathematics (e.g., perceived usefulness of mathematics) or seemed likely to affect such effort (e.g., goal orientation), interview questions were drafted. When composing questions, it was difficult to know how much detail to include in each question. Overly detailed questions tend to cue respondents to give the answer they think the researcher wants to hear. Questions that lack specificity often lead to answers that lack depth or detail. One of the major advantages of an interview as opposed to a survey or questionnaire is that when questions fail to produce detailed answers, the researcher can probe further. For this interview instrument, initial questions were designed so as not to lead respondents toward specific answers. Follow-up questions and prompts were provided for students who did not provide detailed initial answers.

The initial interview questions were arranged in a single document so that related questions were asked in sequential order whenever possible. This ordering made it relatively easy for interviewers to cross reference responses when writing down a student's comments. Using input on wording provided by eight individuals with expertise in mathematics education, an instrument was generated that contained 56 questions, most of which included follow-up questions to be used as needed.

The instrument was administered to 56 high school student volunteers in the mid-western United States. The students were selected from four schools and were enrolled in mathematics courses ranging from general mathematics (review of computational algorithms) to beginning calculus. Ten students were in ninth grade, 13 were in tenth grade, 15 were in eleventh grade, and 16 were in twelfth grade. In other words, the students interviewed represented a cross section of students enrolled in four-year high schools in the United States.

Most interviews took place during students' study periods. The interviews took two 45 to 50 minute periods and responses were noted in writing and audio taped. In addition, after each interview session, the interviewer wrote down a summary of (a) the student's overall level of motivation, and (b) those factors that seemed to be most influential in determining the student's motivation. All interviewers were experienced teachers who were used to talking to students about issues relating to education.

When all interviews were completed, questions were reassessed individually to determine if other ways of wording them might result in better information, if some could be combined or eliminated, and if new questions should be added where responses were ambiguous. Data were also scrutinized for unexpected answers, particularly unexpected answers that came up a number of times. As a result of this scrutiny, several minor changes in wording and content took place for the instrument. The final form of the instrument contained 51 questions and is included in Appendix A.

2.1. Interview Categories and Questions

In the final version of the instrument, the 51 questions were spread across 12 major categories. Although it could be argued that the diversity of the categories resulted in questions that did not fit well together, unifying theories of motivation in mathematics are not available in the literature (Carr, 1996). Thus, having variety in the categories provided the opportunity to determine if some perspectives were more effective than others in understanding students' thoughts about motivational issues. Following are the question categories for the interview instrument.

2.1.1. General Background (questions 1-5)

Questions here focused on previous mathematics courses taken and plans for the future. The rationale for these questions was that background information helps in understanding a student's success and failure pattern in mathematics (Stage & Kloosterman, 1995) and in school in general (Graham & Weiner, 1996). Plans for the future help to understand when a student might be willing to work hard to achieve specific career goals (Maple & Stage, 1991; Reyes, 1984).

2.1.2. Indiana Mathematics Belief Scales

After answering the first five questions, students were asked to complete the 36 items of the *Indiana Mathematics Belief Scales* (available in Kloosterman & Stage, 1992). These Likert-type scales assess students' perceptions of (a) their self-confidence in solving mathematics word problems, (b) the extent to which mathematics is always a step-by-step process, (c) the usefulness of mathematics, (d) the importance of conceptual understanding in mathematics, (e) the importance of textbook word problems, and (f) the relation between effort and ability in mathematics. For example, one of the textbook word problem items is "Word problems are not a very important part of mathematics" and one of the effort and ability items is "Ability in math increases when one studies hard." Because the scales are quantitative, they do not give the type of insight that an interview can give but they do provide baseline data on key student beliefs.

2.1.3. Feelings About School and About Mathematics (questions 6-9)
Questions in this category dealt with general like and dislike of school and of mathematics. Data from these questions are helpful in determining the extent to which motivation in mathematics is indicative of a broader pattern of motivation in all school subjects. Clearly, there is sentiment that some students have very different attitudinal and motivational patterns in mathematics as compared to other subjects (e.g., Kloosterman, 1996; National Research Council, 1989).

2.1.4. Effort in Mathematics (questions 10-13)
On the assumption that a student's effort can vary with the task at hand (Blumenfeld, 1992; Kloosterman, 1996), students were asked about their level of effort in mathematics class, on homework, and in other school subjects. They were also asked directly about what motivated them in mathematics. Data from these questions provided background for interpreting later responses about how specific content and classroom factors influence motivation.

2.1.5. Non-School Influences on Motivation (questions 14-21)
Non-school factors, including parents' comments and actions, can be very influential in the motivation of students (Eccles et al., 1985; Shernoff & Schneider, 2000). However, as a teacher, school administrator, or researcher, one must be very careful not to ask questions that violate rights to privacy. Questions about non-school influences were thus intentionally general and dealt with how parents use mathematics themselves, amount of parental support for learning mathematics, perceptions of how mathematics is used in the workplace, and whether peers view learning mathematics as useful.

2.1.6. Self-confidence in Mathematics (questions 22-23)
Self perceptions of ability have been tied to motivation and achievement in general academic settings (Alderman, 1999; Covington, 1992; Graham & Weiner, 1996; Schunk, 1991) and in mathematics (Kloosterman, 1988, 1996; Reyes, 1984). Because student confidence can vary by topic area within mathematics (Kloosterman & Cougan, 1994), a question about variation in confidence by mathematics topic was included (#23).

2.1.7. Natural Ability in Mathematics (questions 24-26)
When an individual believes that ability is fixed and that he or she does not have a lot of ability, motivation and performance are negatively affected (Alderman, 1999; Graham & Weiner, 1996). Given the perception of many adults in the United States that one either has a "math mind" or one does not (National Research Council, 1989), self perceptions of ability in mathematics seem particularly important for understanding student motivation in mathematics.

2.1.8. Goal Orientation and Effort (questions 27-30)
Students were asked questions designed to determine the extent to which they were task-oriented, ego-oriented, affiliative-oriented, and work-avoidant-oriented. All of these orientations relate to motivation (Ames, 1992; Blumenfeld, 1992; Elliott & Dweck, 1988; Graham & Weiner, 1996) yet little has been done to look at the orientations within the content area of mathematics.

2.1.9. Study Habits in Mathematics (questions 31-35)
Given the substantial efforts being put into reforming the way students learn mathematics as well as changes in what they need to learn (see Carl, 1995), it is important to know where students put their effort when working in class or at home. Ford (1994), for example, notes that fifth-grade teachers in her study felt that problem solving was primarily just the application of computational skills and thus focused on having students decide which computational algorithm to apply. Such instruction may undermine students' motivation for deeper understanding of how to solve complex problems. Questions in this portion of the instrument dealt with the extent to which students learned from memorization as opposed to building connections and conceptual understanding. Queries involving use of textbooks and reliance on peers were also included. In general, motivation is important but motivation to simply memorize is not sufficient for conceptual understanding. That is, the direction in which students put their efforts needs to be considered (Kloosterman, 1996).

2.1.10. Mathematics Content (questions 36-40)
This category is related to the previous one in that studying the "wrong" mathematics, like studying mathematics using inefficient strategies, can result in wasted effort. The first question in this category (#36) was designed to get a sense of how students viewed mathematics content. If mathematics is just a set of rules, then students are likely to try to memorize the rules (Kloosterman, 1996) or try to make inappropriate generalizations from the rules (Tsamir & Tirosh, this volume). A question was asked about activities in mathematics class on the assumption that such activities affect students' views of the nature of mathematics (Yackel & Rasmussen, this volume). In brief, reforms in mathematics education call for a significant change in views of the discipline of mathematics (Steen & Forman, 1995). For motivation to be most effective, it needs to be aimed toward learning the mathematics that will be needed in the present and the future rather than the mathematics that was used in the past.

2.1.11. Assessment Practices (questions 41-43)
A major tenet of the efforts to reform assessment in school mathematics is that students will work to learn the mathematics on which they are assessed (Lester & Kroll, 1991). In addition, grading "on a curve" is a competitive assessment system that promotes ego-orientation much more so than task-orientation (Alderman, 1999; Nicholls, 1984). Questions in this category dealt with types of assessment and

grading systems students had experienced, and the extent to which they felt that grading practices influenced their motivation.

2.1.12. Students' Expectations of Teachers (questions 44-50)

The final category of questions in the instrument dealt with what students expected their teachers to do in mathematics class. This issue is connected to motivation because when teachers use pedagogical techniques that students are not used to seeing, the students may question the competence of the teacher and thus not bother to try to learn what the teacher is teaching (Yackel & Rasmussen, this volume). For example, when a teacher withholds help as a means of getting students to reflect on ways to solve a problem without help, some students believe the teacher is abdicating his or her responsibility and they quit working (see Kloosterman, 1996).

3. REFLECTIONS ON USE OF THE INSTRUMENT

After using the instrument with 56 students and then making final revisions to it, three general issues emerged. First of all, although it was not a surprise, there was significant variability in how much secondary school students had to say about what they like to do and why they like to do it. Some students were quite verbal but others, especially the third of the sample who had taken or were taking pre-algebra or other remedial courses, had very little to say about what they liked and disliked about school or what motivated them. Cognitive theories of motivation assume that students make conscious choices about when to work and when not to work (Alderman, 1999; Graham & Weiner, 1996). A number of the students, especially those in the lower-level classes, seemed to be coasting through school without really thinking about where they were heading or what they might do in the future. Another way to say this is that they were concerned about what they were doing after class or on the coming weekend but not when they finished high school. If these students really do not think about having to support themselves or about the purpose of their education, then current frameworks for understanding academic motivation may need to be revised.

Second, in a finding similar to that of DeGroot (2000) who used interviews to understand more general motivational factors, the instrument proved to be an effective tool for constructing a picture of the interest and effort for the students interviewed. The interview format proved to have all the virtues and limitations described by Clement (2000). It was labor intensive and analysis of the data was somewhat subjective. However, the utensil provided much more detail than a survey, especially for students who needed prompting to be open about their feelings.

Third, as anticipated, the instrument usually took about two 45-minute periods to complete. Near the end, the interview questions sometimes became redundant because students had talked about the issues brought out in those questions in their responses to earlier questions. However, the length of the interviews was almost never a problem for the students.

4. GENERAL FINDINGS FROM THE INTERVIEWS

Detailed analysis of student responses for the full instrument is beyond the scope of this chapter, although summaries of the findings for six different students can be found in Kloosterman (1998). However, because others using the instrument are likely to wonder what types of responses to expect, the following summary of the types of comments that students gave in each category is provided. No effort was made to quantify the number of each type of response because the comments were often hard to put into specific categories and because, as will be explained later in the chapter, it was common for students to have responses to questions that were contradictory to responses to other questions. All generalities noted below were true for at least 25 percent of the sample unless otherwise stated. As previously noted, there was considerable variation across students in the level of detail of their answers.

4.1. General Background

The questions in this category were good for getting students to be at ease and for getting a sense of the achievement level and aspirations of each individual. Question 4, which dealt with post-graduation plans, was most effective with juniors and seniors. That is, students getting near graduation had thought more about plans after graduation and thus their answers tended to be better formulated than the answers of some of the younger students. Question 5, about extra-curricular activities and things done just for fun in school, proved useful in identifying those students who were minimally involved in the non-academic aspects of school.

4.2. Indiana Mathematics Belief Scales

Data from the *Indiana Mathematics Belief Scales* proved useful for identifying students who were outliers in terms of how they viewed mathematics, the utility of mathematics, and their views of themselves as mathematics learners (see Kloosterman, 1998).

4.3. Feelings About School and About Mathematics

Question 6, "What words would you use to describe school?" resulted in a variety of answers. Some students described enjoyable and frustrating aspects of school, others described the reasons that school is important, and many described the social aspects of school. The primary advantage of this question was that it allowed students to provide the first ideas that came to their minds and thus gave the interviewer insight into the priorities of the students. The responses, however, were so diverse that making generalizations from them was difficult. Questions 7 through 9 asked about students' enjoyment, and perceived usefulness, of school and mathematics. As expected (see Kloosterman & Cougan, 1994), many students reported that their views of mathematics changed as they moved from middle school to algebra and then to geometry and more advanced courses. That is, students who enjoyed algebra often disliked geometry while those who disliked algebra sometimes found geometry more interesting. (N.B., In the United States, college-bound students take algebra, geometry, a second year of algebra and then pre-calculus. Because these are taught as separate courses, the distinction between enjoyment of algebra as opposed to geometry is less likely to be found in other countries.)

4.4. Effort in Mathematics

Questions about effort in mathematics were necessary given that a primary purpose of the interview was to determine what aspects of school and mathematics were most motivating. Although students seemed quite honest in their responses, it was apparent that questions about how hard one works meant different things to different individuals. For some, working hard simply meant working until an assignment was done with little regard for how correctly and completely an assignment was completed. Some students noted that directions given by teachers were vague and thus they were not sure exactly what they were supposed to do. For example, teachers would remind students to study for tests but gave them little guidance on how to do that. Without that guidance, the students did not really know if they had "worked hard" or not (although most assumed they probably did). It was hard from these questions to determine the extent to which students worked hard to "understand" as opposed to simply get an answer. Comments about why students liked or disliked school and mathematics proved to be quite interesting. Like elementary students interviewed by Kloosterman and Cougan (1994), a number mentioned liking challenge. Also mentioned was what teachers did to make mathematics understandable – a key issue for many of them. Finally, interviewers were asked to note students' comments relating to goal orientation (question 12) but such comments turned out to be quite rare.

4.5. Non-school Influences on Motivation

Questions 20 and 21 dealt with peer influences on motivation in mathematics. Students were quite willing to talk about their peers and provided a number of examples of things that peers did that those being interviewed thought was inappropriate. On question 20 for example, a student who was indifferent about school in general but saw mathematics as worth learning, said "My friends don't do nothing in math, they just copy off me. They might see math as useful but they don't show it in class."

4.6. Self-confidence in Mathematics

Similar to elementary students (Kloosterman & Cougan, 1994), students in this study usually mentioned grades and teacher comments as evidence that they were doing well or poorly in mathematics. When answering question 9 about enjoyment of mathematics, many students reported greater self-confidence in some areas of mathematics than others. For example, one senior indicated that "I'm probably best at algebra and I'm worst at memorization.... If I like have a note card with the formulas on it for the test ... I can usually figure out the stuff [algebra]."

4.7. Natural Ability in Mathematics

Kloosterman and Cougan (1994) found almost all the elementary students in their study felt that anyone who tried could learn mathematics. At the high school level, most students felt that effort made a difference and that making mistakes in mathematics was to be expected (question 25). However, a number also felt that lack of ability is a significant impediment for some students ("Some people will get it, some not"). As noted earlier, those who believe they have low ability are often less willing to try than those who know that they can learn (Graham & Wiener, 1996). A majority of students felt that memorization skills were important for learning mathematics (question 26). One would assume that believing that memorization is a key to learning mathematics has significant impact on where a student puts her or his efforts.

4.8. Goal Orientation and Effort

The goal orientation questions were effective in getting a sense of the extent to which students worked to learn the material (task orientation) as opposed to out-perform their peers (ego orientation). Similar to the findings of Seegers and Boekarts (1993) who reported positive correlations between task orientation and ego orientation in mathematics, many students in this study reported both wanting to learn for the sake of learning, and to do well in relation to their peers. A number stated that looking good in the eyes of the teacher was much more important to them than looking good to their peers. Twenty percent fit the definition of "work avoidant" in that they stated flatly that they did the least amount that they could to get by (question 30).

4.9. Study Habits in Mathematics

Questions 31 and 32 proved difficult for students to answer in that the students seemed used to doing what the teacher asked without considering whether their time was being used efficiently. Like the students of Yackel and Rasmussen (this volume), the norm for classroom culture was usually one of doing what the teacher asked without regard for the purpose of an assignment. With the current emphasis on conceptual understanding in mathematics (Carl, 1995; NCTM, 2000), questions about the use of study time are important but those who ask them need to be aware that students may have a hard time answering them. Questions 34 and 35 are appropriate when trying to determine if students will put effort into group activities. These questions, along with question 33 about using a textbook, were easy for students to answer.

4.10. Mathematics Content

Question 36 was written with the hope of getting a sense of what students thought it meant to learn or do mathematics. As will be explained later in this chapter, a few students did give some very detailed responses about how they saw mathematics, what they did in mathematics class, and how they used mathematics. Few said anything about the conceptual as opposed to rule-based aspects of mathematics. Most answered by saying again whether they liked or disliked mathematics. Responses to questions 37-40 provided more detail on the types of activities that students were exposed to in mathematics and thus where they felt they should be putting their effort when doing mathematics.

4.11. Assessment Practices

Student answers to question 41 indicated that tests and homework were common in most classes. Grades were strong motivators for almost all students interviewed, with comments ranging from working to get the best grade in mathematics one could to working just hard enough to get a passing grade. Question 43, which focused on how a teacher's grading system affected effort, resulted in a wide variety of answers. Some students claimed they worked very hard to learn exactly what the teacher graded on and even that they worked much harder in classes where the teacher had a tough grading scale. Others claimed that the exact mechanisms for evaluation had little to do with their effort (in seeming contradiction to their desire to get good grades). Only a handful of students made comments about learning because of interest or enjoyment in the material.

4.12. Students' Expectations of Teachers

Students provided a number of examples of what they considered good and bad teaching in response to questions 44-50. In particular, many students were like those of Yackel and Rasmussen (this volume) in their desire to have teachers explain clearly what they were supposed to do. From a motivational perspective, the belief that the teacher should explain everything can cause conflict and lack of student effort in a classroom where the teacher expects students to take an active role in learning. Teachers who expect students to be active problem solvers are likely to have to confront the issue of teachers' role to avoid having students be too discouraged (Kloosterman & Mau, 1997; Yackel & Rasmussen, this volume). Question 49, which dealt with differing expectations by teachers for different students, brought near unanimous responses that all students should be expected to complete the same work. Such opinions could have significant impact on student motivation in a classroom where a teacher tried to account for individual differences by varying expectations or assignments as a method of dealing with individual differences in background or ability.

5. BELIEFS ABOUT THE NATURE OF MATHEMATICS AND MATHEMATICS LEARNING

Although beliefs about the nature of mathematics and mathematics learning were only part of the interview study, findings with respect to these beliefs are particularly relevant to this volume and thus will be reported here. In essence, analysis of student responses to interview questions 26 and 36 (Appendix A) can help in answering the research question "What do students think mathematics is and how does one learn mathematics?" Written notes summarizing responses to several other interview questions were also considered when they helped in answering this question.

After reading through the transcripts and interview notes several times, three primary themes related to the research question emerged. Once the themes were specified, the data were reviewed one more time for comments that were contrary to the themes. None of the themes was true for every student, but this final review of the data confirmed that they were valid in most situations. (Exact numbers of students with specific beliefs will be noted only in situations where student comments were clear enough to categorize each student as having a specific belief.)

The first two themes are related and are thus considered together. To support the validity of the themes, transcriptions of student responses are provided.

Theme 1: The nature of mathematics as a discipline is not an issue that United States high school students think about.

Theme 2: When students are pressed to talk about the nature of mathematics, they mention that mathematics can be used to solve a variety of problems and that it involves numbers. They often mention the procedural nature of mathematics. They sometimes mention the logical nature of mathematics but almost never mention deduction or proof.

Because the nature of mathematics should become apparent after years of exposure to the subject, it was assumed that when questioned, students would be able to provide their views about what comprises mathematics. Theme one reflects the fact that this turned out not to be the case. When asked indirectly about the nature of mathematics (question 36), most students talked about whether they liked mathematics, why one needed to learn mathematics, and the daily procedures in mathematics classes (grading homework, listening to an explanation, etc.).

Lana: (female junior, second-year algebra) I'd probably explain the benefits of math and tell the alien that everything you do basically has to do with math [The term alien comes from the wording of question 36]. The words I'd use to describe it? Probably the same words I'd used before – interesting, difficult at times but sometimes it's really easy too. I'd probably tell the alien that about what math class I'm in right now and describe it to him or it or whatever it is. And tell it the progress I'm making.

Interviewer (I): How would you describe the math that you're in right now?

Lana: Advanced, difficult, a lot of application, that's it.

Sid: (male junior, first-year algebra) It's fun, the teachers I have ... makes it fun. It ain't dull. They [teachers] do extra stuff to show you more how to do it [and] let you learn better.

I: What kind of extra stuff do you mean when you say extra stuff.

Sid: Like projects and stuff. Like extra sometimes it's extra credit assignments to do and helps you get better.

Noah: (male senior, pre-calculus) I would say to the alien to only take what you need to take because from my point of view a lot of it after a certain point is purposeless except if you're going to go on to college and that's the only other time you're going to use a lot of the extended math.

I: Okay.

Noah: If I could elaborate a little.... I believe everyone should know how to do simple equations, word problems for example. But when you get into this cosine and sine and I don't even know how to describe all this stuff. I have a hard time seeing when it's going to be used unless you're going to teach trig or something and go on to be a mathematician. I have a hard time finding use for this, so I'd say I'd tell them that definitely take math. It's very important but to an extent.

I: Okay. You say you had a hard time seeing how things in trig for example are going to be used. Do you have to do word problems or any things like that?

Noah: Very rarely.

Fifteen of the 56 students mentioned specifically that mathematics involved steps, procedures, or formulas.

Lisa: (female junior, pre-calculus) Basic at first, just like steps and skills, and of course as you progress in math it gets more challenging. Can be difficult at times but it can always be solved. Basically I don't think there isn't ever an answer. I mean there probably is.

I: You said it was basic at first – steps and skills. Does that change?

Lisa: Yes. The steps get more complex and more involved in harder classes.

I: Is math still a set of steps? Is that the way you think of it even at the higher levels?

Lisa: Basically yeah. I think math problems have a lot of formulas. I do math in steps instead of just being able to, you know, you [I] kind of have to look at things and make sure I know what I'm doing.

Seven students spoke of mathematics as a way of thinking or a logical system. One of the students, who was in his last year of high school, spoke specifically about the structure of mathematics. However, he was like many other students in that it was difficult to understand exactly what his comments referred to.

Arlan: (male senior, calculus) It's like a structure, math is like the study of a structure.

I: Math is study of a structure?

Arlan: It's the study of an imaginary structure that you can use to analyze all kinds of problems and we use to label things. Because the structure fits to a label, we label objects by using math.

I: Okay.

Arlan: It's hard to kind of explain.

Only one student mentioned proof as an important aspect of mathematics. This student was taking geometry at the time she was interviewed and "formal two-column proof" was a major topic in her course.

Allie: (female junior, geometry) How would I describe math here in Indiana?

I: Yeah. Just the idea is you're describing it to somebody who doesn't have any conception of what math is. How do you get this thing to understand what math is?

Allie: Dealing with a lot of numbers, problem solving, proofs. I guess that's all....

I: When you say problem solving, what do you mean by that?

Allie: Like word problems.

Overall, most of the students interviewed assumed that "math is math" and that on the surface, it is the same for everyone. Almost all students found it difficult to describe the nature of mathematics, even with significant prompting. The fact that 15 of the 56 students mentioned the stepwise nature of mathematics without any prompting indicates that many students do feel mathematics is a set of rules to be mastered.

Upon further reflection, these findings seem quite reasonable. Rather than asking what math is, parents and peers ask students what they like about a class, or what they are learning in a class. In fact, a sizable number of those interviewed seemed to interpret the question about the nature of mathematics as a question about the nature of mathematics *class*. In short, because students are not regularly asked about the nature of mathematics, they do not form opinions on this issue. I will return to the implications of this finding later in the chapter.

Theme 3: Students tend to feel that memorization, and the ability to memorize procedures, is an important part of being successful in mathematics. On the other hand, they also feel that students who are not good at memorizing can still learn mathematics if they work hard enough.

In the United States, there is a common perception that one either has a "math mind" or one does not (National Research Council, 1989). The 1996 National Assessment of Educational Progress (NAEP) found that 89% of grade 4 students and 73% of grade 8 students agreed with the statement "Everyone can do well in mathematics if they try." By grade 12, however, only 50% agreed, with 29% disagreeing and 21% undecided (Mitchell, Hawkins, Jakwerth, Stancavage, & Dossey, 1999). Mathematics has traditionally been taught as a series of procedures to be applied to specific types of problems rather than as the ability to build and analyze models of complex real-life situations. As such, memorization of procedures can be important to learning mathematics. From a motivation perspective, those students who have doubts about their ability to memorize formulas and procedures have good reason to question whether they have ability to do mathematics and thus whether they should bother to try.

Like the question about the nature of mathematics, the question about memorization in mathematics caught a number of students by surprise. They understood what they were being asked and many had very strong opinions. However, many of the comments they had seemed contradictory in that they would claim that memorization was very important and then they would also claim that students who did not memorize well could still be high achievers in mathematics.

Others would say that memorization was not that important and then talk about how difficult it was to know mathematics if you did not remember the procedures. Overall, 20 of the 56 students used the terms very important or really important when first asked about the need to memorize. Another 19 said memorization was important but were not emphatic about it. Fifteen said that it was somewhat important or that it was important in some ways but not in others. One said that it was not important and one said it might be important but that she really did not know.

Although there was overwhelming agreement that memorization was important in some if not all aspects of mathematics, the nature of the specific comments proved to be more interesting than the actual numbers in each category. For example, Beth, a female sophomore taking first-year algebra, was asked about the importance of memorization when trying to learn mathematics [question 26]. Beth's comments were quite typical in that (a) she felt memorization was important, (b) even though she said memorization was very important, she felt that someone who did not memorize well could be good at mathematics, and (c) the interviewer had to probe to get her to say why she felt the way she did.

Beth: [Memorization is] real important. Because if you don't remember for example, your basic adding and subtracting, then how [are] you really going to do anything? And your multiplication, you'll have to go back to really, you know, if you don't remember it then, then it'll take you forever to do anything.

I. So... it sounds like here you're giving examples of basic facts.... addition, subtraction, multiplication. Is there any other place where you need memorization ... or is it always just a matter of memorizing these isolated pieces of math?

Beth: Yeah, like I don't know. (Giggle.) Just, I mean, if you don't remember it then you'll just have to always have somebody tell you how to do it and explain it again and again. And it helps a lot to remember so you don't have to go over it and over it.....

I: And does that ... apply to, for instance, the kind of math you're doing now in algebra?

Beth: Yeah.

I: And what kinds of things do you memorize in there?

Beth: Like, just basically how to do it and how to like take one paragraph, or one parentheses and split it into two and like, like stuff like that.

I: Those are like steps or procedures?

Beth: Yeah.

I: Okay. Are you good at memorizing?

Beth: Well, math I am. Yeah.

I: Can someone who's not very good at memorizing be good at math?

Beth: Yeah. I mean, they'd have to have it all written down or have to have someone really explain it to 'em. But, they can still do good in math.

6. DISCUSSION

The instrument and findings described in this chapter are intended to better establish the link between students' beliefs and mathematics achievement. The instrument is an example of how general psychological theories of motivation should be applied to mathematics where the content, pedagogy, and attitudes are in many ways unique to the discipline. It is assumed that as additional research is done on motivation and beliefs in mathematics, a better understanding will ensue of how motivation and beliefs in mathematics differ from those in other content areas.

With respect to the instrument itself, several general comments are appropriate. First of all, the structured interview format proved to be more effective than Likert scales (e.g., Fennema & Sherman, 1976; Kloosterman & Stage, 1992) in assessing students' beliefs, attitudes, and overall motivation. In a finding consistent with that in general qualitative motivation research (DeGroot, 2000), the instrument described here provides a clearer picture of the factors that influence how hard a student works than would be found with a quantitative survey or scale. One could argue that this instrument is too structured, that an even more open-ended format would reveal greater insight. More open-ended measures might pick up on factors not mentioned in this instrument but such measures also have the risk of not getting to some of the key factors mentioned in the literature (e.g., self-confidence, peer influence, goal orientation).

One factor that needs to be considered with respect to instruments developed to measure motivation is administration of the instrument. Administration of a survey is relatively easy. All three individuals who were involved in the testing of the interview instrument (Appendix A) were experienced classroom teachers who were adept at getting students to talk openly. This is essential in an interview study. The more open ended the interview, the more crucial the training and background of the interviewer becomes.

An important motivational variable that is not well addressed by this instrument is how task difficulty affects motivation. A number of researchers (e.g., Blumenfeld, 1992; Paris & Turner, 1994; Stipek, 1996) note the importance of providing students with moderately difficult tasks. Tasks that are too difficult frustrate students, while tasks that are too easy can lead to boredom (Kloosterman & Cougan, 1994) or the feeling that the teacher has low expectations (Nicholls, 1984). Given NCTM's (2000) challenges to have students involved in mathematics that requires deeper thinking than is currently expected in most classrooms, more needs to be done to see how students react to challenge. (Student reaction to challenge was an underlying principle in the development of many of the questions in this instrument. However, because students were exposed to predominantly traditional curricula with predominantly routine tasks, it became obvious in the interviews that it would not be possible to determine how complex mathematical tasks affected their attitudes and motivation.)

An obvious question to ask with respect to the instrument is the extent to which it can be condensed or expanded. In interview research, changing questions is always an option and thus researchers interested in specific aspects of motivation in mathematics could easily add or subtract items as needed. Our experience indicated that by the time we got to questions 41 to 50 on classroom and grading practices, most students had already touched on most of what we were going to ask in their responses to earlier questions – particularly those about what they liked about mathematics and what caused them to work hard in the subject (questions 6 to 13). For those students who did not address grading and instructional issues earlier, questions 41 to 50 were important. Moreover, looking at responses to several related questions often helped in identifying apparent inconsistencies in what students said (e.g., comments about importance of ability as outlined in Theme 3).

With respect to the data reported on beliefs about mathematics, it is clear that the question "what is mathematics" is difficult for many students to answer, probably because they are never asked to think about the nature of mathematics. Responses to the alien question (number 36) as well as other questions throughout the interview indicated that students felt there was a significant procedural aspect to mathematics but most also felt that some people could be successful without having to memorize everything. The inherent contradiction between believing that memory was very important, and believing that people who cannot memorize can still do mathematics, was typical of many student comments. That is, logical inconsistencies in students' beliefs were common and seemed to indicate that students were often doing mathematics assignments without thinking about what it takes to learn or why knowing mathematics is important. Note that while the types of questions asked in this study were very different from the questions about dividing by zero asked by Tsamir and Tirosh (this volume), both studies show that what mathematics education researchers view as inconsistency is often not seen as inconsistency by students. Moreover, although DeGroot (2000) was looking at general academic rather than mathematics motivation, she also found that data from student interviews indicated many things adults would find as illogical seemed quite logical to children. Examples of inconsistency can also be found in the more general psychological literature. Kelley (1973) for example, described a framework for interpreting how individuals are likely to act when faced with competing attributions (beliefs about success or failure). His framework makes it clear that even adults are likely to make decisions that other adults would question.

Going back to the issue of reform mathematics, there is still concern that when students are (a) expected to complete mathematics tasks that require far more than procedural understanding, (b) work cooperatively with peers, and (c) solve problems when the teacher will not explain the steps to get the answer, they will hit a roadblock. There is evidence that with time and with adequate rationale for doing mathematics differently, students can thrive in a reform mathematics environment (e.g., Hiebert et al., 1997; Huntley, Rasmussen, Villarubi, Sangtong, & Fey, 2000; Yackel & Rasmussen, this volume). Because the students we interviewed had not experienced reform mathematics instruction, it was not possible to determine their beliefs about such instruction. It appears that reform instruction goes against many of their intuitions about what mathematics should be but it also appears that they simply do what they are told. One important goal of reform instruction is that students are actually expected to see mathematics as a discipline for dealing with complex questions and problems (NCTM, 2000). It seems that many of the issues addressed in the instrument are ones that should be made more explicit to students. Is mathematics more than procedures? How should one study for a math quiz? Should the teacher explain everything? Should one know how to do math with paper and pencil if the calculator breaks? Discussion of such issues is likely to help students know more about how mathematics is learned and thus how they can learn more quickly.

In short, this chapter describes a variety of important beliefs as seen from the perspective of cognitive theories of motivation. Pedagogy and content in mathematics have many unique characteristics, which, in addition to societal attitudes about the difficulty of mathematics, make studying motivation in mathematics unique from the study of motivation in general. Specifically, the chapter describes beliefs that should be moved from implicit to explicit when implementing innovative mathematics curricula.

With respect to beliefs about mathematics and mathematics learning, data presented in this chapter indicate that (a) the nature of mathematics is an issue that students seldom think about, (b) students have a hard time separating the activities in a mathematics class from the nature of the discipline itself, and (c) students tend to see memorization as an important aspect of learning mathematics while also believing that students who do not memorize facts easily can still learn mathematics. It would be interesting to look at similar issues in other content areas. Do students see chemistry, for example, as more than formulas for building molecules and compounds? Do they have a sense of what chemists do and can they describe the discipline beyond what happens in a chemistry class? Can a student who has trouble memorizing chemical formulas still do well in chemistry?

Finally, it is important to reiterate the connection between beliefs and motivation. The two are not often dealt with concurrently in instructional or psychological theories yet the connection between them is strong. What can be said about the connection? The data collected from the instrument in general and from the nature of mathematics questions in particular indicate that students hold a complex web of beliefs and that what makes sense to one individual may not make sense to someone else. The data also indicate that some of the beliefs that students act on are well formed (e.g., beliefs about one-self as a mathematics learner), while others are not (e.g., beliefs about ability as a limiting factor in mathematics learning). In this chapter, I have worked to identify key variables in belief webs under the assumption that knowing those variables helps to understand what motivates students. There are no easy answers to the problem of motivating students in mathematics but the beliefs and motivation framework described here should provide an important piece of the motivation puzzle in mathematics.

7. REFERENCES

Alderman, M. K. (1999). *Motivation for achievement*. Mahwah, NJ: Lawrence Erlbaum.

Ames, C. (1992). Classrooms: Goals, structures, and student motivation. *Journal of Educational Psychology. 84,* 261-271.

Atkinson, J. W. (1957). Motivational determinants of risk-taking behavior. *Psychological Review, 64,* 359-372.

Bandura, A. (1997). *Self-efficacy: The exercise of control.* New York: Wiley.

Bartsch, K., & Wellman, H. (1989). Young children's attribution of action to beliefs and desires. *Child Development, 60,* 946-964.

Blumenfeld, P. C. (1992). Classroom learning and motivation: Clarifying and expanding goal theory. *Journal of Educational Psychology, 84,* 272-281.

Carl, I. M. (Ed.) (1995). *Prospects for school mathematics.* Reston, VA: National Council of Teachers of Mathematics.

Carr. M. (1996). Afterward. In M. Carr (Ed.), *Motivation in mathematics* (pp. 173-177). Cresskill, NJ: Hampton Press.

Clement, J. (2000). Analysis of clinical interviews: Foundations and model viability. In A. E. Kelly & R. A. Lesh (Eds.), *Handbook of research design in mathematics and science education* (pp. 547-589). Mahwah, NJ: Lawrence Erlbaum.

Covington, M. V. (1992). *Making the grade: A self-worth perspective on motivation and school reform.* Cambridge, England: Cambridge University Press.

DeGroot, E. (2000, April). *In their own words: Adolescents and their teachers talk about learning and schooling*. Paper presented at the Annual Meeting of the American Educational Research Association, New Orleans.

Eccles, J., Adler, T. F., Futterman, R., Goff, S. B., Kaczala, C. M., Meece, J. L., & Midgley, C. (1985). Self-perceptions, task perceptions, socializing influences, and the decision to enroll in mathematics. In S. F. Chipman, L. R. Brush, & D. M. Wilson (Eds.), *Women and mathematics: Balancing the equation* (pp. 95-121). Hillsdale NJ: Erlbaum.

Elliott, E. S., & Dweck, C. S. (1988). Goals: An approach to motivation and achievement. *Journal of Personality and Social Psychology, 54,* 5-12.

Fennema, E., & Sherman, J. A. (1976). Fennema-Sherman Mathematics Attitude Scales. *JSAS: Catalog of Selected Documents in Psychology, 6*(1). (Ms. No. 1225).

Ford. M. I. (1994). Teachers' beliefs about mathematical problem solving in the elementary school. *School Science and Mathematics, 94,* 314-322.

Graham, S., & Weiner, B. (1996). Theories and principles of motivation. In D. C. Berliner & R. C. Calfee (Eds.), *Handbook of Educational Psychology* (pp. 63-84). New York: Macmillan.

Hiebert, J., Carpenter, T. P., Fennema, E., Fuson, K. C., Wearne, D., Murray, H., Olivier, A., & Human, P. (1997). *Making sense: Teaching and learning mathematics with understanding*. Portsmouth, NH: Heinemann.

Huntley, M. A., Rasmussen, C. L., Villarubi, R. S., Sangtong, R. S., Sangtong, J., & Fey, J. T. (2000). Effects of *Standards*-based mathematics education: A study of the Core-Plus Mathematics Project algebra and functions strand. *Journal for Research in Mathematics Education, 31,* 328-361.

Kelley, H. H. (1973). The process of causal attribution. *American Psychologist, 28*, 107-128.

Kloosterman, P. (1988). Self-confidence and motivation in mathematics. *Journal of Educational Psychology, 80*, 345-351.

Kloosterman, P. (1996). Students' beliefs about knowing and learning mathematics: Implications for motivation. In M. Carr (Ed.), *Motivation in mathematics* (pp. 131-156). Cresskill, NJ: Hampton.

Kloosterman, P. (1998, April). *How hard do you work in mathematics? Motivational profiles of six high school students.* Paper presented at the annual meeting of the American Educational Research Association. San Diego. Retrieved May 30, 2002 from:
http://www.indiana.edu/~pwkwww/AERA98.html

Kloosterman, P., & Cougan, M. C. (1994). Students' beliefs about learning school mathematics. *Elementary School Journal, 94,* 375-388.

Kloosterman, P., & Mau, S. T. (1997). Is this really mathematics? Challenging the beliefs of preservice primary teachers. In D. Fernandes, F. Lester, A. Borralho, & I. Vale (Eds.), (1997). *Resolução de problemas na formacão inicial de professores de matemática: Multiplos contextos e perspectivas* (*Solving problems in the preparation of mathematics teachers: Multiple contexts and perspectives* (pp. 217 - 248). Aveiro, Portugal: Grupo de Investigação em Resolução de Problemas.

Kloosterman, P., & Stage, F. K. (1992). Measuring beliefs about mathematical problem solving. *School Science and Mathematics, 92,* 109-115.

Lester, F. K., & Kroll, D. L. (1991). Evaluation: A new vision. *Mathematics Teacher, 84,* 276-284.

Maple, S. A., & Stage, F. K. (1991). Influences on the choice of math/science major by gender and ethnicity. *American Educational Research Journal, 28*, 37-60.

Mitchell, J. H., Hawkins, E. F., Jakwerth, P. M., Stancavage, F. B., & Dossey, J. A. (1999). Student work and teacher practices in mathematics. *Washington, DC: National Center for Education Statistics.*

National Council of Teachers of Mathematics (2000). *Principles and standards for school mathematics.* Reston, VA: Author.

National Research Council (1989). *Everybody counts: A report to the nation on the future of mathematics education.* Washington, DC: National Academy Press.

Nicholls, J. G. (1984). Conceptions of ability and achievement motivation. In R. E. Ames & C. Ames (Eds.), *Research on motivation in education* Vol. 1: *Student motivation* (pp. 39-73). Orlando, Academic Press.

Paris, S. G., Turner, J. C., (1994). Situated motivation. In P. R. Pintrich, D. R. Brown, & C. E. Weinstein (Eds.), *Student motivation, cognition, and learning: Essays in honor of Wilbert J. McKeachie* (pp. 213-237). Hillsdale, NJ: Lawrence Erlbaum.

Pintrich, P. R., & Schrauben, B. (1992). Students' motivational beliefs and their cognitive engagement in classroom academic tasks. In. D. H. Schunk & J. L. Meece (Eds.), *Student perceptions in the classroom* (pp. 149-183). Hillsdale, NJ: Lawrence Erlbaum.

Reyes, L. H. (1984). Affective variables and mathematics education. *Elementary School Journal, 18,* 207-218.

Schunk, D. H. (1991). Self-efficacy and academic motivation. *Educational Psychologist, 26,* 207-231.

Seegers, G., & Boekarts, M. (1993). Task motivation and mathematics achievement in actual task situations. *Learning and Instruction, 3,* 133-150.

Shernoff, D., & Schneider, B. (2000, April). *Engagement in math and science among high school students: Examining school, individual, and family influences.* Paper presented at the Annual Meeting of the American Educational Research Association, New Orleans.

Stage, F. K., & Kloosterman, P. (1995). Gender, beliefs, and achievement in remedial college-level mathematics. *Journal of Higher Education, 66,* 294-311.

Steen, L. A., & Forman, S. L. (1995). Mathematics for work and life. In I. M. Carl. (Ed.), *Prospects for school mathematics* (pp. 219-241). Reston, VA: National Council of Teachers of Mathematics.

Stipek, D. J., (1996). Motivation and instruction. In D. C. Berliner & R. C. Calfee (Eds.), *Handbook of Educational* Psychology (pp. 85-113). New York: Macmillan.

Weiner, B. (1984). Principles for a theory of student motivation and their application within an attributional framework. In R. E. Ames & C. Ames (Eds.), *Research on motivation in education* Vol. 1: *Student motivation* (pp. 15-38). Orlando, FL: Academic Press.

APPENDIX A: STUDENT INTERVIEW INSTRUMENT

Reminders for Interviewer

Whenever you prompt a student to get a response, put a "P" before writing that response. In many cases, it will not be possible to write down exactly what a student says. However, whenever you have enough time to include verbatim comments, please do so (put exact wording in quotes). Notes about student non-verbal responses are also appropriate (e.g., nod of the head can be interpreted as a yes or no, confused look should be noted to indicate student didn't understand question, etc.). Make sure you write summary comments while the interview is fresh in your mind.

Initial Instructions for Student

Interviewer introduces him/herself and says "I am interested in knowing your thoughts and feelings about some of the kinds of school work you do, especially in math. I'll be asking you questions and I hope that you will be at ease and give honest responses. There are no right or wrong answers for any questions, I would just like to know how you personally think and feel. If there is a question that you don't want to answer, you don't have to. Just say that you don't want to answer it. I won't tell your teacher, or parents, or friends what you say but if you want to talk to them about it, you can. Do you have any questions about what we are going to do?" (If necessary, make additional "small talk" to put student at ease.)

General Background

1. What other high school math courses have you taken? What grades did you get in those math courses?
2. What other math courses do you think you'll take in high school?
3. What grades do you tend to get in other courses (English, history, PE, etc.)?
4. What plans do you have when you leave high school (4-year college, 2-year college, job, join the army, etc.)? Be as specific as you can.

5. Are there things you do in school just for the fun of it? (play sports, go to the computer lab, participate in clubs, etc.)

Indiana Mathematics Beliefs Scales
Give student IMBS (from Kloosterman & Stage, 1992) and pencil. Fill in the student's name and tell him/her that some of the questions are similar but they all are different and ask the student to answer all of them truthfully. If the student is not sure about what to put, ask him/her to mark his/her initial reaction.

Feelings About School and About Mathematics

6. What words would you use to describe school?
7. On a scale of 1 to 10, with 10 being the highest, how much do you like school? What specific things do you like and dislike about school?
8. On a scale of 1 to 10, with 10 being the highest, how useful do you think school is for the things you want to do? Are there some things you do in school that are more useful than others?
9. On a scale of 1 to 10, with 10 being the highest, how much do you like math? Are there some parts of math you like and some you don't? Please explain. (Look for topics the student likes, such as likes fractions but dislikes algebra. Also look for level of challenge student prefers -- are textbook exercises boring? -- are story problems too hard?)

Effort in Mathematics

10. How hard do you work in math class? Do you always do everything the teacher assigns?
11. Do you always do your homework? How much time do you spend on homework each day? If you don't do your homework, why don't you do it?
12. In general, what influences you to work hard in math? Is there anything that causes you to work very hard? (Although this issue comes up again later, if there is any evidence of *task orientation, ability orientation,* or any type of *social orientation*, make sure it is noted.)
13. How do you like math in comparison to other subjects? Are the factors that make you work hard in other subjects different from the ones that make you work hard in math?

Non-school Influences on Motivation

14. Do you sense that your parents use mathematics very much? How do they use it? (Look for examples of job related or any other parental uses of mathematics.)
15. Do your parents want you to do well in school? How much support do they give you? Do they ever help with homework? Do they ask you about school?
16. Do you have, or have you had, paying jobs -- anything from baby-sitting and lawn mowing to working at McDonalds. Did you like those jobs? Did you use any math in those jobs?
17. Do your parents do anything special to encourage (or discourage) you in math as compared to other subjects?
18. What do you do in your spare time at home? (How much time is spent watching TV, visiting friends, doing homework, at a job, etc.)
19. I've already asked about math in jobs you have had. Do you use math in other things you do? (Probe for examples)
20. What do your friends think about math? Do they like it? Do they see it as useful? Do they work hard?
21. Are you influenced by what your friends think about math and about school? How do they influence you?

Self-confidence in Mathematics

22. How good are you in math? How do you know? (Look for how judgments are made and what it means to student to be good in math -- if needed, probe to see if student compares him/herself to other students. Also, are judgments made on basis of grades? Teacher comments? Is there any evidence of internal factors such as "I know I'm good because I understand math?")

23. Are you better at some kinds of math than others? For example, are you better at long division than you are at fractions or are you better at computations than problems that require a lot of thinking?

Natural Ability in Math

24. Do you think it takes special talent to do well in mathematics? Do you have such talent? Can people do OK in math even without special talent?
25. When someone makes mistakes in mathematics, does it mean that person is dumb in math? (probe for explanation - particularly if student feels making mistakes is part of the learning process in math)
26. How important is memorization in mathematics? Are you good at memorizing? Can someone who is not very good at memorizing be good in mathematics (or even "OK" in math?)

Goal Orientation and Effort

27. How often do you work hard in math just to learn the material? (look for evidence of task-orientation [motivation just to learn the material or accomplish the task])
28. Do you care about what your classmates and teacher think of your skills in math? Is it important to you to look like a good math student or poor math student to your friends and teacher? (look for evidence of an ego-orientation [trying to do better than others])
29. How often do you work hard in math so you can help your friends, or at least work with them (look for evidence of affiliative-orientation [desire to share work with peers])
30. How often do you do the least amount of work you can to get by? (look for evidence of work avoidant-orientation)

Study Habits in Mathematics

31. How do you study for a test or quiz in math? Has anyone taught you special skills for studying in math? What does your teacher say about how to study?
32. When learning a new topic in mathematics, do you try to see how it relates to other math topics or topics you have learned before? Is relating new topics to other topics something your teacher stresses? (Try to see how topics are related. Do you just add one step to the process learned the day before or is there a real attempt at deep understanding of mathematical structure?)
33. How important is your textbook in helping you learn math? Do you read the text or just do problems?
34. Do you ever work with others in math? Does your teacher (or past teachers) encourage working with others? What kinds of things do you do with others in math?
35. Do you think working with others in math is a good idea? (Probe for when it is appropriate and when it isn't.)

Mathematics Content

36. Suppose an alien from outer space landed in your back yard and started asking you what math was like in Indiana. What would you tell him? What words best describe mathematics? (Try to get a sense of how much the student sees math as "rule-based" and believes that math involves complex problems as opposed to textbook exercises.)
37. Describe what you normally do in math class and how this compares to previous math courses. Give examples of the types of problems, activities, and projects you have done in math.
38. Of the activities you mentioned in your response to the last question, are there any that were particularly enjoyable or interesting? Are there any that were particularly dull?
39. Do you use calculators or computers in math class or on math homework? What do you use them for? Is it ever "cheating" to use a calculator in math class?
40. Are there ever problems in math that can be solved more than one way or that don't have an exact answer? (give examples)

Assessment Practices

41. How are you graded in math and how does it compare to grading procedures in previous math classes or in subjects other than math?

42. How important is it to you to get a good grade in math and other courses?
43. How much does your teacher's grading system affect what you try to learn? Do you ever try to learn things that you know you won't be graded on?

Students' Expectations of Teachers

44. Have you had any particularly good or bad math teachers? (don't ask for names) In general, have your math teachers done a good job of getting you to work hard in math? What did they do to get you to work hard? What could they have done to get you to work harder?
45. Does your teacher always tell you if you are right or wrong in math? Is this something a teacher should do?
46. Should a teacher always explain everything in math or are there times when it is better for a teacher to let students figure some things out for themselves? (probe for examples)
47. Do you ever learn better if you have to figure something out for yourself in math? Do you like to figure things out for yourself?
48. Does your math teacher pass back test papers in order of scores or do other things to let you know who is doing the best in the class? Do you think teachers should let all students know who is doing well and who isn't?
49. Does your teacher ever let some students "get by" doing less work than other students (for example, if a few students don't finish their work, is that OK with the teacher)? Does the teacher ever expect more from any really good students?
50. Does (or would) the practice of letting some students do more or less work than others affect how hard you work in the class where this happens? (Probe for examples).

Final Instructions to Student

51. We are now at the end of this set of interviews. Your responses have been very useful in helping me understand what you think about mathematics. Is there anything that you think is important about learning mathematics that you haven't said?

Thank you very much for your responses. They will be very helpful to me in understanding what you like and don't like about mathematics, and thus what we might do to make mathematics teaching better. Do you have any questions?

CHAPTER 16

BRIAN GREER, LIEVEN VERSCHAFFEL, AND
ERIK DE CORTE

"THE ANSWER IS REALLY 4.5": BELIEFS ABOUT WORD PROBLEMS

Abstract. In the course of a program of research into how students respond to typical word problems, it quickly became clear that patterns in their responses showing an apparent willingness to suspend sense making could not be explained in cognitive terms alone. Rather, it is necessary to consider the culture of the mathematics classroom and, in particular, the set of beliefs underlying the "Word Problem Game" that, largely implicitly, governs classroom practice. Findings from systematic research studies with students and teachers-in-training are reported that cohere with others in the literature and anecdotal evidence to elucidate the nature of practices surrounding word problems. Further, initial teaching experiments are reported which suggest that it is possible to change beliefs about word problems. However, these beliefs cannot be considered in isolation. Rather, they form part of more general beliefs about the nature of mathematics and its relation to the real world. Moreover, beliefs about word problems shaped by classroom culture are embedded within the nested and interacting contexts of school culture, the educational system, and society in general. We argue for the reconceptualization of word problems as a vehicle for promoting early awareness of the relationship between mathematics and aspects of reality that it models. This proposal reflects our own beliefs about the nature of mathematics and the proper goals of mathematics education.

In this chapter, we examine a program of research on how students and teachers respond to word problems and propose a theoretical explanation in terms of the nature of practices surrounding word problems as typically presented in mathematics classrooms, and in particular the beliefs that are engendered by these practices. From the results of our research and analysis we argue that word problems provide a compelling example of how school mathematics is shaped by beliefs – not just at the level of the classroom, but in the broader contexts of the school, the educational system, and, ultimately, society itself.

1. THE NATURE OF WORD PROBLEMS

By "word problem" is meant a text (typically containing quantitative information) that describes a situation assumed familiar to the reader and poses a quantitative question, an answer to which can be derived by mathematical operations performed on the data provided in the text, or otherwise inferred.

G. C. Leder, E. Pehkonen, & G. Törner (Eds.), Beliefs: A Hidden Variable in Mathematics Education? 271-292.

1.1. A Historical Perspective on the Purpose of Word Problems

Word problems constitute a very recognizable genre showing a remarkable homogeneity over thousands of years and across many cultures. The following examples come from a 13th century Chinese text and a 15th century Italian text (the Treviso Arithmetic), respectively:

> Thirty feet of a foundation are built by 7 men in 2 days. How many men are required for building 3,570 feet in 3 days? (Libbrecht, 1973, p. 94)

> If 17 men build 2 houses in 9 days, how many days will it take 20 men to build 5 houses? (Swetz, 1987, p. 163)

There is extensive documentation of such problems but there is relatively little discussion of the central question of what are the purposes of word problems. Indeed, the continuity of the genre suggests that word problems have been handed on across generations and between cultures as traditional forms with little questioning of their role in mathematical education (Gerofsky, 1996; Lave, 1992).

We may, however, suggest two main purposes (for more detailed discussion, see Verschaffel, Greer, & De Corte (2000), chapters 8 and 9; Galbraith & Stillman, 2001). The first main purpose is practising the solution of classes of problems that represent applications of mathematics to physical and social phenomena, reflecting the central role of mathematics as a means of describing, and drawing inferences about, aspects of reality.

The second is as vehicles for reasoning about mathematical structures, using imaginable but often not realistic "stories". The following example is from a 12th century Indian classic:

> In the interior of a forest, a number of apes equal to the square of 1/8th of their total number are playing with enthusiasm. The remaining 12 apes are on a hill. The echo of their shrieks by the surrounding hills rouses their fury. What is the total number of apes? (Srinivasiengar, 1967, p. 86)

This scenario could hardly be interpreted as the posing of a problem that would arise in anybody's experience, yet it is imaginable and can be translated into mathematical terms and solved algebraically. The purpose of problems of this type is "to convey a mathematical meaning, that is the use of suitable concrete objects to represent or reify abstract mathematical notions" (Toom, 1999, p. 37), in other words to act as "mental manipulatives".

The distinction between the two purposes is observable in the Treviso Arithmetic. As well as exercises in many types of standard business transactions, there are what Swetz (1987, p. 288) terms "intellectual exercises", an example being:

> A hare is 150 paces ahead of a hound, which pursues him. The hare covers 6 paces while the hound covers 10. In how many paces will the hound overtake the hare? (Swetz, 1987, pp. 245-246)

Swetz (1987, p. 245) describes this as "a standard exercise in European arithmetic books for centuries" and it is but one of numerous examples of classic problems whose variants span thousands of years and many civilizations (see also Wells, 1992).

1.2. Word Problems and the Modeling of Reality

While acknowledging the importance of the second type of word problem for mathematics education, our focus is on word problems of the first type, those that, at least ostensibly, describe practical situations arising in people's lives. Such application problems may be considered as exercises in mathematical modeling, the core elements of which process are the translation of the situation into mathematical relationships, the manipulation of those relationships to yield derivations, and the interpretation of those derivations within the original context.

Mathematical modeling raises fundamental questions about the relationship between the real world and mathematics. Until relatively recently, against a background of general belief that mathematics (for example, Euclidean geometry and Newtonian mechanics) provided exact descriptions of reality, the predominant belief has been that situations described in word problems could be unproblematically mapped on to mathematical structures. For example, the problem cited above from the Treviso Arithmetic about men building houses has been analyzed by Säljö (1991) in terms of the assumptions underlying the stated answer, 19 days and 3 hours. Apart from assumptions that all men work equally hard, that the time taken is directly proportional to the number of men, and so on, Säljö argues (pp. 262-263) that the solution implies a working day of 24 hours and concluded that "Arguments rooted in an external reality, in which people do not work for 24 h[ours] a day can – and have to – be temporarily disregarded; if they are not the problem becomes difficult to handle".

Freudenthal (1991, p. 32) commented as follows:

> Mathematics has always been applied in nature and society, but for a long time it was too tightly entangled with its applications for it to stimulate thinking on the way it is applied and the reason why this works
>
> ... money changers, merchants and ointment mixtures behaved as if proportionality were a self-evident feature of nature and society ...
>
> ... Modeling is a modern feature. Until modern times the application of rigorous mathematics to fuzzy nature and environment boiled down to more or less consciously ignoring all of what had appeared to be inessential perturbations spoiling the ideal case.

2. LEARNING TO PLAY THE WORD PROBLEM GAME: STUDENT BELIEFS

The lack of a critical attitude towards the modeling of situations, as just described in historical context, is echoed in the way in which word problems are typically presented in mathematics classrooms to this day. De Corte and Verschaffel (1985, p. 7) introduced the term "Word Problem Schema" to refer to the system of beliefs about word problems shared by students and teachers, involving the perceived intent of word problems, interpretation of stereotyped semantic structure (e.g., Nesher, 1980; Reusser, 1988), and a complex network of implicit rules and expectations that govern playing of the "Word Problem Game" (De Corte & Verschaffel, 1985).

Beliefs may be defined as the implicitly or explicitly held conceptions (understandings, premises, propositions, expectations,...) subjects hold to be true (Op t'Eynde, De Corte, & Verschaffel, this volume). These beliefs are organized in complex belief systems, which are characterized by cluster structure, quasi-logicalness, and psychological centrality (Op t'Eynde et al., this volume). They serve subjects' needs, desires and goals, such as self-protection and personal and social control, and cause biases in perception, judgment and action in social situations as a result. The student beliefs involved in the Word Problem Schema can be considered as a part of that (sub)system of Op 't Eynde et al.'s taxonomy of mathematics-related beliefs termed "beliefs about mathematical problem solving, teaching and learning". As argued later in the chapter, students develop their beliefs about what constitutes a word problem, what it means to do mathematical word problem solving, and what they and others are expected to do in a lesson on mathematical word problem solving from their experiences and interactions during the classroom activities in which they engage. Putting it another way, students' beliefs develop from their perceptions and interpretations of the socio-mathematical norms (Yackel & Cobb, 1996; Yackel & Rasmussen, this volume) that determine – explicitly to some extent, but mainly implicitly – how to behave in a mathematics class, how to think, how to communicate with the teacher, and so on.

Examples of beliefs specific to word problems that many authors (e.g., Gerofsky, 1996; Lave, 1992; Puchalska & Semadeni, 1987; Schoenfeld, 1991) have drawn attention to are the following:

- Any word problem presented by the teacher or in a textbook is solvable and makes sense.
- There is a single, correct, and precise numerical answer.
- This single answer must be obtained by performing one or more mathematical operations with the numbers embedded in the text (a manifestation related to the "perform-the-operation" belief analyzed by Tsamir and Tirosh (this volume)).
- The task can be achieved by applying familiar mathematical procedures.
- The text contains all the information needed and no extraneous information may be sought.
- Violations of your knowledge or intuitions about the everyday world should be ignored.

2.1. Apparent Suspension of Sense-Making: Examples from the Literature

Circumstantial evidence for the existence of these beliefs has been provided by a large number of studies. The most spectacular case is that of the French and German researchers (Institut de Recherche sur l'Enseignement des Mathématiques de Grenoble, 1980; Radatz, 1983, 1984) who posed nonsensical problems such as "There are 26 sheep and 10 goats on a ship. How old is the captain?" and found that many children supplied answers produced by arithmetical operations on the numbers in the text. These findings and many others suggest, indeed, that many children in elementary school develop the very specific beliefs that a word problem should be solved by adding, subtracting, multiplying, or dividing the two numbers supplied, and that the choice may be guided by superficial cues, such as the presence of "key words" in the text. As Gravemeijer (1997, p. 391) put it, word problems are reduced to "decorated exercises in the four basic operations".

2.2. *Initial Research and Replications*

Stimulated by the "How old is the captain?" research and many other examples in the literature, we carried out preliminary pencil-and-paper studies (Greer, 1993; Verschaffel, De Corte, & Lasure, 1994) using a set of problems including those listed in Table 1 (provided with verbal labels for ease of reference).

Table 1. Examples of P-items used in Greer (1993) and Verschaffel et al. (1994).

Balloons	Grandfather gives his 4 grandchildren a box containing 18 balloons, which they share equally. How many balloons does each child get?
Planks	Steve has bought 4 planks each 2.5 meters long. How many planks 1 meter long can be sawn from these planks?
Runner	John's best time to run 100 meters is 17 seconds. How long will it take him to run 1 kilometer?
Rope	A man wants to have a rope long enough to stretch between two poles 12 meters apart, but he has only pieces of rope 1.5 meters long. How many of these pieces would he need to tie together to stretch between the poles?
Flask	This flask is being filled from a tap at a constant rate. If the depth of the water is 4 cm after 10 seconds, how deep will it be after 30 seconds? (Accompanied by a drawing of water from a tap filling a cone-shaped flask)

We termed each of these items "problematic" in the sense that they require (from our point of view) the application of judgment based on real-world knowledge and assumptions rather than the routine application of one or more simple arithmetical operations. Each such P-item was paired with an S-item (for "standard") in which the "obvious" calculation is (we would argue) appropriate. For example, the S-item paired with the first P-item listed above was:

> Steve has bought 5 planks each 2 meters long. How many planks 1 meter long can be sawn from these planks?

The distinction between S-items and P-items is a matter of judgment, and a matter of degree rather than absolute. For example, the answer "10" to the item just cited could (and has been) criticized on the grounds that it ignores the wood lost as sawdust in the process of sawing. In fact, such considerations draw attention to a key aspect that should be part of what students learn about the modeling process, namely the level of precision appropriate to the circumstances and goals of the modeling act, which always involves simplification (see Gravemeijer, 1997, p. 393).

For each item, the students, as well as recording an answer, were invited to comment on the problem and their response. A response to a P-item was classified as a "realistic reaction" (RR) if *either* the answer given indicated that realistic considerations had been taken into account *or* if a comment indicated that the student was aware that the problem was not straightforward. For example, an RR classification for the Planks P-item would be given to a student who gave the answer "8" or who made a comment such as "Steve would have a hard time putting together the remaining pieces of 0.5 meters".

Table 2. Percentages of Realistic Reactions (RRs) on selected P-items in the studies of Greer (1993) (G), Verschaffel et al. (1994) (V1), Caldwell (1995) (C), Hidalgo (1997) (H), Reusser and Stebler (1997) (RS), Yoshida et al. (1997) (Y), Renkl (1999) (R), and Verschaffel et al. (1999) (V2).

	Country	Age/grade	N	Ballons	Planks	Runner	Rope	Flask
G	N.I.	13-14	100	85	-	6	8	2
V1	Belg.	10-11	75	59	14	3	0	4
C	N.I.	10-11	75	81	-	0	1	-
H	Ven.	Grade 5	119	55	16	0	0	0
RS	Switz	13	439	75	14	5	6	0
Y	Japan	Grade 5	45	52	0	7	2	4
R	Germ.	Grade 4	93	51	21	2	0	-
V2	Belg.	Grade 5	64	-	17	1	5	5

These initial studies were subsequently replicated in Northern Ireland (Caldwell, 1995), Venezuela (Hidalgo, 1997), Switzerland (Reusser & Stebler, 1997), Japan (Yoshida, Verschaffel, & De Corte, 1997), Germany (Renkl, 1999) and Belgium (Verschaffel, De Corte, & Lasure, 1999), using similar methodologies and, to a considerable extent, the same items (subject, of course, to differences due to translation). The results, shown in Table 2, are clear-cut (for more details, see Verschaffel et al., 2000 and the specific papers listed with Table 2). By our criteria, none of the questions listed was answered in a realistic fashion by more than a small percentage of students, and the results were strikingly consistent across many countries. These findings, and many others in the literature, constitute strong evidence for the postulated student belief that their knowledge or intuitions about the everyday world should be ignored when solving word problems in mathematics class.

2.3. What Students Say about Word Problems

As well as inferring students' beliefs on the basis of their responses to word problems, it is, of course, possible to ask them directly, and here we present some illustrative findings using this approach. Interviews carried out by Caldwell (1995) and Hidalgo (1997) suggest that, while unfamiliarity with the contexts involved in the problems and lack of appropriate heuristic and metacognitive skills may provide contributory explanations, it is the more or less conscious (mis)beliefs about school arithmetic word problems that constitute the major reason why so many pupils solve the P-items in a non-realistic way.

Some students indicated that they were aware of the nature of the Word Problem Game. For instance, a 10-year-old interviewed by Caldwell (1995, p. 39) commented as follows in response to the interviewer's question as to why she did not make use of realistic considerations when solving the P-items in the context of the written test:

> I know all these things, but I would never think to include them in a maths problem. Maths isn't about things like that. It's about getting sums right and you don't need to know outside things to get sums right.

As part of a study in which various interventions were introduced to see whether students could be induced to pay more attention to realistic considerations, Reusser and Stebler (1997) used two items, one being the Runner item (see Table 1), and attempted to "sow the seeds of doubt" (p. 316) by asking 18 pairs of students first if they were sure that their solution (10 x 17 = 170 seconds) was correct and then to imagine themselves in the role of the runner. As a result of this intervention, the number of pairs giving RRs went from 3 to 11 (a similar result was found for the other item). A whole-class discussion was then initiated in which the students were asked to comment on the following:

- Why were the problems solved without anyone wondering whether they could be solved?
- Why did many students think about the difficulties but not mention them?

Among the responses were the following:

> I did think about the difficulty, but then just calculated it the usual way. (Why?) Because I just had to find some sort of solution to the problem and that was the only way it worked. I've got to have a solution, haven't I?

> It would never have crossed my mind to ask whether this task can be solved at all.

The contrast between these two responses (echoed in others) suggests that students differ in the degree to which they experience a conflict between the demands of the classroom context and their need to make sense in terms of their life experiences. How aware are students and teachers, in general, of the rules of the word problem game, and do students develop an ability to adapt their behavior accordingly? Investigation of individual differences in this regard – why, in the words of Freudenthal (1991, p. 70) "children's immunity against this mental deformation is so varied" – is one appropriate focus for further research.

In an unpublished study, Mukhopadhyay and Greer (2000) used similar P-items in pencil-and-paper tests and interviews with a contrasting population, namely adult students in a pre-algebra class taken as a basic prerequisite at a Community College in the US. It might be thought that the much more extensive and widely varying life experience of these students would be reflected in more realistic responses. On the other hand, it could be conjectured that they share the school students' beliefs and expectations about the appropriate reaction to word problems and mathematics in general. Bearing in mind that the sample size of 13 was small and more extensive research is necessary to draw any firm conclusions, the percentages of realistic reactions were not much higher, generally, than for school students, as illustrated in Table 3.

Table 3. Percentages of Realistic Reactions (RRs) on selected P-items in the study of Community College students by Mukhopadhyay and Greer (2000) (N = 13).

Balloons	Planks	Runner	Rope	Flask
54	54	8	0	8

Comments by several of these students were interesting. Part of the title of this chapter comes from what one said about the balloons problem. The answer given was "4", but this was followed by the comment that "the answer is really 4.5, but you can't have half a balloon".

3. TEACHERS' BELIEFS

It is not simply the students' beliefs that are implicated, of course. Teachers are also participants in the Word Problem Game (as are other players in the instructional environment, notably text-book authors, test developers, and constructors of examinations).

3.1. Anecdotal Evidence

While there is limited systematic evidence regarding the beliefs of teachers, our strong impression is that most teachers share the belief that realistic considerations should not complicate the "real" mathematics that word problems are intended to evoke. There are many anecdotes that vividly reflect this impression. For example, Keitel (1989, p. 7) observed a teacher using a problem about mixing paint to illustrate proportionality. When a boy objected, on the basis of practical knowledge due to his father being a painter, that paint mixed exactly in proportion does not reliably look like a sample, the teacher responded by saying "Sorry, my dear, we are doing ratio and proportion".

Another anecdote, related to one of us (LV), concerned a discussion between a teacher and the mother of one of his students. The mother and son rejected their initial answer to a problem about the price of a loaf of bread on the grounds that it was unreasonably low, and calculated a different answer. The teacher revealed that the problem had been taken from an old textbook and added:

> Of course, we all know that nowadays a loaf of bread costs considerably more than 21.5 francs. But, after all, that's not what students have to worry about when doing algebra problems. It's the construction and execution of the mathematical expression that counts, all the rest is decor (Van der Spiegel, personal communication, 1997).

3.2. A Study of Student-Teachers' Beliefs about Word Problems

One systematic investigation that has been carried out (Verschaffel, De Corte, & Borghart, 1997) analyzed future elementary school teachers' conceptions and beliefs about word problems, specifically the role of real-world knowledge about the problem contexts, as indicated by:

1. Responses to a set of "problematic" word problems.
2. Evaluations of (imaginary) student answers to these problems taking or not taking into account relevant real-world knowledge.

The participants were 332 preservice elementary school teachers from three teacher-training institutes in Flanders. About two-thirds were preservice teachers who had just started their first year of training, the remainder were third-year students who had almost completed their preservice training. A paper-and-pencil test was constructed consisting of 14 word problems: 7 standard items (S-items) and 7 problematic items (P-items). The seven problematic items (P-items) selected included the Planks, Runner, Rope, and Flask items (see Table 1).

The test was given twice to all preservice teachers with different instructions. For the first task (Test 1), the student teachers simply answered the word problems themselves. Calculations and comments could be written down in a "comments box" below the "answer box". Immediately after they had finished and handed in this test, they were given the second (Test 2), in which they were asked to score four different answers from pupils to the same word problems as presented in the first test, awarding each answer 1 point (absolutely correct), 1/2 point (partly correct), or 0 points (completely incorrect). The four constructed answers to each P-item belonged to different categories, namely a non-realistic answer (NA), a realistic answer (RA), a technical error (TE) and another answer (OA) derived by using the wrong operation or giving one of the numbers in the problem. At the bottom of each problem, there was a box for writing explanations and/or comments.

For Test 1, the student teachers' answers to the P-items were categorized as realistic reactions (RRs) and non-realistic reactions (NRs), on the same basis as described earlier for tests with school students. For Test 2, analysis focused on the scores (1, 1/2 or 0) given to realistic answers (RAs) and non-realistic answers (NAs) for each of the 7 P-problems.

The main findings for Test 1 were as follows:

1. The overall percentage of RRs for all the participants, averaged across 7 problems, was 48%. These problems included the Planks, Runner, Rope, and Flask items, results for which are shown in Table 4. The percentages of RRs are considerably higher than in the studies with upper elementary and lower secondary school pupils reported earlier (Table 2) but are still remarkably low (from our perspective).

Table 4. Percentages of Realistic Reactions (RRs) on selected P-items in the study of teachers in training by Verschaffel et al. (1997) (N = 332).

Planks	Runner	Rope	Flask
64	31	37	39

2. There was a statistically significant, but rather small, difference in the overall number of RRs between the first-year (45%) and the third-year (54%) student teachers. The size of the difference between the two years was dissimilar for the three teacher-training institutes involved in the study, suggesting that student-teachers' disposition toward realistic modeling of arithmetic word problems is at least partially influenced by the courses on mathematics education received during their preservice training (and further suggesting that a study of the beliefs about word problems held by the teachers' instructors would be an appropriate extension of the research).

The main results for Test 2 were as follows:

1. The evaluations of realistic and non-realistic answers were as shown in Table 5. Thus, student teachers' overall evaluations of the non-realistic answers to the P-items were considerably more positive than for the realistic answers.

Table 5. Percentages of scores for realistic and non-realistic answers on P-items for the first-year and third-year student teachers (Verschaffel et al., 1997).

Year/N	Realistic answer			Non-realistic answer		
	1	½	0	1	½	0
1 (N = 228)	44	7	49	58	25	17
3 (N = 104)	33	5	42	52	26	22
Total	47	6	47	56	26	18

2. There was, again, a statistically significant difference between the first-year and third-year student teachers, with third-year students evaluating RAs higher and NAs lower. As for Test 1, the size of the difference between first-year and third-year students varied noticeably among the three teacher-training institutes.

The data were further analyzed for the relationship between responses on Test 1 and Test 2, showing a strong relationship between the NR on P-items during Test 1, and the evaluations of the corresponding RAs and NAs during Test 2. The cases for which an NR was given to a P-item during Test 1 were followed by scores of 1 for the corresponding NA in 89% of cases and scores of 0 for the corresponding RA in 83% of cases. Thus, in a high percentage of such cases, the non-realistic answer was positively evaluated, presumably because this answer corresponded to the non-realistic answer the student teacher had given on the same item during Test 1. Moreover, in a high percentage of such cases, the realistic answer was negatively evaluated, presumably because the student teacher did not consider valid the context-based considerations underlying this answer.

Ten percent of the NRs to a P-item during Test 1 were followed by a score of 1 for the RA during Test 2. This suggests that in those cases the confrontation with the realistic answer during Test 2 had functioned as a scaffold toward (more) realistic modeling. However, the finding that only 10% of the scorings following a non-realistic response yielded evidence for the scaffolding effect of the confrontation with the realistic answer can be interpreted as additional evidence for the strength of the tendency among student-teachers not to take realistic considerations into account, and to resist being influenced in that direction.

The congruence between the RRs on Test 1 and the scorings of corresponding RAs and the NA during Test 2 was less straightforward. The evaluations of the RA were, indeed, generally in line with the reactions on Test 1, with 85% of the RRs on Test 1 being followed by a score of 1 for the RA on Test 2. However, the scorings for the NA were rather surprising – in only 34% of the cases where a subject reacted in a realistic way to a P-item during Test 1 was the corresponding NA given a score of 0. A detailed analysis implies that in many instances where student teachers reacted themselves to a P-item in a realistic manner, they were nevertheless tolerant towards elementary school pupils who interpreted and solved these P-items in a non-realistic fashion. According to their written explanations in the comments box, they thought that it would be unfair to punish a fifth-grader for solving the P-item in a stereotyped, non-realistic manner. This is illustrated by the following comment in relation to the Runner problem: "I scored alternative D (the RA, which was "It is impossible to answer precisely what John's best time on 1 kilometer will be") with 1 because the pupil who gave this answer knows that is not realistic to assume that John will be able to run at his record speed for 1 kilometer. But I also gave 1 for alternative A (the NA, which was "17 x 10 = 170. John's best time to run 1 kilometer is 170 seconds") because from a purely computational point of view this is the correct answer."

3.3. General Conclusions from the Study

This study provides some insight into one of the instructional factors considered responsible for the development of the tendency among children to disregard realistic considerations, namely the teachers' own conceptions and beliefs about the importance of real-world knowledge in arithmetic word problem solving. While the findings convincingly demonstrate that many future teachers have knowledge and beliefs about teaching and learning arithmetic word problems that are detrimental according to our beliefs, it does, of course, still not yield direct evidence that these teachers' conceptions and beliefs are responsible for children's strong tendency to exclude real-world knowledge from their problem solving endeavors. However, based on the literature on the relationship between teachers' mathematical beliefs and behavior in the mathematics lessons, on the one hand, and students' mathematics learning, on the other hand (Fennema & Loef, 1992; Thompson, 1992), there is good reason to assume that teachers' cognitions and beliefs about the role of real-world knowledge in the interpretation and solution of school arithmetic word problems may have strong impact on their actual teaching behavior and, consequently, on their students' learning processes and outcomes.

We should stress that we do not want to imply that teachers are to blame. As with the students, the (student-)teachers in the study of Verschaffel et al. (1997) were (sometimes deliberately, but mostly unconsciously) behaving "rationally" in accordance with the requirements they have to observe as partners in the didactical contract (Brousseau, 1997) for arithmetic word problem solving. Nevertheless, we are concerned about the apparent lack of sensitivity to the complexity of the mathematization of situations described in word problems among the majority of (future) teachers, and we recommend that more attention should be given to this fundamental topic in teacher training.

4. EXAMINING AND CHANGING BELIEFS ABOUT WORD PROBLEMS

Current school practice in relation to word problems, insofar as it results in a "suspension of sense-making" (Schoenfeld, 1991, p. 316) on the part of students as documented in many "disaster studies", has been widely criticized (e.g.,. Gerofsky, 1996; Greer, 1993; Lave, 1992; Nesher, 1980; Puchalska & Semadeni, 1987; Reusser, 1988; Schoenfeld, 1991; Verschaffel et al., 1994). Given that a major part of this criticism relates to the beliefs that students develop about word problems, we may ask whether and how such beliefs may be changed. A prerequisite is to be more reflective about the purposes of word problems in school mathematics, purposes which have tended to remain unanalyzed (Gerofsky, 1996, Lave, 1992).

Based on our own beliefs about mathematics education, we have put forward a radical proposal whereby word problems become reconceptualized as exercises in mathematical modeling (Verschaffel et al., 2000). In order to effect such a shift, a comprehensive change in socio-mathematical norms (Cobb, 1996; Yackel & Cobb, 1996) would be required.

4.1. Changing Beliefs: An Intervention Study

As a first move beyond ascertaining studies (i.e., studies that primarily document the state of affairs as it exists) to intervention studies (i.e., studies that seek to intervene and thereby improve the state of affairs), Verschaffel and De Corte (1997) set up a small-scale teaching experiment in which they strove to change students' beliefs about the role of real-world knowledge in responding to word problems, and to develop a disposition towards more realistic mathematical modeling. Students were immersed in a radically different classroom culture in which word problems were explicitly treated as modeling problems. Three classes were included in the study, of which two acted as controls. The students in the experimental class participated in a program of realistic modeling based on five teaching/learning units of about 3 hours each, while those in the control classes followed the regular mathematics curriculum.

For the experimental class, the impoverished and stereotyped diet of standard word problems typical of traditional mathematics classrooms was replaced by a variety of non-routine problems designed to elicit thinking about the complexities involved in taking into account realistic considerations, and to provoke discrimination between realistic and stereotyped solutions. Each teaching/learning unit focused on a prototypical problem type, such as interpreting the outcome of division problems with remainder or modeling the union or separation of sets with joint elements.

Secondly, a varied set of highly interactive instructional techniques was deployed, such as small-group collaborative work followed by whole-class discussion.

Thirdly, an attempt was made to establish a different classroom culture by explicitly negotiating social norms about the role of the teacher and the students in the classroom, and new socio-mathematical norms about what counted as a good mathematical word problem, a good solution, and a good response (Yackel & Cobb, 1996; Schoenfeld, 1991). By way of specific example, a sustained attempt was made to replace the belief cited earlier (p. 274) that "there is a single, correct, precise numerical answer" by the understanding that a problem might legitimately elicit more than one answer, or an approximate answer. In short, this study instantiated the recommendation by Yackel and Rasmussen (this volume, p. 328) that "one way to give explicit attention to student beliefs in the mathematics classroom is to be deliberate about initiating the negotiation of classroom norms".

Using similar tests and classification scheme for responses as described earlier, the percentage of RRs for the experimental class was found to rise considerably (and statistically significantly) between a pre-test and a post-test, from 7% to 51%. By contrast, the control classes showed small (and statistically non-significant) rises, from 20% to 34%, and from 18% to 23%. Further, results on far-transfer items and a retention test were also positive for the experimental group (for more details see Verschaffel & De Corte, 1997). A replication of this teaching experiment with German students by Renkl (1999) yielded similarly promising results. Given the relatively short span of the intervention, these results suggest that it is feasible to alter students' beliefs about the role of real-world knowledge in solving word problems and so develop a disposition towards mathematical modeling that takes into account realistic considerations.

However, a local intervention of this nature is unlikely to achieve a lasting change on its own given that beliefs about word problems form part of a system of beliefs about mathematics in general with parallel issues arising in other branches, the most obvious example being algebra word problems. Moreover, as outlined in the next section, what happens in the classroom is determined to a degree by school policy, which in turn is affected by the educational system and, at the most general level, beliefs about the nature of mathematics and goals of mathematics education prevalent in society.

5. SHAPING OF SCHOOL MATHEMATICS BY BELIEFS: THE EXAMPLE OF WORD PROBLEMS

The case of word problems strikingly illustrates how the practice of mathematics in school, involving both students and teachers, is shaped by beliefs. Indeed, Lave (1992, p. 74) characterized word problems as "a microcosm of theories of learning".

5.1. Effects of Classroom Culture

In the course of this research, we quickly became aware that the results we were observing could not be accounted for in narrowly cognitive terms; rather, it became clear that the explanation must be sought in the way in which word problems are embedded in classroom culture. As Boaler (1999, pp. 264-265) commented recently:

> Within the mathematics classroom ... students do not only learn mathematical concepts and procedures; they learn how to interact in the classroom; they learn particular sets of beliefs and practices and they learn the appropriate way to behave in the mathematics classroom...

Specifically, as stated by Lave (1992, p. 77):

> There is a discourse of word problems – a set of things everyone knows how to say about word problems or that can be expressed in "word-problemese", issues and questions that come up when people begin to talk about them; and things that are not and cannot be said within this framework.

From the ascertaining studies that we and our colleagues, and others, have carried out we have built up a theoretical explanation in terms of children and teachers learning to play the "Word Problem Game" (De Corte & Verschaffel, 1985), in the course of which numerous beliefs about word problems become firmly established. This theory specifically relating to word problems dovetails naturally within more general theoretical frameworks, notably the concepts of "didactical contract" (Brousseau, 1997) and sociomathematical norms (Cobb, 1996; Yackel & Cobb, 1996).

A limitation of our work to date is that we have paid scant attention to individual differences, in particular to what extent students are differentially aware of the implicit rules of the game, as hinted at by some responses in interviews cited earlier. Further, in our research to date we have not addressed issues of gender and class raised by the work of, amongst others, Cooper and Dunne (1998, 2000) and Boaler (1994, 1997a).

An implication of the above discussion is that improving the teaching of word problems is not simply a matter of improving the quality and design of the problems themselves and teaching a better (in some sense) set of procedures for solving them. Rather, as exemplified in the teaching experiment described earlier, it implies changing beliefs about appropriate responses to word problems, renegotiating the didactical contract, establishing different sociomathematical norms.

5.2. Beyond the Classroom: Effects of the School Environment

While it is an appropriate starting point to take a classroom as the unit of analysis, from a wider perspective, students' and teachers' beliefs about word problems (as for mathematics in general) are embedded in the school setting, the educational system, and most broadly in society, as represented in outline in Figure 1 and as illustrated by the examples that follow.

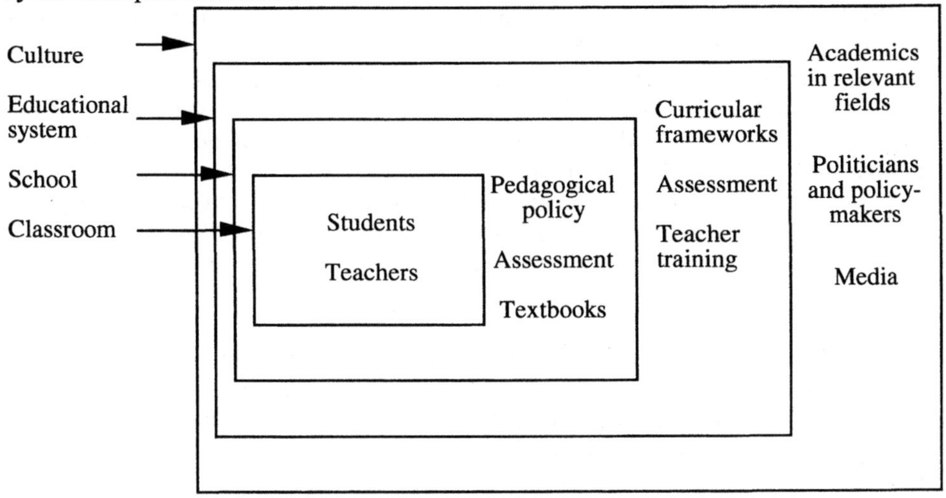

Figure1. Factors shaping beliefs about word problems

Teachers and students do not operate in isolation, and their beliefs are shaped by both the immediate and wider instructional environments in which they work. However, as Boaler (1999, p. 259) has pointed out: "The influence of the community upon knowledge development and use has, until recently, received relatively little attention in the field of education". Boaler (1997b, 1999) carried out an in-depth study of two schools in England teaching mathematics in accordance with strikingly different pedagogical approaches, characterized as follows (Boaler, 1999, pp. 259-260):

> One school used traditional, demonstration and practice methods of teaching. These methods encouraged students to develop mathematical beliefs and practices that were effective in the mathematics classroom, but remarkably ineffective in most other places. The other school used project-based methods of teaching that were more consistent with the demands of the classroom and the 'real world'.

Of particular relevance to our work is Boaler's analysis of how students' responses to mathematical tasks in relation to their real-world experience are differentially shaped. In what she characterizes as "the artificial school" (Boaler, 1999, p. 264), for example:

> If a question required some real world knowledge, or non-mathematical knowledge – for example a question I observed in which the students had to say why there are more females than males in the population -- students would stop and ask for help. They

> would be able to answer the question if prompted, they would probably be able to answer the question if they were in a science classroom, or if they were at home, but their expectation of the knowledge they should use in a mathematics classroom stopped them from answering such questions. (pp. 265-266)

Moreover, there are strong gender effects whereby "girls are more likely than boys to underachieve in contexts which present real world variables but do not allow the variables to be taken into account" (Boaler, 1994, p. 551).

While we do not deal with this aspect in any detail here, it is relevant also to point out that the phenomena observed in relation to students doing word problems are paralleled in other parts of the mathematics curriculum. Most obviously, at a later stage, students encounter algebraic word problems, which share many of the same characteristics as arithmetic word problems in terms of ignoring aspects of reality. More generally, it can be argued that school mathematics pays insufficient attention to the nature of mathematical modeling as the link between mathematical structures and aspects of the real world (Verschaffel et al., 2000, Chapter 7).

5.3. Beyond the School: Effects of the Educational System

Schools, in turn, are situated within a wider context of an educational system. An in-depth analysis by Cooper and Dunne (2000) provides an example of how decisions taken at a national level have implications for the shaping of beliefs about word problems, in particular the extent to which realistic aspects should be taken into account in answering them. In brief, the circumstances they studied arose from a combination of a shift in emphasis towards applications of mathematics (the Cockcroft Report (1982) was highly influential in this regard) and politically imposed forms of high-stakes assessment using paper and pencil tests. As a result, "a situation had arisen by the early 1990s in which children's knowledge and understanding were being assessed primarily by 'realistically' conceptualized items in two timed test papers given at several points in their school career" (Cooper & Dunne, 2000, p. 194). Their work represents (p. 204) "a contribution to our general understanding of the relations between socio-cultural background and cognitive processes and products".

Besides incisive criticisms of many of the items and associated marking schemes used, Cooper and Dunne (2000) analyzed "differences between children in their predispositions to import their everyday knowledge and experience into the context of problem-solving in school" (p. 194), and differences linked to class and gender. They found that working- and intermediate-class children performed relatively poorly (as judged by the marking schemes used) on "realistic" items because they tended to bring in more real-world considerations than the authors of the items envisaged – in other words, they were less well attuned to the rules of the version of the Word Problem Game represented by the assessment. Cooper and Dunne (2000) also found gender differences that were smaller, but still noticeable, with girls performing less well on "realistic" items for similar reasons (see also Boaler, 1994, 1997a).

As illustrated by this example, at both the level of the school and that of the educational system, assessment is a major agent of belief shaping. Assessment impacts instruction because it transmits powerful signals conveying the goals of instruction, what counts as competence in mathematics, and what forms of mathematical performance are valued. Typically, written forms of assessment are closed "in terms of time, in terms of information, in terms of activity, in terms of social interaction, in terms of communication" (Verschaffel et al., 2000, p. 72) and, as such, ill-fitted to asking questions that imply the activation of real-world experience and knowledge. The same comments apply to textbooks, which have been severely criticized in relation to how they present word problems (Verschaffel et al., 2000).

5.4. Beyond the Educational System: Effects of Society

At the most general level, the beliefs about word problems prevalent in society and culture are part of the picture. (For present purposes, we will take "society" to mean "the totality of social relationships among organized groups of human beings" and "culture" to mean "the total of the inherited beliefs, values and knowledge which constitute the shared bases of social action" (definitions from Collins English Dictionary, 1998)).

In our work, we did not set out to study cultural differences; by no means does this imply that we regard such differences as unimportant. Indeed, with reference to the general question of cross-cultural differences, it might be expected that specific cultural beliefs would strongly influence response to word problems. The evidence that does exist, as far as we are aware, relates mainly to contrasts between schooled and unschooled groups of which a classic example is the work of Luria (1976). Cooper and Dunne (2000) suggest that the lack of attention to culture:

> ... may have resulted partly from a working assumption that children's cognitive development followed similar patterns everywhere. But, in the case of mathematics, it may also have been partly dependent on various tacit assumptions concerning the universality of a discipline that many appear to believe to be above and beyond 'everyday' cultural differences. (p. 4)

We may conjecture that there is homogeneity of school mathematics teaching that masks cultural differences insofar as students' behavior is classroom-situated. The highly consistent results from replications of our research carried out in Northern Ireland, Belgium, Switzerland, Germany, Japan, and Venezuela, as illustrated in Table 2, provides support for this conjecture. Historical analysis, moreover, shows a striking continuity in word problems across cultures and centuries, as exemplified by examples cited earlier.

5.5. Contrasting Beliefs of Cognitive Psychologists and Mathematics Educators

Within a given culture, many groups may be identified that have an influence on shaping beliefs about mathematics learning in general and word problems in particular. These include those associated with several relevant academic fields – mathematics educators, mathematicians, philosophers of mathematics, cognitive psychologists with an interest in mathematics, and social scientists (Ernest, 1991).

We will not attempt here to discuss all the complex interrelationships among these groups; rather, we focus on the particular relationship between the beliefs of cognitive psychologists with an interest in mathematics and mathematics educators as exemplified in the contrast between two recent books on word problems (Reed, 1999; Verschaffel et al., 2000).

Reed's work, summarized in his book, fits within a historical pattern whereby psychologists who have taken an interest in mathematics have tended to exploit it as grist for their own theoretical mills and have had different perceptions, goals, beliefs, and, ultimately, values in relation to mathematics education than mathematics educators (De Corte, Greer, & Verschaffel, 1996). Thus, Reed (p. ix) commented that, originally, "Algebra word problems just happened to be a convenient source of problems for my research". The most fundamental doubt that mathematics educators raise relates to the applicability of traditional cognitive psychological research, the argument being that "although the extraction of general principles might be valued in certain academic circles, it is not clear that it contributes to the improvement of educational practice" (Cobb & Bowers, 1999, p. 13). With commendable honesty, and some puzzlement, Reed (1999, p. ix) admitted that "my work has had little influence on mathematics education, and work in mathematics education has had little influence on my work". Stimulated by his personal experience as a parent in California faced by debate over school mathematics education, he wrote his book with the aim of bridging the gap. In the book, he refers to our early work (Greer, 1993; Verschaffel et al., 1994) and recently (Reed, 2001) has written a review of our book (Verschaffel et al., 2000) in which he acknowledges, with an open mind, our point of view. Nevertheless, his position on word problem instruction is summarized as follows:

> ... instructors should begin with routine problems but do a better job in teaching students how to solve these problems. It seems to me overly idealistic to expect students to solve more challenging problems if they are still struggling with routine problems. If this first goal can be reached, then instructors will be in a better position to teach students when the assumptions for creating routine solutions are unreasonable, how to relate problems that belong to different categories, and how to combine information from several routine problems to solve the big problems. (pp. 164-165)

It should be clear from this chapter how this position differs from ours.

5.6. Philosophical Views on the Relationship of Mathematics to Reality

Influencing teachers' and students' beliefs about word problems at the most general level of all are pervasive views about the nature of mathematics and its relation to reality. Mathematics, in our view, has a dual nature:

> On the one hand, mathematics is rooted in the perception and description of the ordering of events in time and the arrangement of objects in space, and so on ... and in the solution of practical problems. On the other hand, out of this activity emerge symbolically represented structures that can become objects of reflection and elaboration, independent of their real-world roots. (De Corte et al., 1996, p. 500)

In our work (Verschaffel et al., 2000), we have elaborated our own belief that word problems (insofar as they ostensibly describe authentic situations and are not intended as "mental manipulatives" (Toom, 1999)) should be reconceptualized as exercises in mathematical modeling, whereas in standard practice the assumption is that they map situations unproblematically on to arithmetic operations. The relationship between statements of arithmetic and aspects of reality which they may or may not model has been clarified in our view by Hersh's (1997) suggestion that a statement such as "2 + 2 = 4" has two meanings. On the one hand, it is a statement about "mathematical objects", objects that are neither physical nor mental, but social constructions. On the other, it can be taken as a potential model of reality, for example as a statement about physics, that "two discrete, reasonably permanent, noninteracting objects collected with two others make four such objects" (Hersh, 1997, p. 15).

Hersh (1997, pp. 103-104) quoted the declaration by St. Augustine that:

> Seven and three are ten, not only now but always; nor was there a time when seven and three were not ten, nor will there ever be a time when seven and three will not be ten. I say, therefore, that this incorruptible truth of number is common to me and to any reasoning person whatsoever...

In our judgment, this conception remains the common view of arithmetical statements. At the general cultural level, the "folk perception" of mathematics, as reflected in the media and in popular culture, is still predominantly that mathematics is about precise computational procedures (as in the students studied by Kloosterman (this volume)). The complexity of the relationship between physical and social phenomena and the use of mathematical structures to model these aspects of reality is not part of popular consciousness.

5.7. Interactions between Levels

In analyzing the nested contexts (Figure 1) within which beliefs are situated, it is important also to take into account their interactions. For example, those with political power over education (which in many countries is increasing) tend to share the folk perception of mathematics. As suggested in De Corte et al. (1996, p. 534) "[P]erhaps the enlightenment of political decision-makers, and other groups such as parents, administrators, and the public in general ... is the biggest educational challenge facing reformers". Thus, it is possible to trace the lines of influence from a political/philosophical view on education at the level of policy-makers through to everyday practice in the classroom.

Tracing effects in the opposite direction, it is our view that the traditional and prevalent approach to teaching word problems helps to establish in children what we characterize as suspension of sense-making, and associated maladaptive beliefs about mathematics – what Freudenthal (1991, p. 70) called "an anti-mathematical attitude". Moreover, these beliefs are reinforced by the teaching of algebra and other branches of mathematics without appropriate attention to the nature of modeling. Approaching word problems from the perspective of modeling, in our view, offers a way of introducing students early to this fundamental aspect of mathematics and thereby fostering more balanced beliefs about the nature of mathematics as both an abstract construction and a way of modeling physical and social phenomena (Verschaffel, 2002).

6. REFERENCES

Boaler, J. (1994). When do girls prefer football to fashion? An analysis of female underachievement in relation to "realistic" mathematical contexts. *British Educational Research Journal, 20*, 551-564.

Boaler, J. (1997a). Equity, empowerment and different ways of knowing. *Mathematics Education Research Journal, 9*, 325-342.

Boaler, J. (1997b). *Experiencing school mathematics: Teaching styles, sex and setting*. Buckingham: Open University Press.

Boaler, J. (1999). Participation, knowledge and beliefs: A community perspective on mathematics learning. *Educational Studies in Mathematics, 40*, 259-281.

Brousseau, G. (1997). *Theory of didactical situations in mathematics*. (Edited and translated by N. Balacheff, M. Cooper, R. Sutherland, & V. Warfield). Dordrecht: Kluwer.

Caldwell, L. (1995). *Contextual considerations in the solution of children's multiplication and division word problems*. (Master's thesis). Belfast, Northern Ireland: Queen's University, Belfast.

Cobb, P. (1996). Accounting for mathematical learning in the social context of the classroom. In C. Alsina, J. M. Alvarez, B. Hodgson, C. Laborde, & A. Perez (Eds.), *Eighth International Congress on Mathematical Education: Selected Lectures* (pp. 85-99). Sevilla, Spain: S. A. E. M. Thales.

Cobb, P., & Bowers, J. (1999). Cognitive and situated learning perspectives in theory and practice. *Educational Researcher, 28*(2), 4-15.

Cockcroft, W. H. (1982). *Mathematics Counts* (Report of the Committee of Inquiry into the Teaching of Mathematics in Schools). London: Her Majesty's Stationery Office.

Collins English Dictionary, 4th Edition. (1998). Glasgow: Harper Collins.

Cooper, B., & Dunne, M. (1998). Anyone for tennis? Social class differences in children's responses to National Curriculum mathematics testing. *Sociological Review, 46*, 115-148.

Cooper, B., & Dunne, M. (2000). *Assessing children's mathematical knowledge: Social class, sex and problem solving*. Philadelphia: Open University Press.

De Corte, E., Greer, B., & Verschaffel, L. (1996). Learning and teaching mathematics. In D. Berliner & R. Calfee (Eds.), *Handbook of educational psychology* (pp. 491-549). New York: Macmillan.

De Corte, E., & Verschaffel, L. (1985). Beginning first graders' initial representation of arithmetic word problems. *Journal of Mathematical Behavior, 4*, 3-21.

Ernest, P. (1991). *The philosophy of mathematics education*. Basingstoke: Falmer Press.

Fennema, E., & Loef, M. (1992). Teachers' knowledge and its impact. In D. Grouws (Ed.), *Handbook of research on learning and teaching mathematics*, (pp.147-164). Reston, VA: National Council of Teachers of Mathematics. New York: Macmillan.

Freudenthal, H. (1991). *Revisiting mathematics education*. Dordrecht, The Netherlands: Kluwer.

Galbraith, P., & Stillman, G. (2001). Assumptions and context: Pursuing their role in modeling activity. In J. F. Matos, W. Blum, S. K. Houston, & S. P. Carreira (Eds.), *Modeling and mathematics education, ICTMA9: Applications in science and technology* (pp. 300-310). Chichester, England: Horwood.

Gerofsky, S. (1996). A linguistic and narrative view of word problems in mathematics education. *For the Learning of Mathematics, 16*(2), 36-45.

Gravemeijer, K. (1997). Solving word problems: a case of modeling? *Learning and Instruction, 7*, 389-397.

Greer, B. (1993). The modeling perspective on wor(l)d problems. *Journal of Mathematical Behavior, 12*, 239-250.

Hersh, R. (1997). *What is mathematics, really?* New York: Oxford University Press.

Hidalgo, M. C. (1997). *L'activation des connaissances à propos du monde réel dans la résolution de problèmes verbaux en arithmétique*. (Unpublished doctoral dissertation). Quebec, Canada: Université Laval.

Institut de Reserche sur l' Ensiegnement des Mathématiques (IREM) de Grenoble (1980). *Bulletin de l' Association des professeurs de Mathématique de l' Ensiegnement Public, no. 323*, 235-243.

Keitel, C. (1989). Mathematics education and technology. *For the Learning of Mathematics, 9*(1), 7-13.

Lave, J. (1992). Word problems: A microcosm of theories of learning. P. Light & G. Butterworth (Eds), *Context and cognition: Ways of learning and knowing* (pp. 74-92). New York: Harvester Wheatsheaf.

Libbrecht, U. (1973). *Chinese mathematics in the thirteenth century: The Shu-shu chiu-chang of Ch'in Chiu-shao*. Cambridge, MA: MIT Press.

Luria, A. R. (1976). *Cognitive development: Its cultural and social foundations*. Cambridge, MA: Harvard University Press.

Mukhopadhyay, S., & Greer, B. (2000). *Community College students' perceptions of word problems*. Unpublished study.

Nesher, P. (1980). The stereotyped nature of school word problems. *For the Learning of Mathematics, 1*(1), 41-48.

Puchalska, E., & Semadeni, Z. (1987). Children's reactions to verbal arithmetical problems with missing, surplus or contradictory data. *For the Learning of Mathematics, 7*(3), 9-16.

Radatz, H. (1983). Untersuchungen zum Lösen eingekleideter Aufgaben. *Zeitschrift fur Mathematik-Didaktik, 4*(3), 205-217.

Radatz, H. (1984). Schwierigkeiten der Anwendung arithmetischer Wissen am Beispiel des Sachrechnens. In: *Untersuchungen zum Mathematik Unterricht* (Band 10) (pp 17-29). Bielefeld, Germany: Institut fur Didaktik der Mathematik, Universitat Bielefeld.

Reed, S. (1999). *Word problems: Research and curriculum reform*. Mahwah, NJ: Lawrence Erlbaum Associates.

Reed, S. (2001). Review of "Making sense of word problems". *Mathematical Thinking and Learning, 3*(1), 87-91.

Renkl, A. (1999, August). *The gap between school and everyday knowledge in mathematics*. Paper presented at the Eighth European Conference for Research on Learning and Instruction, Göteborg, Sweden.

Reusser, K. (1988). Problem solving beyond the logic of things: Contextual effects on understanding and solving word problems. *Instructional Science, 17*, 309-338.

Reusser, K., & Stebler, R. (1997, August). *Realistic mathematical modeling through the solving of performance tasks*. Paper presented at the 7th European Conference on Learning and Instruction, Athens, Greece.

Säljö, R. (1991). Learning and mediation: Fitting reality into a table. *Learning and Instruction, 1*, 261-273.

Schoenfeld, A. H. (1991). On mathematics as sense-making: An informal attack on the unfortunate divorce of formal and informal mathematics. In J. F. Voss, D. N. Perkins & J. W. Segal (Eds.), *Informal reasoning and education* (pp. 311-343). Hillsdale, NJ: Erlbaum.

Srinivasiengar, C. N. (1967). *The history of ancient Indian mathematics*. Calcutta, India: The World Press.

Swetz, F. J. (1987). *Capitalism and arithmetic: The new math of the 15th century*. La Salle, IL: Open Court.

Thompson, A. (1992). Teachers' beliefs and conceptions: a synthesis of the research. In D. Grouws (Ed.), *Handbook of research on learning and teaching mathematics* (pp. 127-146). Reston, VA: National Council of Teachers of Mathematics. New York: Macmillan.

Toom, A. (1999). Word problems: Applications or mental manipulatives *For the Learning of Mathematics, 19*(1), 36-38.

Verschaffel, L. (2002, July). *Taking the modeling perspective seriously at the elementary school level: Promises and pitfalls*. Plenary address at the 26th Annual Meeting of the International Group for the Psychology of Mathematics Education, University of East Anglia, England.

Verschaffel, L., & De Corte, E. (1997). Teaching realistic mathematical modeling and problem solving in the elementary school. A teaching experiment with fifth graders. *Journal for Research in Mathematics Education, 28*, 577-601.

Verschaffel, L., De Corte, E., & Borghart, I. (1997). Pre-service teachers' conceptions and beliefs about the role of real-world knowledge in mathematical modeling of school word problems. *Learning and Instruction, 4*, 339-359.

Verschaffel, L., De Corte, E., & Lasure, S. (1994). Realistic considerations in mathematical modeling of school arithmetic word problems. *Learning and Instruction, 4*, 273-294.

Verschaffel, L., De Corte, E., & Lasure, S. (1999). Children's conceptions about the role of real-world knowledge in mathematical modeling of school word problems. In W. Schnotz, S. Vosniadou & M. Carretero (Eds.), *New perspectives on conceptual change* (pp 175-189). Oxford: Elsevier.

Verschaffel, L., Greer, B., & De Corte, E. (2000). *Making sense of word problems*. Lisse, The Netherlands: Swets & Zeitlinger.

Wells, D. (1992). *The Penguin book of curious and interesting puzzles*. London: Penguin.

Yackel, E., & Cobb, P. (1996). Sociomathematical norms, argumentation, and autonomy in mathematics. *Journal for Research in Mathematics Education*, *27*, 458-477.

Yoshida, H., Verschaffel, L., & De Corte, E. (1997). Realistic considerations in solving problematic word problems: Do Japanese and Belgian children have the same difficulties? *Learning and Instruction*, *7*, 329-338.

CHAPTER 17

NORMA PRESMEG

BELIEFS ABOUT THE NATURE OF MATHEMATICS IN THE BRIDGING OF EVERYDAY AND SCHOOL MATHEMATICAL PRACTICES

Abstract. Evidence from two research projects, one with high school students and the other with graduate students in an Ethnomathematics course, is presented to suggest that the beliefs students hold about the nature of mathematics both enables and constrains their ability to construct conceptual bridges between familiar everyday practices and mathematical concepts taught in school or university. Using a semiotic theoretical framework, graduate students learned to construct chains of signifiers linking a cultural practice that was personally meaningful to them, with abstract and general mathematical ideas. In the process, a majority of students broadened their conceptions, both of the nature of mathematics and of its relationship with cultural practices.

1. WHAT IS MEANT BY "BELIEFS"?

It is clear that the notion of beliefs is endowed with various meanings according to the needs of different researchers (Thompson, 1992). A belief may be regarded as "the multiply encoded cognitive configuration to which the holder attributes a high value, usually a truth value, including associated warrants" (Goldin, 1999). Resonating with this definition, in this chapter beliefs are taken in a slightly broader sense to have cognitive and affective components, and also to confer a propensity to act in certain ways. If actions are taken to be indicators of beliefs, and beliefs in turn confer the propensity to act, there is a certain methodological circularity. The circularity is avoided, however, if people are asked about their beliefs as well as observed in action. Professed beliefs do not always resonate with actions (Cooney, 1985; Thompson, 1984, 1992), but this does not necessarily indicate that people are being dishonest in what they say. This last point is consonant with Cooney's (1999) view of belief, following Scheffler, as "a cluster of dispositions to do various things under various associated circumstances". This definition allows for the common phenomenon that different circumstances may evoke different clusters of beliefs, and that these clusters may appear to be contradictory and may remain unexamined by the person holding them (Rokeach, 1960; Presmeg, 1988). In the projects described in this chapter, students were asked to write or speak about their beliefs.

In this research, the beliefs of high school students and graduate students about the ontological nature of mathematics could be characterized as what Thompson

G. C. Leder, E. Pehkonen, & G. Törner (Eds.), Beliefs: A Hidden Variable in Mathematics Education? 293-312.

(1992) called "conceptions". These terms (beliefs and conceptions) will be taken as interchangeable in the context of the students' views of what mathematics is. Thompson (1992, p. 130) considered conceptions to be more general mental structures that encompass, among other things, beliefs, meanings, concepts, mental images, propositions, rules, and preferences. The grain of the research reported in this chapter was not fine enough to distinguish between beliefs and more general conceptions about the nature of mathematics. When students were asked, "What is mathematics?" their replies were taken to indicate a professed view of the nature of this subject, which could be called a belief or a conception.

In this chapter, students' views of the nature of mathematics will be shown to enable or to constrain the making of links of certain kinds, in the process of constructing connections between home or cultural activities and school or college mathematics topics. The need for such connections, and the importance of beliefs about the nature of mathematics in making these links, became apparent to me over a number of years in several research projects. After a brief section detailing the evolution of my own beliefs in this regard, the two research projects that attempted to examine mathematical ontological beliefs of students, and the possibility of changes in such beliefs as a result of certain mathematical experiences, are described.

2. AN INTELLECTUAL "BELIEFS" AUTOBIOGRAPHY

In the late 1970s, after I had been teaching high school mathematics for about twelve years, I became interested in Albert Einstein's thinking and the imagery that he considered to lie at the root of his exceptional creativity (Presmeg, 1980). Visual imagery and diagrams are an important component of doing mathematics for most people; but an intensive three-year study of the roles of such imagery in teaching and learning mathematics revealed large individual differences in the need for visualization among high school teachers and their students (Presmeg, 1985, 1986). Further, certain forms of imagery that allowed for mathematical generalization were powerful tools in the mathematical cognition of the successful visualizers in the study. And even more strikingly, all the mathematical difficulties experienced by the 54 visualizers in the study were related in one way or another to problems of generalization (Presmeg, 1997). The importance of mathematical generalization became abundantly clear in my own belief system.

The next episode in my intellectual "beliefs" journey concerned a research project with prospective high school mathematics teachers in Florida in the early 1990s, to investigate their changing images of classroom practice (Presmeg, 1993). My colleagues had listed three main components as necessary prerequisites when teachers change their pedagogy. These were the identification of some perturbing elements in their current practice, a vision of what their mathematics classroom might look like, and a commitment to change classroom practice in the direction of the vision (Shaw & Jakubowski, 1991). In my research, changes in practice entailed changes in students' beliefs about the nature of mathematics as well as pedagogy, resonating with the conclusions of other researchers in this regard (Thompson, 1992).

Resonating with the realistic mathematics education that developed in The Netherlands (Treffers, 1993), a third strand of this intellectual journey had been developing during these years. Consonant with my belief in the need to link generalized mathematical principles and lived experiences of students, this strand concerned a desire to find ways of using real-life experiences and cultural practices of students in their construction of mathematical ideas in school learning of mathematics at all levels, for purposes of equity in the teaching and learning of mathematics. In fall of 1993 I taught a graduate course, *Mathematics and Culture*, for prospective and practicing teachers, which later became a regularly taught course called *Ethnomathematics*. In 1995-96, the same principles used in this course were tried in a yearlong research project with seven students in an Algebra II class in the County's high school for immigrant and foreign students. Most of these Algebra students believed that mathematics is "a bunch of numbers" and a way of solving "word problems". A discrepancy was observed between beliefs of these seven students, who could not at first describe connections between school mathematics and their daily lives or cultural practices, and those of the graduate students whose broader ontological conceptions of mathematics enabled them to construct such connections in the projects they were required to do in the *Ethnomathematics* course (Presmeg, 1998a). This discrepancy led to my strong conjecture (elaborated later) that beliefs about the nature of mathematics were enabling or constraining the bridging process between everyday practices and school mathematics.

Finally, semiotic chaining was seen as a potentially valuable tool to enable the building of links between home and school mathematical practices in a way that allowed for mathematical generalization (Presmeg, 1997, 1998b). Without such chaining of representations, in a direction from everyday practice to classroom mathematical concepts, it had been found that students' bridging was inclined to lead them to quite trivial mathematics in many instances. Requiring that they construct a chain with several links encouraged more generalized mathematical principles, leading to wider potential links with the mathematics of the school or college curriculum at various levels. Research that investigated students' beliefs or conceptions of mathematics and their construction of such chains is described in the following sections after a more detailed account of some of the earlier studies mentioned in this section.

This beliefs autobiography has described some of the rationale behind my belief that it is important to link everyday and school mathematical practices in such a way that, firstly, structure is preserved in the move towards mathematical abstraction and generalization, and, secondly, that some of the meanings of the everyday practice are retained in the transition to school mathematical practices.

3. HIGH SCHOOL STUDENTS' BELIEFS ABOUT THE NATURE OF MATHEMATICS

The aim of the high school project, conducted with the help of two research assistants (Stephen Sproule and Bineeta Chatterjee), was to investigate ways in which the lived experiences of students, including their cultural practices, could be used in the classroom learning of mathematics at high school level. Consonant with the exploratory nature of the study, the research methodology was qualitative and interpretive, involving regular tape- and video-recorded interviews with students chosen from those in a multicultural high school Algebra II class, as well as classroom observation of lessons. (The research is described more fully in Presmeg, 1998a.)

I want to offer here a word about the methodology of research on beliefs. In chapter 20 (this volume), Lester refers to the pitfall of research that relies on students' actions to infer their beliefs, while at the same time predicating those actions on the beliefs in a circular argument. He is dismissive of interview and other self-report methods (e.g., journal writing) for collecting data about beliefs. He is perhaps overlooking that even open or Likert-type questionnaires involve self-report (Kloosterman, chapter 15, this volume). Thus, like most of the authors in this section (e.g., Greer et al., chapter 16) I shall rely on self-report methods to substantiate my claims, thus avoiding the pitfall of Lester's circular scenario. As far as possible, the students' own words will be provided verbatim, to enable the reader to judge whether these serve as evidence for the interpreted meanings.

The seven students in the high school project came from varied cultural backgrounds, but their conceptions of mathematics were remarkably similar, and although these views did change in the course of the research, these changes were of a superficial nature, illustrating the tenacity of these deeply held conceptions of what mathematics is (Table 1). The students were asked "What is mathematics?" in the first interview of the series, and again in the final one.

Table 1. Student beliefs concerning the nature of mathematics

Mathematics is ...

Student	First interview: October 31, 1995	Final interview: April 9, 1996
Erica	using numbers to solve problems.	solving problems, complex problems with numbers and equations.
Ajeng	word problems.	like solving problems; a whole bunch of numbers put together to solve something.
Keri	numbers, equations, triangles, word problems ("I can't stand them!")	(moved)
Big Al	learning how to solve problems.	more than just solving problems.

Student	First interview: October 31, 1995	Final interview: April 9, 1996
April	word problems.	a problem solving strategy.
Kristie	adding, subtracting, multiplying and dividing.	the study of numbers.
Jessica	confusing problems.	a bunch of problems you have to solve.

It is likely that the students' classroom experiences of mathematics were influential in these characterizations. Indeed, given the beliefs of the students and their teacher about what classroom learning of mathematics entailed, we were unable to implement successfully the cultural mathematical activities we had planned for this class (Presmeg, 1998a). However, it was encouraging that in a series of interviews the students described a wide range of their own activities that the researchers considered to have potential for the development of school mathematical concepts, even if the students did not themselves recognize this potential at first. These activities could be classified under four headings that emerged in our analysis. These categories encompassed the past, the present, and the future: cultural Heritage, Hobbies, Hopes for the future, and Homeland in the cases of immigrant students (the "four Hs"). After describing their home activities, the students were asked whether mathematics was used in these lived experiences. They were also asked what work their parents did and whether mathematics was involved in this, their achievement in and feelings towards school mathematics, and whether they saw any links between mathematics and any other subjects they were studying at school (Table 2).

Table 2. Activities described by students (series of interviews)

Student	Activities
Erica	coin collecting (Korea, Italy, France, Germany, England, Denmark, USA - countries in which Erica had lived); collecting dolls and postcards; reading fiction and non-fiction; listening to music; attending theatre; games from Korea
Ajeng	housework; listening to music; swimming; camping (Girls Scouts, in Indonesia)
Keri	barrel racing with horses (for seven years, on a farm in Central Florida)
Big Al	farming (horses, cows, hogs, chickens, greens, potatoes); football; basketball; Gospel music
April	cheerleading; volleyball; softball; writing poetry
Kristie	basketball; house plastering (father's work); designing and making furniture (father's hobby)
Jessica	housework; tennis; volleyball; modeling clothes (Puerto Rico)

By the final interviews in April, 1996, when these questions were revisited, the students were describing where they could see uses for mathematical connections, in other subjects of the school curriculum, and in their everyday activities. It is possible that despite the lack of deep change in their "formal" responses to the question, "What is mathematics?" there might have been some broadening in their conceptions of what mathematics is and how it is used, as evidenced by their comments in this final interview (Table 3).

Table 3. Other activities in which potential for mathematics was seen by students (final interview)

Student	Activities
Erica	music; English poetry ("Now I see math in almost everything.")
Ajeng	working as a cashier ("in the numbers, in the change").
Keri	(moved)
Big Al	counting money (paying for gas, fund raising as in car washes, buying clothes)
April	television (rhythm and patterns, making connections with problems in mathematics class)
Kristie	biology (lab; "measuring stuff in cups")
Jessica	music (lower and higher tones); cooking (recipes, "Like you are trying to make your food from other countries, you have to make sure you know the ingredients and how much you have to put in."); chemistry (periodic table, lab experiments)

Despite these changes in beliefs about where mathematics is used or "found", resonating with psychological literature (Rokeach, 1960) we became convinced that mathematical ontological beliefs do not change easily, and also that for changed practices in mathematics classrooms it is necessary that participants - both students and teachers - broaden their conceptions of mathematics as well as their beliefs about what is entailed in the learning of mathematics. These interpretations from our research data resonate with much of the literature on teachers' beliefs and conceptions (Thompson, 1992). We concluded that the bridging of home and school mathematical practices could be facilitated by teachers if they valued such bridging, and that a good place to start introducing such issues is in preservice and inservice courses such as our Ethnomathematics course. Accordingly, a year later, in this course, a cohort of 119 teachers of mathematics, mostly in elementary and middle schools in Miami, Florida, provided more overt evidence of changed beliefs, in this case about the relationships between mathematics and culture, as described in the following section.

4. CHANGES IN BELIEFS CONCERNING CULTURE AND MATHEMATICS

In summer, 1996, we started a distance learning project with 250 teachers in Miami, Dade County, in which these students would study for Master's and Specialist's degrees in either mathematics or science education[1]. One of the courses for teachers specializing in mathematics was the *Ethnomathematics* course, taught in summer of

1997. At the end of the course, the 119 students were asked to respond in writing to the question, "Have your beliefs about the relationships between mathematics and culture changed this summer? If so, in what ways?" 95% of the students reported that their beliefs had changed in one way or another, sometimes dramatically. There was no negotiation of the meaning of the term "beliefs", thus naïve, intuitive, or colloquial meanings can be inferred. However, their comments illuminate salient issues. In the following I include written comments by many of the students to illustrate the extent of the belief changes they experienced. Even amongst those who reported no change in response to the focus of the question, there were those who described an enlarged or heightened awareness of mathematical elements around them. For instance, Amy wrote as follows:

> Well I don't really think my beliefs changed, but I do think that I am looking at mathematics differently. I never realized how much math is in our cultures. Math is everywhere really. Math is in everything. We are constantly on a schedule, what time do we do what. Time is mathematics! It is interesting to see how much mathematics we can find in our lives, in our culture, and within ourselves.

In like vein, Jennifer reported as follows:

> My beliefs about Math & Culture have not really changed, because I had never really thought about it before this class. What has happened, however, is I am now aware of how important and useful it is to become mathematically enculturated. There are so many wonderful opportunities at both our reach and our students' reach, if we incorporate cultural experiences into the learning of math. Not only will we learn from each other, but we learn that math has value and is literally all around the world (her emphasis).

Amongst those who responded that their beliefs had changed, a similar focus of the reports was this awareness of mathematics in everyday activities. The following subheadings document categories of the changes the students reported.

4.1. Seeing the World "with Mathematical Eyes"

In his description of the structure of mathematical abilities, Krutetskii (1976) included a "mathematical cast of mind" in the characteristics of the mathematically "capable" students whose cognition he studied. Such a cast of mind enabled Krutetskii's students to identify mathematical ideas in all their activities, that is, to view their worlds "with mathematical eyes". In the summer Ethnomathematics course, several students who reported changed beliefs also expressed an increased facility in this regard:

Maria S: Prior to my experiences in this course I never considered culture to play a part in mathematics. In fact, I thought of them as two separate entities. I now realize that culture is a major part of mathematics and mathematics is a major part of culture. After all, mathematics is in *everything*! After doing my project I realized that I had so much to offer others from my own culture. Students, too, have so much to offer from their cultures. Therefore it is pertinent that we allow our students to share their ideas with one another. Culture is who we are, and by sharing it with each other we create new learning experiences (her emphasis).

Shirley: I find myself more cognizant of issues related to math in a lot of ways. As I ride down streets, I'm naming the shapes and designs that I see. As situations occur, I'm thinking about an available math lesson that could be found in the situation.

Ann: Above all, I don't think I will ever view math as an isolated discipline or subject. I will now see it as an intrinsic part of our daily lives and culture.

Maria B: Like I have mentioned in my project, I now see much more connection between mathematics and culture than ever before. I look at just about everything with mathematician's eyes. I used to be a bit skeptical about how to incorporate mathematics through a whole language approach. But now I feel confident enough to say and share that we "hear, smell, touch, taste, and even breathe mathematics daily". Therefore I'm better prepared to be an effective mathematics teacher.

Maria B's project was a powerful analysis of mathematical elements in the construction of Cuban rafts from inner tubes of tires, and boards, and ropes. Facing incredible danger and hardship, some of Maria's compatriots succeeded in reaching the shores of Florida on such makeshift "boats". Certainly this topic was personally meaningful to Maria. The themes of personal meaningfulness and of improved mathematical pedagogy were recurrent ones.

4.2. "Math is Becoming More of a Personal Thing."

For many of the students, their personal cultural research projects were an opportunity for them to take ownership of the mathematical ideas they were constructing, sometimes for the first time.

Ana: I did my summer research on the musical history of my native country (Cuba). I found the research to take me to two other countries. Aside from this, I found out just how much math is in music. Even though I have loved and studied music for years, I never associated the fun of music with the horror of mathematics! It was refreshing to find a link between something that I love and something that I have never found appreciation for. Hence it was quite fulfilling and worthwhile.

Michele: The research project (Mayan mathematics in Michele's case) helped me to be aware of how important it is that we know our culture background. As for mathematics and culture being related, yes they are; it gave me great joy this summer to be able to discover this.

Clara: Ethnomathematics has opened a door that once had been closed, both in my mind and in my heart. ... Learning about my own culture and the cultures of others is truly a wonderful opportunity. I learned that there's math in everything, and everywhere. I also learned how to allow my students to explore and examine the culture of mathematics and its connection to the world around them.

Twyla: My beliefs about mathematics and cultures have changed this summer, because looking around at things that we see everyday and take for granted are things that can enhance our mathematical skills and techniques, whether it is a student or a teacher. Both individuals can learn from these aspects. ... Math is becoming more of a personal thing.

4.3. The "Cultural Significance" Required in Teaching Mathematics

For many of these teachers, reflection on their own practice was a natural outcome of their ownership and heightened awareness of mathematical constructs "frozen" in cultural practices (Gerdes, 1998).

Mayra: Ethnomathematics has changed my view as a learner. Going back to when I was in school, I was never taught that *there's mathematics everywhere*. I have now realized that mathematics is universal. Students should be exposed to the different mathematics from other parts of the country and be more knowledgeable. Ethnomathematics has really made a great impact; as a teacher I plan to allow my students to explore and discover the mathematics around us. I believe that students' attitudes towards the mathematics would be more positive and that they would be more confident about themselves. By experimenting and generalizing the mathematics around us, students will become better problem solvers and be able to apply their skills in the real world (her emphasis).

Susan: I feel awareness has been the greatest enlightenment. I am so aware of math as a culture and how culture affects math. I think how I visualize myself teaching math is different now as a result of how I've been exposed or taught how to look at things. ... The world's a very good and enormous place.

Several teachers expressed wonder at the richness of possible mathematical constructs in cultural practice. Shabana expressed it well:

Shabana: In researching Islamic Art and the mathematics involved in it I feel a deeper appreciation for the arts of all cultures. I was always aware of the mathematical components involved in art; however, I found the extent of those mathematical elements to be astounding. Also, with the various activities done in class with the instructors, mathematical light was shed on even the most ordinary occurrences of many cultures. Similar activities would be a great way to answer the question, "When am I ever going to need this?" or "Why do we have to learn this?"

Clova: I now can experience a better appreciation for the songs, customs, and way of life of people from different cultures. Firstly is my attempt to research a folk song (Jamaican) that I have been singing since childhood days and never stopped to think of. [Knowing] the background experience that goes with it, I now can sing it with greater meaning. This led me to different aspects of my own culture that I was not aware of. Secondly, I now realize that if I could approach mathematics from a cultural point of view, the students in my class would develop a deeper love for the subject and find the learning of it fun. I am amazed to find out that there is so much mathematics in the different aspects of one's culture, namely, the songs, flag, travel, and how they used to count in their own way. I will make every effort to change my approach in teaching mathematics.

5. THE NEED FOR MATHEMATICAL GENERALIZATION

The foregoing are just some of the reported changes in beliefs that these practicing teachers described. These personal comments about cultural practices and the learning and teaching of mathematics provide some insight into why changes in beliefs about using cultural diversity in the classroom may enhance the mathematical learning that takes place in these classrooms. However, the conception that "mathematics is everywhere" needs to be tempered with the realization that the patterns and relationships of cultural practices can be symbolized in various ways, and that it is the systematization of these relationships that allows for mathematical generalization and enables them to be linked with the formal required mathematics of school curricula (Presmeg, 1998a, 1998b; Civil, 1998).

These considerations led to my interest in semiotics as a tool for deepening mathematical generalizations, because of its potential to link seemingly disparate domains of human experience, such as everyday home practices and classroom mathematics, through a process of chaining of signifiers that may preserve elements of structure and meaning.

5.1. Purposeful Semiotic Chaining

Whitson (1997, p. 99) has pointed out that semiosis refers to activity with signs whereas semiotics properly refers to the study of semiosis. In a dyadic model such as the one used in this study, a sign consists of a signifier and what it signifies. A promising analytic tool for building connections between everyday practices and classroom mathematics is the chaining of signifiers in a semiotic system based on Lacan's inversion of Saussure's model (Whitson, 1997; Presmeg, 1997, 1998b). The relevance of semiotic chaining in this chapter lies in the suggestion that students' conceptions of the nature of mathematics influenced the chains they were able to construct between cultural activities and mathematical concepts. Since it is not central to this chapter, the theory behind semiotic chaining will not be reported in detail here (see Whitson, 1997, Walkerdine, 1988). The kernel of the theory is that the signifier from one sign or link in the chain may become signified in the next link, with the new sign encompassing all that went before, in a "sliding under" effect. The new signifier stands for all that went before. The direction of development of the chain of signifiers is towards more abstract and general forms. Symbolism of structure and relationships may be considered to lie at the heart of doing mathematics (Pimm, 1995). Thus the chaining of signifiers (often mathematical symbols) is eminently suited to the connections that may help students to link everyday practices with mathematical systems at many levels. Because interpretation of the signs is involved, I believe that there is a need in the theory to incorporate a triadic model such as that of Peirce (1998), but there is not room to do that here.

For my purposes, it was the need to encourage students to systematize and deepen mathematical relationships that they identified in their own cultural practices, that led to the request that they symbolize these relationships in semiotic chains (Presmeg, 1998b). In the summer of 1998, 53% of the 97 students in Miami who made such chains were judged to have attained some degree of mathematical generalization. When the course was taught in Tallahassee in the fall of 1999, with 22 students in the class, the percentage of successful chains constructed by students was higher: 77%, or 17 students were judged to have created successful bridges between cultural practices authentic to them, and school or college mathematical topics. A few examples follow, after an examination of some of the mathematical ontological beliefs of the students in the 1999 class.

5.2. Beliefs of Graduate Students about Mathematics and Culture

As shown on Table 4 (see the summary at the end), 13 of the 22 graduate students in this class started the course with a predominantly numerical view of the nature of mathematics. They were asked the same questions again at the end of the course, and again responded in writing. Consonant with the changes reported by the Miami teachers in the previous section, most of these students also reported not so much a change in their beliefs about the nature of mathematics, but a broadening in their perceptions of the links between everyday practices and mathematics. The following table shows the initial perceptions, together with a final column that indicates the broad range of personally meaningful projects that the students chose, and from which they constructed mathematical ideas.

Table 4. Beliefs of students about the nature of mathematics: Fall, 1999

Questions (and replies given in August, 1999):
1. What is mathematics?
2. Is there a relationship between mathematics and culture? If so, describe it.

Name Project & type of mathematics involved	**Math is ...** (primary definition)	**Plus ...** (extra elements)	**Relationship between math. and culture**
Jan Karate: spatial patterns	numbers & symbols	method of describing universe	None
Mark Tossing cowrie shells	study of numbers		Yes: learning styles
Judy US flag: space & numbers	numerical & geometrical concepts	working & reasoning with these	Yes: approach to numerical reasoning
Melanie Needlework: spatial, number & abstractions	numbers & abstraction	study of these	Yes: different teaching practices
Bev Flute: sound waves (grade 2 students)	numbers, relationships & patterns	computation & problem solving, etc	Yes: teaching, but universal elements too
Veon Maypole dancing: units, geom., trig.	language/science of numbers	how these relate to the physical world	Yes: teaching, but techniques & formulas are general
Angel Mancala: strategies in playing the game	numbers, concepts, ideas	reasoning & problem solving	Yes: problem solving methods (contradiction)
Margelet African American hand games: modulus, num. analysis, & circle	numbers & their significance	understanding of these	Yes: different understandings of math
Stacie Board games: num. analysis of moves	number, quantity & orientation	tool for applying science	Yes, of course: way of writing, creating math

Robert H Baseball: num. & graph. analysis of field	using numbers to understand world	how to count & organize objects, data, even people	Yes: in use of math, problem solving, types of problems
Rod Billiards: geometry, num. measurement & scaling, coordinates	language that uses numbers & symbols	problem solving	Yes! In some cultures math symbols are written differently
Datasha African names: descriptive statistics, & length of names	using numbers & symbols to calculate	everyday life; critical thinking & problem solving	Yes: derived from many cultures; but universal language too
Robert W Techno-kaleidoscope: numbers & geometry	study of numbers and space	logic	Yes: passed on by ancestors
Cathie Tap dancing: 3 notations, digraph & tables	patterns, observation, deductions, etc.	problem solving, fun, standards, calc, alg, graph theory, chaos	Of course: math influences culture & vice versa, in what is taught
Derek Tennis: table of symmetries, group theory	study of patterns	discipline that develops anal thinking	Yes: different cultures affect the math produced
Heather US quilting: symmetry, scaling, measurement & geometry	problem solving & pattern finding	in everyday life	Yes: approach to problems, and in spreading ideas
Jackie Western wall: squares, table, number patterns, formulas	unique discipline of both art & science	developed by people for utility & aesthetic value	Yes: since math is developed by people
Erdogan "Okey": rules of game, scoring, strategies	language of the universe	numbers, with a special meaning	Yes: culture contains math & vice versa
Baran Oriental carpets, patterns, symmetries, isometries	positive science branch	derive solutions from parameters by formulas	Yes: math is a unique part of culture
Onder Islamic & Jalaali calendars: number &	description without words		Yes

moon relationships			
Hasan "Ihale" card game: system of trumps, tricks, & limits	part of life & of being human	basis of almost everything	Yes: cannot think of math & culture separately
Suleyman Turkish green houses: geom.. & numerical analysis	system for calculation		Yes: history

Summary of characterizations of mathematics given by 22 students

Numbers	Patterns	Art & science	Other
13	3	1	5

Three of these cases (Melanie, Veon, and Derek) will be examined in more depth, to illustrate the consonance between the students' ontological beliefs and the mathematics they were able to "see" in their projects. Thus I shall concentrate on their responses (before and after the course) to the first question, "What is mathematics?" These three cases were chosen because these students constructed particularly successful semiotic chains linking the practices embedded in their cultural projects with more generalized and systematized mathematical concepts. Melanie and Veon were Masters degree students; Derek was in the Ph.D. program.

5.2.1. Melanie

From the beginning of the course, Melanie's definition of mathematics included elements of abstraction and generalization. She wrote,

> Mathematics is often the study of numbers and abstract representations, relationships between such, and theory and proof behind such numerical and abstract principles and formalizations (8/30/99).

At the end of the course, she characterized mathematics as follows:

> I don't think that my idea of what mathematics is has changed. However, I think that I may better be able to express it. Mathematics is the study of patterns and the relationships between such. It is often a method of organization in practical life ideas, which can then easily be generalized into abstractions of such patterns and numerical systems (11/29/99).

Melanie chose to examine "Mathematical ideas in needlework." After an historical introduction, her first mathematical focus was an analysis of the isometric symmetries in patterns of blackwork embroidery. In ten charts, she identified symmetries of translation, horizontal reflection, glide reflection, vertical reflection, and rotation through various angles. She also analyzed a further three patterns from the point of view of tessellations.

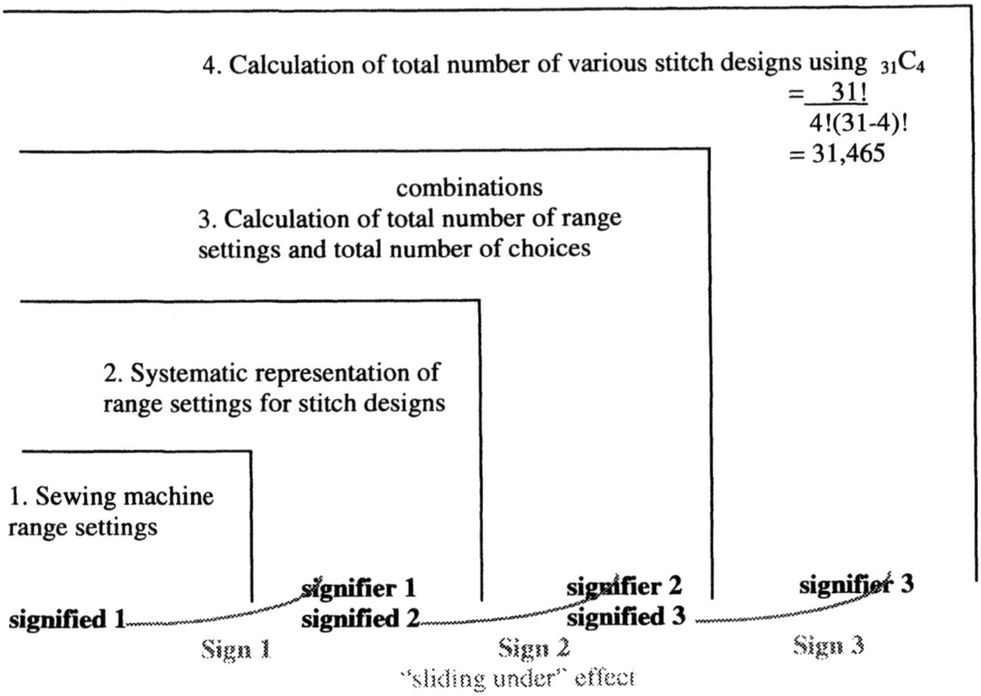

Notes:

- Sign 2 encompasses sign 1.
- Sign 3 encompasses both sign 2 and sign 1.
- Within each sign, the signifier "stands for" all the links that precede it in the chain.

Figure 1: Melanie's semiotic chain for combinations of various stitch designs

Melanie's next focus was on the mathematical aspects of sewing machine stitch designs, taking into account the tensions of upper and lower threads, stitch width and length (four "effects", as she called them), and range settings. She described her analysis as follows:

> With a wide variety of range settings, and with so many factors affecting stitch design, I began to question how one might find the total number of various stitch designs one could have. I first began to systematically chart each of the four effects, trying to match each one with the other, so as to get all combinations of settings. I began with the upper and lower tensions set equal at a balanced and ideal range of four, since the balanced stitch is the most ideal, and strongest stitch. After only a few entries and combinations, I realized this work would become too tedious, and remembered probability and the calculation of combinations. From here, I decided to total the number of choices for range settings, of which there were 31, and choose four of these each time. After careful computation, I found that there are 31, 465 different stitch designs which can be

> produced by a machine with the given range settings. WOW, that is a lot of different designs!

Melanie's semiotic chain for this part of her analysis is given in Figure 1.

Melanie also decided to explore the mathematical aspects related to working with repeat designs in needlepoint, and she presented a second semiotic chain illustrating this aspect of her work. Both of her chains were consistent with the nature of her stated beliefs about mathematics as "the study of patterns and the relationships between such", including "numerical and abstract principles and formulations". There are spatial and numerical aspects to the systematic way she set out to construct mathematical relationships; abstraction and generalization are implicit.

In the next case, Veon also initially stressed the numerical aspect of mathematics, but this conception broadened in her later definition.

5.2.2. Veon

Veon's definitions of mathematics were as follows:

> Mathematics is the language/science of numbers and how these numbers relate to the physical world (8/30/99).

> Mathematics is the study of patterns and the way these patterns relate to everyday living. It involves more than the manipulation of numbers, formulae, and stated facts. It includes the way one thinks and the techniques of doing even the simplest task (11/29/99).

As Veon explained, Maypole dancing has been a part of her country Jamaica's heritage ever since the 18th century. It is often performed by school children at festivals and school fairs. The musical accompaniment is usually mento, one of the traditional Jamaican dances, but Veon wrote that "it is not unusual to have groups perform this [Maypole] dance to our popular reggae music." When it is performed during October, the cultural heritage month, folk dancing groups compete for prizes. The poles vary from eight to about 20 feet in height, and the number of ribbons (each one- and-a half times the height of the pole) should be a multiple of four. There are two kinds of Maypole dance, the Closed Plait, where the pattern is wound round the pole, and the Open Plait, made by the outer ribbons winding round the inner ones, away from the pole. Veon described and analyzed examples of both kinds, barber's pole and spider's web respectively, suggesting ways these dances could be used in school mathematics classrooms. She wrote, "These two dances are chosen because they distinctly show mathematical concepts that can be used to reinforce the teaching of trigonometrical and geometrical topics in the mathematics classroom especially in middle and high school." The links in her two chains were as follows.

<u>Barber's pole dance</u>

1. Barber's pole dance.
2. Analysis of height of pole and number of dancers.
3. Calculation of length of a ribbon, and total amount of ribbon needed.
4. Diagrams showing ribbon, pole and dancer, and circular path of dancers.

5. Calculation of distances using Pythagorean theorem, and use of formulae for finding the area and circumference of a circle.
6. Tables showing circumference and area of circles with varying radii. Deduction of ratios.

Spider's web dance

1. Spider's web dance.
2. Height of pole and number of dancers.
3. Length of ribbon and total amount.
4. Diagram showing ribbon, pole, and dancer.
5. Calculation of angles of depression and elevation, volumes and slant height of varying radii.
6. Table showing volumes and slant heights of cones.

Both Melanie and Veon were able to demonstrate beautiful examples of the crafts or dances that they analyzed in their projects: Veon showed a video of an actual Maypole dance in her country. Certainly the patterns in her second definition of mathematics were elements in her analysis; mathematics "involves more than the manipulation of numbers, formulae, and stated facts."

Patterns were also central in Derek's definition of mathematics.

5.2.3. Derek

> Mathematics can be thought of as a study of patterns. I also like to characterize math as a discipline that develops analytical thinking skills (8/30/99).

> Mathematics encompasses several factors. One is the detection and analysis of patterns. It is important to not only be able to recognize patterns, but also discover properties related to those patterns. This ties in to the second characteristic of mathematics – the ability to generalize. While the patterns themselves may be viewed as building blocks, attention should also be given to the ways these patterns can be extended to include a broader class of examples. It is through patterns that we both understand and build mathematics (11/29/99).

As in the cases of Melanie and Veon, Derek's project, "The mathematics of tennis" also resonated well with his definition, the patterns he identified being suitable for a college mathematics course. He analyzed the dimensions of a tennis court and the scoring system, not only in their present-day forms, but also in their historical evolution. Then he turned his attention to the symmetries of the court, summarizing the vertical, horizontal, and rotational symmetries in an operation table that evidences group properties. He concluded that the set of symmetries of a tennis court forms an Abelian group of order four, or D2, the dihedral group of order 4. He continued,

> But the dihedral group D_n usually stands for the set of symmetries of a regular n-gon. Therefore it is customary to speak of D_n for $n \geq 3$. In this case, D_2 is isomorphic (i.e., has the same group structure) as the direct product of the integers modulo two, represented as $Z_2 \times Z_2$. In my opinion, this notation is more appropriate for this context, as it suggests the structure of a doubles match: two players versus two players, with four

players total on the court. Also, $Z_2 \times Z_2$ is a non-cyclic group, meaning there does not exist one element which can generate the entire group. The analogy carries over to a doubles match, where it is nearly impossible to have one person dictate the flow of the match. Rather, a team with two players working together will almost always be victorious against a team with one dominant player. As it turns out, $Z_2 \times Z_2$ is generated by two elements, (0,1) and (1,0)!

Derek was apparently having fun in his combination of two loves of his life, group theory and tennis! His semiotic chain in illustration of this is given in Figure 2.

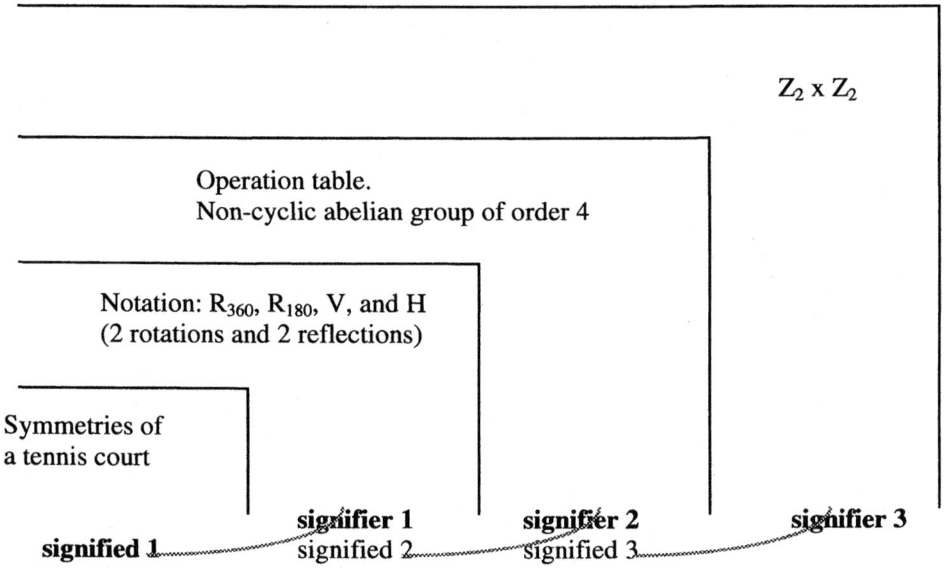

Figure 2: Derek's semiotic chain for the symmetries of a tennis court

The data from the case studies of Melanie, Veon, and Derek were presented to illustrate a certain consonance between their conceptions of mathematics, including base knowledge from mathematics courses they had taken, and the way they constructed their semiotic chains linking personally meaningful cultural practices and generalized mathematical principles.

In both of the research projects described in this chapter, it appeared that students' views of the nature of mathematics either constrained their construction of links between home and cultural practices and school mathematical practices, or enabled such links to be constructed. These themes are continued in the concluding section.

6. CONSTRAINTS AND AFFORDANCES

It seems clear from the data presented as illustration in this chapter, that we see what we believe, perhaps more surely than we believe what we see! What I mean by this is that the kinds of mathematics that the students could recognize in the structures of their everyday practices hinged on their ontological beliefs about the nature of mathematics. These beliefs were thus constraints and affordances, in many cases simultaneously limiting and facilitating the semiotic chains they constructed.

The implications for mathematics education are profound, because it is clear that classroom teaching and learning of mathematics cannot change in the direction of building bridges between home activities and classroom practices without changes in the beliefs that underlie these practices (Civil, 1998). In this sense, beliefs about the nature of mathematics and its teaching are fundamental in attempts to understand what happens in classroom learning of mathematics, and thus in any attempts to change these practices. Such beliefs may constrain what happens there (as we experienced in our high school algebra project), or beliefs may provide affordances, as illustrated in the narratives of the Miami teachers in describing their broadened views of mathematics and where it is seen, and their ideas of how they could use the rich cultural diversity of their students as resources for the teaching and learning of mathematics.

In this chapter, I have concentrated on ontological beliefs of students (high school and graduate) about what mathematics is and what it entails, particularly with regard to its implicit links with practices in everyday life. I have not reported data from the fall, 1999 project that suggest that courses such as the Ethnomathematics course do help to broaden participants' views about how rich everyday activities of students may be linked with their classroom learning of mathematics. The evidence shows that this is indeed the case. I have not examined at all, beliefs about what constitutes teaching and learning of mathematics. But from the student beliefs that have been presented in this chapter concerning the nature of mathematics, and ways in which those beliefs may broaden based on experiences that students have, I am convinced in my own belief system that beliefs of preservice and inservice teachers are an important determinant of what happens in mathematics classrooms, and a key to effecting change in those practices.

Teaching preservice teachers to use chaining of signifiers in their lesson preparation is one way that we are currently investigating, of facilitating the use of school students' own practices in the teaching and learning of classroom mathematics. The rationale for making these connections has been developed over more than a decade in the ongoing curriculum work of the realistic mathematics education team of the Freudenthal Institute (Streefland, 1988; de Lange. 1993). My belief in the need for systematization and progressive generalization in such endeavors resonates with what Freudenthal called vertical mathematization, complementary to the horizontal mathematization that makes connections between activities from students' lived realities and the mathematical ideas of the classroom. Systematic chaining of signifiers has the potential to facilitate both kinds of mathematization. The evidence presented in this chapter suggests that a broadening of beliefs or conceptions about the nature of mathematics and of how it is related to everyday activities accompanies both kinds of mathematization, and in turn facilitates the construction of these links in a reflexive process. Without such a broadening of ontological beliefs, the process of building links may be constrained, as it was in the project with high school students, who were not engaged in either form of mathematization in their regular classroom activities. Conceptions of what mathematics is, and related beliefs of how it is involved in all aspects of life, appear to be key elements in the endeavor to use students' lived realities in classroom learning of mathematics.

7. NOTES

[1] This project was funded by Dade County under National Science Foundation Grant number 945-3669, awarded under the Urban Systemic Initiative program. Opinions expressed are not necessarily those of the NSF.

8. REFERENCES

Civil, M. (1998, April). *Bridging in-school mathematics and out-of-school mathematics.* Paper presented at the Annual Meeting of the American Educational Research Association, San Francisco, California.

Cooney, T. J. (1985). A beginning teacher's view of problem solving. *Journal for Research in Mathematics Education, 16*, 324-336.

Cooney, T. J. (1999, November). Examining what we believe about beliefs. In E. Pehkonen & G. Törner (Eds.), *Mathematical beliefs and their impact on teaching and learning of mathematics* (pp.18-23). Proceedings of the Workshop in Oberwolfach.

de Lange, J. (1993). Between end and beginning. *Educational Studies in Mathematics, 25*, 137-160.

Gerdes, P. (1998). On culture and mathematics teacher education. *Journal of Mathematics Teacher Education, 1*(1), 33-53.

Goldin, G. A. (1999, November). Affect, meta-affect, and mathematical belief structures. In E. Pehkonen & G. Törner (Eds.), *Mathematical beliefs and their impact on teaching and learning of mathematics* pp. 37-42. Proceedings of the Workshop in Oberwolfach.

Krutetskii, V. A. (1976). *The psychology of mathematical abilities in schoolchildren.* Chicago: University of Chicago Press.

Peirce, C. S. (1998). *The essential Peirce: Selected philosophical writing* Vol. 2 (pp.1893-1912). Bloomington, Indiana: Indiana University Press.

Pimm, D. (1995). *Symbols and meanings in school mathematics.* New York: Routledge.

Presmeg, N. C. (1980). *Albert Einstein's thought, creativity, and mathematics education.* Unpublished M.Ed. dissertation, University of Natal, South Africa.

Presmeg, N. C. (1985). *Visually mediated processes in high school mathematics: A classroom investigation.* Unpublished Ph.D. dissertation, University of Cambridge, England.

Presmeg, N. C. (1986). Visualisation in high school mathematics. *For the Learning of Mathematics, 6*(3), 42-46.

Presmeg, N. C. (1988). School mathematics in culture-conflict situations. *Educational Studies in Mathematics, 19*(2), 163-177.

Presmeg, N. C. (1993, October). Changing visions in mathematics pre-service teacher education. In J. R. Becker & B. J. Pence (Eds.), *Proceedings of the Fifteenth Annual Meeting of the North American Chapter of the International Group for the Psychology of Mathematics Education* Vol. 2 (pp. 210-216). Asilomar, California.

Presmeg, N. C. (1997). Generalization using imagery in mathematics. In L. D. English (Ed.), *Mathematical reasoning: Analogies, metaphors and images* (pp. 299-312). Hillsdale, New Jersey: Lawrence Erlbaum Associates.

Presmeg, N. C. (1997). A semiotic framework for linking cultural practice and classroom mathematics. In J. A. Dossey, J. O. Swafford, M. Parmantie, & A. E. Dossey (Eds.), *Proceedings of the Nineteenth Annual Meeting of the North American Chapter of the International Group for the Psychology of*

Mathematics Education Vol. 1 (pp. 151-156). Columbus, Ohio: ERIC Clearinghouse for Science, Mathematics, and Environmental Education.

Presmeg, N. C. (1998a). Ethnomathematics in teacher education. *Journal of Mathematics Teacher Education, 1*(3), 317-339.

Presmeg, N. C. (1998b). A semiotic analysis of students' own cultural mathematics. Research Forum Report, in A. Olivier & K. Newstead (Eds.), *Proceedings of the 22nd Conference of the International Group for the Psychology of Mathematics Education* Vol. 1 (pp. 136-151).

Rokeach, M. (1960). *The open and closed mind.* New York: Basic Books.

Shaw, K. L., & Jakubowski, E. H. (1991). Teachers changing for changing times. *Focus on Learning Problems in Mathematics, 13*(4), 13-20.

Streefland, L. (1988). Reconstructive learning. In A. Borbás (Ed.), *Proceedings of the 12th Annual Conference of the International Group for the Psychology of Mathematics Education* (pp. 75-89).

Thompson, A. G. (1984). The relationship of teachers' conceptions of mathematics teaching to instructional practice. *Educational Studies in Mathematics, 15*, 125-127.

Thompson, A. G. (1992). Teachers' beliefs and conceptions: A synthesis of the research. In D. A. Grouws (Ed.), *Handbook of research on mathematics teaching and learning* (pp. 127-146). New York: Macmillan.

Treffers, A. (1993). Wiskobas and Freudenthal: Realistic mathematics education. *Educational Studies in Mathematics, 25*, 89-108.

Walkerdine, V. (1988). *The mastery of reason: Cognitive development and the production of rationality.* New York: Routledge.

Whitson, J. A. (1997). Cognition as a semiosic process: From situated mediation to critical reflective transcendence. In D. Kirshner & J. A. Whitson (Eds.), *Situated cognition: Social, semiotic, and psychological perspectives* (pp. 97-149). Mahwah, New Jersey: Lawrence Erlbaum Associates.

CHAPTER 18

ERNA YACKEL AND CHRIS RASMUSSEN

BELIEFS AND NORMS IN THE MATHEMATICS CLASSROOM

Abstract: The central purpose of this chapter is to demonstrate that by coordinating sociological and psychological perspectives we can explain how changes in beliefs might be initiated and fostered in mathematics classrooms. In particular, we examine: 1) the coordination of students' beliefs about mathematical activity and classroom social norms and 2) the coordination of specifically mathematical beliefs and classroom sociomathematical norms. Examples from a university level differential equations class are used for purposes of clarification and illustration.

1. INTRODUCTION

For more than a decade we and our colleagues[1] have collaborated to study students' mathematical learning in the context of the classroom. In the process of doing so, we have developed an interpretive framework (see Table 1) for analyzing classrooms that coordinates both individual (psychological) and collective (sociological) perspectives. In this work we were strongly influenced by Bauersfeld, Krummheuer, and Voigt's (Bauersfeld, 1988; Bauersfeld, Krummheuer, & Voigt, 1988) long standing work in advancing symbolic interactionism[2] as a theoretical framework for investigating mathematics teaching and learning. The central thesis of this chapter is that by coordinating sociological and psychological perspectives it is possible to develop ways to explain how changes in beliefs might be initiated and fostered in mathematics classrooms. The purpose of this chapter is to develop this thesis. In particular, we discuss those aspects of the interpretive framework that relate to student beliefs and the corresponding classroom norms. The beliefs we consider in this chapter are beliefs about one's role, others' roles, and the general nature of mathematical activity in school and specifically mathematical beliefs and values. We use a university level differential equations class as an example to clarify and illustrate these constructs within the framework. The example demonstrates both the normative aspects of the classroom and the corresponding student beliefs. In each of the classrooms we have studied over the past years, from elementary school mathematics to university level differential equations, student mathematical beliefs changed dramatically over the course of the teaching experiment. In this chapter we demonstrate how the theoretical constructs of the interpretive framework can be used to explain this change. The significance of this work is that it begins to address

G. C. Leder, E. Pehkonen, & G. Törner (Eds.), Beliefs: A Hidden Variable in Mathematics Education? 313-330.

a major challenge in working with beliefs, namely, the initiation of changes in students' beliefs about mathematics and mathematics instruction (also see Greer, Verschaffel, & De Corte, this volume; Tsamir & Tirosh, this volume).

2. BACKGROUND

Following on the seminal work of Erlwanger[3] (1973), a number of mathematics educators have argued for the need to consider students' beliefs about mathematics when attempting to make sense of their mathematical behavior. For example, Cobb (1985) demonstrated that the mathematical activity of the young children who participated in an extended teaching experiment could not be accounted for solely in terms of their mathematical conceptions. However, by complementing a conceptual analysis with an analysis of the children's beliefs it was possible to explain the radically different behavior of children to whom similar concepts were attributed. At the same time, Schoenfeld's work with university level students led to similar conclusions (Schoenfeld, 1983).

As early as 1986, Cobb conjectured that mathematics instruction, as a socialization process, influences student beliefs (see also Greer, Verschaffel, & De Corte, this volume). This conjecture, which was based on working with children in one-on-one settings, was confirmed in a classroom teaching experiment[4] we conducted in 1986-87 in one second-grade classroom. As we have reported elsewhere (Cobb, Yackel, & Wood, 1989), student beliefs at the beginning of the school year were compatible with a "school mathematics tradition," but as the year progressed their beliefs became compatible with an "inquiry mathematics tradition"[5]. Initially,

> [T]he teacher's expectations that the children should [attempt to construct their own solutions to problems and] verbalize how they actually interpreted and attempted to solve the instructional activities ran counter to their prior experiences of mathematics instruction in school (Wood, Cobb, & Yackel, 1988). The teacher, therefore, had to exert her authority in order to help the children reconceptualize their beliefs about both their own roles as students and her role as the teacher during mathematics instruction. She and the children initially negotiated obligations and expectations at the beginning of the school year which made possible the subsequent smooth functioning of the classroom. Once established, this mutually constructed network of obligations and expectations constrained classroom social interactions in the course of which the children constructed mathematical meanings (Blumer, 1969). The patterns of discourse served not to transmit knowledge (Mehan, 1979; Voigt, 1985) but to provide opportunities for children to articulate and reflect on their own and others' mathematical activities. (Cobb et al., 1989, p. 126)

As we will explain below, in order to investigate how it was that student beliefs were influenced by the socialization process, we sought to analyze the social (participation) structure of the classroom. The sociological perspective we followed was that of symbolic interactionism because of its compatibility with psychological constructivism (Voigt, 1996; Yackel & Cobb, 1996)[6]. In the same way that attention to student beliefs is not a logical necessity but proves pragmatically useful because it helps to account for aspects of students' mathematical activity that otherwise are not explainable, taking a sociological perspective is not a logical necessity. However, taking such a perspective proves pragmatically useful because doing so provides means to analyze and ultimately explicate aspects of the teaching and learning of mathematics in the classroom setting that otherwise defy explanation.

3. THE INTERPRETIVE FRAMEWORK

In this section we give a brief overview of the constructs in the interpretive framework that are relevant to this chapter. A more extensive discussion of the framework together with clarifying examples can be found in Cobb and Yackel (1996). First, we wish to emphasize that the interpretive framework is not the result of an a priori theoretical analysis but rather grew out of extensive classroom-based research. It evolved from our attempts to make sense of students' learning in the classroom across several yearlong classroom teaching experiments in elementary school mathematics classes. Our initial efforts included considerable attention to unraveling the complexity of the classroom by focusing on the classroom social norms and later on the sociomathematical norms (Cobb et al., 1989; Yackel, Cobb, & Wood, 1991; Yackel & Cobb, 1996). We use the label social norms to refer to regularities in the interaction patterns that regulate social interactions in the classroom. As such, social norms are expressions of the normative expectancies in the classroom. For example, in a classroom the interaction patterns might be indicative of the expectation that students are to explain their thinking to each other. By contrast, sociomathematical norms refer to regularities in the interaction patterns that are specific to mathematics. For example, in our analyses we have noted that classrooms differ with respect to what becomes normative regarding acceptable mathematical explanations. In this case, the relevant expectations are specifically related to the fact that the subject matter is mathematics, as opposed to history, or literature, or some other subject. Consequently, we have chosen to use the label sociomathematical norms to distinguish these norms from general classroom social norms.

Table1. Interpretive Framework for Analyzing Individual and Collective Activity in Classrooms

Social Perspective	Psychological Perspective
Classroom social norms	Beliefs about one's role, others' roles, and the general nature of mathematical activity in school
Sociomathematical norms	Specifically mathematical beliefs and values
Classroom mathematical practices	Mathematical conceptions

These constructs are sociological in that they refer to the classroom community as a collective group rather than to the individual members of the community. Nevertheless, in attempting to analyze norms, we took the position that there is a reflexive relationship between the individual and the collective[7]. Therefore, our analyses of norms necessarily involved taking account of the corresponding psychological components. As we have noted elsewhere (Cobb, Yackel, & Wood, 1993) we take beliefs to be the psychological correlates of norms. In doing so, we are taking beliefs to be basically cognitive. They are the understandings that an individual uses in appraising a situation. Thus, discussions of norms and discussions of beliefs are intimately intertwined. This interrelationship between beliefs and norms is critical because it provides a means for talking about changes in beliefs. That is, in saying that norms and beliefs are reflexively related we imply that they evolve together as a dynamic system.

Methodologically, both general social norms and sociomathematical norms are inferred by identifying regularities in patterns of social interaction. Thus social norms are identified from the perspective of the observer and indicate an aspect of the social reality of the classroom. However, what becomes normative in a classroom is constrained by the current goals, beliefs, suppositions, and assumptions of the classroom participants. For example, a student's inferred beliefs about his or her own role in the classroom, others' roles, and the general nature of mathematical activity can be thought of as a summarization of the obligations and expectations attributed to the student across a variety of situations. In this sense beliefs can be thought of as an individual's understandings of normative expectancies. Social norms can be thought of as taken-as-shared beliefs that constitute a basis for communication and make possible the smooth flow of classroom interactions (Cobb et al., 1993).

4. AN ILLUSTRATIVE EXAMPLE

To illustrate the reflexive relationship between student beliefs and classroom social and sociomathematical norms, we draw on data from a semester-long classroom teaching experiment conducted in a university level differential equations class. The overall goal of this teaching experiment was to investigate university level students' learning of differential equations in the context of the classroom. We wish to stress that the interpretive framework and the theoretical relationships discussed above were not developed from data taken from this classroom. These theoretical developments resulted from extensive analysis of several yearlong classroom teaching experiments at the elementary school level. However, the differential equations instructor was knowledgeable about research from which the interpretive framework arose and was committed to the goal of developing an inquiry form of instruction[8]. To this end, he gave explicit attention to the constitution of classroom norms that characterize an inquiry mathematics tradition. In particular, the instructor sought to foster the social norms that students are expected to develop personally meaningful solutions, to explain and justify their thinking, to listen to and attempt to make sense of the thinking of others, and to raise questions and challenges when they disagree or do not understand. Within this context, he also sought to promote the sociomathematical norm that explanations should be descriptions of actions on taken-as-shared mathematical objects that are experientially real for the students.

A question of interest to the project research team was the extent to which these social and sociomathematical norms, which had grown out of analysis of elementary school mathematics classrooms, would be equally beneficial at the university level. For this reason, specific attention was given, on a daily basis, to the norms that were operative in the classroom and to their evolution as the semester progressed. Through classroom observations, it was apparent that the norms described above were constituted in the class over the course of the semester. Further, it was apparent that the individual students altered their beliefs about their role, the teacher's role and the nature of mathematical activity. In the following discussion we document these claims. In keeping with the position that beliefs are the cognitive basis that individuals use to interpret situations that arise in the course of social interaction, our methodological approach is to demonstrate an evolution of students' beliefs across the semester by considering the reflexive relationship between beliefs and norms. This approach stands in contrast to approaches that codify student beliefs before and after the instructional period and compare the results.

4.1. The Project Classroom

The project took place in a differential equations class in an American university. The majority of the twelve students in the class[9] were engineering students; the remainder were mathematics majors. The class met for two-hour class sessions twice a week for fifteen weeks. As noted above, the instructor used an inquiry approach to instruction. Further, as part of the overall project, he developed a majority of the instructional activities used in the class[10]. In addition, the class used a reform-oriented textbook[11] for homework problems. Students used a TI-92 graphing calculator with programs specifically designed by members of the project research team to foster their development of a conceptual basis for slope fields and phase portraits.

In addition to regular homework assignments, students submitted weekly electronic journals in which they reflected on their mathematical activity during the prior week. In some instances specific journal prompts were given. For example, the first journal prompt was to "describe at least one idea from the previous week that was most confusing to you and one idea that was clearest to you". Each student also prepared a course portfolio, which was handed in on two occasions, the day of the first exam and at the final exam. The purpose of the portfolio was for students to synthesize their learning across a number of weeks. To prepare a portfolio a student needed to select entries that reflected important instances of learning and, for each entry, write a rationale explaining the insights gained through the instructional activity represented by the entry. An unanticipated benefit of the electronic journals and portfolios was that they provided additional information about students' evolving beliefs.

The data that form the basis for this chapter come from video recordings that were made of every class session and from the students' electronic journals and portfolios. Throughout the teaching experiment the project team met on an ongoing basis to discuss a variety of issues, including classroom norms and students' beliefs. However, no analyses were conduced until after the semester was completed. At that time two members of the research team set out to document the classroom norms and how they were constituted. For this purpose, our field notes and the video recordings were scrutinized for evidence relating specifically to norms and beliefs, with special attention to the first several weeks of class when the negotiation of norms was initiated. We then identified exemplary episodes from our field notes and videotape review notes. Examples were selected on the basis of their clarity for presentation using verbal material only. These examples were then transcribed. As noted above, an unanticipated benefit of the electronic journals and portfolios was that they provided information about students' beliefs. Consequently, these materials were scrutinized to identify evidence related to the students' initial or evolving beliefs. Again, illustrative examples were selected for use in this chapter based on their clarity. This chapter presents the results of our analysis using the selected episodes and examples as the basis for discussion.

4.2. Coordinating Classroom Social Norms and Beliefs about the Nature of Mathematical Activity

4.2.1. Beliefs

Encountering an inquiry approach to mathematics instruction was a novel experience for the students in the differential equations class since all of them had presumably experienced only the school mathematics tradition in their prior grade K-12 and university mathematics instruction. Thus, at the beginning of the semester their classroom mathematical beliefs were based on expectations that the students' role in class is to follow instructions and to solve problems in the way the instructor and/or textbook demonstrate. Similarly, the instructor's role is to explain and demonstrate procedures for the students to follow. These expectations simultaneously clarify the beliefs students held about the nature of mathematical activity in the classroom. Evidence for these claims about initial student beliefs comes primarily from two sources. First, the interaction patterns in the classroom and second, student comments in their electronic journals. In the paragraphs below we give examples of comments from student journals as a means of documenting their beliefs.

To provide a background for making sense of the students' electronic journals we briefly describe the initial class activity of the first class session. The students' task was to attempt to make sense of the spread of an infectious disease. After a brief discussion of the scenario, the students were to sketch graphs of the susceptible population, the infected population, and the recovered populations over time. They had no quantitative information available to them and no procedural solution method to follow. Thus, the initial classroom activity confronted the students' beliefs about what constitutes mathematical activity, about their role, and about the teacher's role. Subsequent activities were of a similar nature. In fact, all instructional activities throughout the semester were designed to focus on sense making and required explaining and justifying one's thinking and mathematical activity.

In the early weeks of the semester a number of students made explicit comments in their electronic journals about the discrepancy between their (prior) beliefs and the expectations in this class and/or their attempts to cope with the shift in expectations. For example, in the first week's journal one student remarked:

> The only thing that I find sort of confusing is the fact that there may not be an exact answer or answers to a specific problem. There are so many variables in some of these problems that the answers that are obtained (if obtained) can only be used to make educated guesses. I'm used to thinking of math as an "exact science" where there is always an exact answer or answers to a problem.

We interpret this comment to mean that the student's belief that mathematics should always yield one exact answer was being challenged. Another student articulated his earlier belief about the teacher's and the students' roles by saying,

> I'm still getting used to the format. I'm more used to the teacher saying everything and not letting the students really have a "voice".

That he is struggling with these new roles is evident from his later remark in the same journal entry,

> You never said exactly how you wanted the homework done.

Another student commented explicitly about the expectations regarding explanation. He wrote in the journal that accompanied his second homework assignment that he was disappointed in his homework score. He had spent many hours completing the assignment yet he earned only half of the possible points. He wrote the following:

> [M]ost of the points lost were due to my failure to explain how I reached my answers. I thought a clear, systematic approach to the math calculations would be sufficient to explain my thought process. I now have a better understanding of the expectations.

This student is articulating that he formerly believed that in mathematics explanation is not required.

Yet another student was clearly struggling with the dissonance between his beliefs about what mathematics instruction should be and his experiences in the first few class sessions. In his second journal entry, he wrote,

> ... the class is interesting, but the problem is that I'm not learning much. ... I won't deny that I enjoy the open discussion that our class has, but let's face it, we aren't learning much. It would be more practical for us to do some examples using the concepts in class.

Some weeks later, in the portfolio he handed in at the first exam, this same student included the following rationale statement for one of his portfolio entries.

> This handout was chosen because it was one of the initial problems that made me think in a way that I was only exposed to in classes based on philosophy. The open discussion in class at this time was foreign to me, especially in a math class. As you may have guessed, we were not sure what to write or what to answer because we were never asked these types of things before. So basicly [sic], I think that this piece was important because it was the introduction of obscure thinking used in this class.

We take this to mean that while the student had, by this time, gained some understanding of the expectations in this class, they were still counter to what he believed mathematics instruction should be. For him, this class was like philosophy.

Yet another student who had taken the same course previously on another of the university's campuses, with unsuccessful results, was able to clearly articulate the differences in the expectations between the two classes. Although he did not indicate what his beliefs are, he linked his instructional preference to these expectations.

> I like the way the class and the book concentrate on practical applications and explanations for differential equations. As you may have noticed from my info [information] card, I have taken this class before at ______ [another campus of the same university] and it was much different. We spent a lot of time trying to memorize all the techniques to solve the equations and learned very few practical ideas. I was lost and disinterested 15 minutes into the first class session. I can honestly say I think I've learned more about differential equations in the first two weeks here than I did in the whole semester there.

This student explicitly linked expectations of memorizing techniques to not learning much about differential equations and expectations of explanation to substantive learning. Other students were in the position of making comparisons of this class to other mathematics classes in their prior experience. This student had the advantage of being able to compare two versions of the "same" class. As a result, he could be much more explicit in stating his views.

The above selections, which are only samples from the entire class, demonstrate that, in general, students' beliefs upon entering the class were consistent with the expectations that mathematics consists of using prescribed rules and procedures to find exact answers to problems (cf. Kloosterman, this volume; Tsamir & Tirosh, this volume). Further, the students' comments indicate that these beliefs were being confronted by the instructional approach they were experiencing in the differential equations class. As we have shown elsewhere (Yackel, 1995a), it is the situation as it is interactively constituted as a social event rather than the social setting per se that is critical in influencing the nature of students' mathematical activity and beliefs.

4.2.2. Classroom Social Norms

The differing expectations of the students and the instructor led to situations of explicit negotiation. For example, on the second day of class the instructor began with a brief statement of the expectations regarding classroom participation. He concluded his remarks by saying, "We had some nice examples of that from Shawn and Natasha last time". We know from prior analysis that one of the ways a teacher can initiate the renegotiation of expectations and obligations is through explicit discussions such as this (Cobb et al., 1989). However, such discussions by themselves are insufficient for establishing a classroom in which routines are regulated by those expectations. The students and instructor must come to act in accordance with the expectations. In this case, the next twenty minutes of the class was devoted to a dual agenda. On one hand, the class was engaged in a mathematical discussion about the rationale behind the rate of change equations they had used in the prior class session to model the spread of an infectious disease. On the other hand, the instructor gave explicit attention throughout to the negotiation of social norms compatible with the expectations for inquiry mathematics listed above. In fact, the majority of the instructor's remarks throughout the episode were (explicitly or implicitly) directed toward the expectations. Only a few of his remarks were explanations related to the mathematical content.

Episode 1

Excerpts from the twenty-minute episode are included here.

Instructor:	Just to sort of recap, last time we were dealing with the spread of a virus. We had [the] elementary school population in Chicago where we had students who were either susceptible to the disease, who were recovered, or who were infected. And we talked about one differential equation. That [was] dR/dt = (1/14)I. Anyone remember why it was one fourteenth? How many people remember? Shawn, why was it 1/14?
Shawn:	Fourteen days from the time you got cured, from the time you got it to the time its over.

In keeping with traditional instruction, Shawn might expect the instructor to evaluate his response as correct or incorrect and then initiate a different question. However, the instructor pursues the same question further. He calls for additional explanation, in particular, he asks how it "makes sense". In doing so, he indicates that students' responses should explain their individual thinking and further, that mathematical thinking is about sense making.

Instructor:	Okay, can you explain to us then why it was 1/14 times I? How did that sort of make sense as a way to express the change in the recovered population?
Shawn:	That it's constant.
Instructor:	Say that again.
Shawn:	That it's constant. The same amount of number of people for each stage.
Instructor:	What do the rest of the people think about that?

With this question the instructor initiates another shift. His question indicates that he expects others to be actively engaged in the discussion. They are to listen to the exchange he and Shawn are having and are to develop their own interpretations about Shawn's response. An implicit expectation is that each student is developing his or her own response to the question even though it was specifically addressed to Shawn initially. Another student offers his thinking.

Jerry:	Each day, there's from day zero to day one (inaudible) from day 14 to day 15, you would see 1/14 of that population recover. And every day thereafter.
Instructor:	(To Shawn.) Is that similar to what you were thinking?

Here again, the instructor does not follow the traditional initiation-response-evaluate pattern (Mehan, 1979). Instead of evaluating Jerry's remark, the instructor indicates that he expects students to listen to and attempt to make sense of each other's contributions. The instructor is attempting to initiate a genuine dialogue between the students. His intention is that they communicate with each other, not only with and through him.

Shawn:	Kinda. Yeah
Instructor:	Kinda. What about the explanation here? Did everyone understand what was said?

Our analysis to this point shows that the instructor is attempting to influence the interpretations the students make of how to engage in the discussion. From this perspective, it might seem that the instructor is the only one in the classroom who contributes to the renegotiation of social norms. However, norms for social interaction are interactively constituted as individuals participate in interaction (Yackel & Cobb, 1996). In this case, as the episode continues students contribute their part to the negotiation of the social norms by increasingly acting in accordance with the expectations. As the discussion progressed, students not only responded to the instructor's questions, they initiated comments of their own that showed that they were beginning to change their understandings of the classroom participation structure. Greg's remarks, as the dialogue continues, are an illustration.

Greg:	I didn't quite understand what he said.
Instructor:	What was that?
Greg:	I didn't quite understand what he said.
Instructor:	So maybe you could rephrase it or say it a little bit louder so people can hear how you're thinking about it.
Jerry:	If every day, we took all the days, we say—, day zero to day one, day one to two, through day 13 to 14, you had population infected, when you got to day 15 we made the assumption that 1/14 of the population was recovered and my understanding is that on day 16 its another 1/14 recovered. It's an assumption that we made.
Instructor:	Anyone want to add to that explanation? Expand on it a little bit? Maybe you still have questions about it.

With these remarks, the instructor continues to emphasize the expectations regarding sense making and explanation. Apparently, if you don't understand, even after extended discussion, you are expected to ask for further elaboration. By asking for further clarification, Greg contributes to the constitution of this expectation as normative in this class.

Greg:	Well, how I understand it is that ... what I don't understand, what I was asking about, whether the—because, initially we said it was a fixed population, whether the fixed population will have some part to play in the formula because that's what I don't understand—if we said (1/14)I, or what, the fixed population or—
Instructor:	A good question, I think I understood what you were saying. Earlier I heard you saying 1/14 of the population. And you kind of said the same thing here. Well, this (points to the differential equation on the chalkboard) is referring to 1/14 of what population?
Jerry:	The infected population.

After further discussion involving Jerry, Shawn, and the instructor, another student initiates a question about the assumptions underlying the use of 1/14 as the scalar multiple of I.

Alicia:	Could you have [just as] easily assumed that dR/dt is 1/15 or 1/16? Is it just the assumption that there is a correlation between the number of days that the disease will run to [sic] the denominator?

An extended discussion ensued in which Alicia, the instructor, Jerry and Shawn continued to make sense of the relevance of 1/14 in the differential equation of interest.

In the above dialogue we see Greg and Alicia acting in accordance with the expectation that classroom activity is about making sense—making sense of the questions posed by the instructor and others, making sense of the explanations offered as part of the discussion, and making sense of the problem scenario. We also see Jerry and Alicia (in the extended discussion not included here) acting in accordance with the expectation of explaining one's thinking to others. We find it encouraging that as early as this point in the second class session students were beginning to act in accordance with the expectations that the instructor was attempting to initiate just a few moments earlier[12]. We maintain that such actions on the part of students are not simply mindless reactions to the instructor's initiations. Rather, these actions indicate the students' interpretations of the instructor's intentions. Smith (1978) has said, "a willingness to act and ... the assumption of some risk and responsibility for action in relation to a belief represent essential indices of actual believing" (p. 24). Accordingly, we would say that as Greg, Alicia, and Jerry act in accordance with these "new" expectations, their beliefs about their role, the instructor's role, and the nature of classroom mathematical activity are evolving. In this case, not only did these students play a critical role in the interactive constitution of the expectations for the classroom, in doing so, they initiated a shift in their individual beliefs about the classroom participation structure. For his part, by maintaining a focus on the negotiation of expectations, the instructor made it possible for the students to reorganize their beliefs in a way that was compatible with the expectations he was attempting to foster.

The effectiveness of the renegotiation of social norms is indicated by considering classroom interactions that became typical later in the semester. As an example, consider the following episode, which occurred toward the end of the semester.

Episode 2

After two students in the class explained how they determined that a particular phase portrait would not have two saddles next to each other, Dave spontaneously added to the discussion with this remark.

Dave: The way I thought about it at first, to make me think that all the points weren't saddles, is that if the next one was a saddle—see how [Bill] has got the one line coming in towards [referring to the phase portrait that Bill had drawn on the blackboard]. Well, if the next one was like that, then you would have to have another point in between those two equilibrium points, like separating, like a source or something. So that's how I started thinking about it. So then $3\pi/2$ might be a source or maybe a saddle point with opposite directions.

Dave's remark elicits the following response from Bill.

Bill: So it's like, you're saying that if there is a saddle, there has to be a source. If there is a sink or a saddle you have to have a, like in this case right here, you would have to have a source in between the saddles in order for it to really make sense.

Shortly thereafter, Bill relates Dave's explanation to their earlier study of autonomous first order differential equations.

Bill: If you draw the phase line with, like, two sinks, one on top of the other, then you would have to have a source between them.

These spontaneous remarks made by Dave and Bill indicate that they have taken seriously the obligations of developing personally-meaningful solutions, of listening to and attempting to make sense of the thinking of others, and of offering explanations and justifications of their mathematical thinking. In the process of acting in accordance with these expectations they are demonstrating their beliefs about their roles and about the nature of classroom mathematical activity. Furthermore, in acting in accordance with the expectations they are simultaneously contributing to their ongoing constitution. In this way, the normative patterns of interaction serve to sustain the expectations and obligations on which they are based and thus to sustain individual participants' beliefs about their role and about what constitutes mathematical activity in this classroom.

4.3. Coordinating Sociomathematical Norms and Specifically Mathematical Beliefs

Earlier we distinguished between classroom social norms and sociomathematical norms as follows. Social norms are regularities in interaction patterns that regulate the social interactions in the classroom. By contrast, sociomathematical norms refer to regularities in the interaction patterns that relate specifically to the fact that the class is a mathematics class. To the extent that interactions involve interpretations or appraisals of a situation, what becomes normative is constrained by the current beliefs of the classroom participants. In the previous section we discussed the beliefs and norms that relate to the participation structure in the class. In this section, we discuss those beliefs that constrain the mathematical aspects of interactions, called specifically mathematical beliefs, and sociomathematical norms[13].

In contrast to social norms, it took much longer to achieve stability with respect to the sociomathematical norms that characterize inquiry mathematics. In this section, we limit the discussion to the sociomathematical norm of what constitutes an acceptable mathematical explanation and related beliefs. In particular, we discuss two aspects of acceptability with respect to mathematical explanation. The first relates to the communication aspect of explanation and the second to the specific expectations that had been established within the classroom in question. In general, an explanation is an individual's attempt to clarify for others aspects of one's thinking that one judges might not be clear (Cobb, Wood, Yackel, & McNeal, 1992). That is, explanation is a form of communication. As such, an explanation is acceptable when it serves a clarifying function. This is one aspect regarding acceptability of explanation. Acceptability, however, cannot be judged apart from those who are attempting to make sense of the explanation. This means that the notion of what constitutes an acceptable mathematical explanation is confounded by the students' developing understanding of the mathematical concepts in the course and the instructor's understanding of students' mathematics. This is where the second aspect of acceptability comes to the fore. For an explanation to be acceptable it has to meet the requirements that have been established through interaction by the participants in the classroom. In general, in mathematics classrooms that follow the inquiry tradition, explanations have to be about actions on mathematical objects that are experientially real for the students. By this we mean that explanations have to be about students' mathematical activity with entities that are part of their mathematical worlds. Descriptions of procedures are typically not acceptable. This is the second aspect of acceptability.

In the following paragraphs we first discuss the general clarifying function of explanation. Next, we discuss the specifically mathematical aspects regarding the acceptability of an explanation. The interactive constitution of what constitutes an acceptable mathematical explanation is closely linked to that of the social norm that explanations are to be given. In this regard, we point to one of the instructor's comments in Episode 1. In response to Greg's remark, "I didn't quite understand what he said," the instructor replied, "So maybe you could rephrase it or say it a little bit louder so people can hear how you're thinking about it". Here the instructor's suggestion to rephrase hints at the communicative function of explanation. If the earlier remark was simply not heard, say it louder. However, if the earlier remark did not serve a clarifying function for the listeners, rephrase it.

There is considerable evidence that by the fourth electronic journal assignment the students were beginning to understand the clarifying function of explanation. In this journal students were asked to explain their understanding of the Existence and Uniqueness Theorem for solutions to differential equations. Many students wrote comments to the effect that their explanations were inadequate. For example, one student ended his journal with this comment, "This is the best way I know to explain it which I know is very lacking". Another wrote, "I am not completely sure I understand this point so I wouldn't try to explain it to someone unless they had some feedback as to what they think it is". We take these comments as indications of the students' belief that the purpose of explanations is to communicate; explanations should clarify one's thinking for others.

The second aspect of adequacy of explanation is specifically mathematical. At the beginning of the semester at least some of the students believed that procedures in the form of calculations were acceptable as explanations. One of the student journal entries we included earlier exemplifies this belief. This was the second journal of the semester.

> [M]ost of the [homework] points lost were due to my failure to explain how I reached my answers. I thought a clear, systematic approach to the math calculations would be sufficient to explain my thought process. I now have a better understanding of the expectations.

Previously we argued that this journal entry shows that the student initially believed that in mathematics explanation is not required. In making that statement, we were using explanation in the sense of providing some insights into one's thinking. From that perspective, the student was not giving an explanation at all. However, from the student's perspective he was giving what he thought counted as an explanation. In terms of what constitutes acceptability, we can say that the student initially believed that systematic calculations constituted acceptable explanations. From the instructor's perspective systematic calculations do not necessarily signify actions on mathematical objects that are meaningful to the student. They may simply be manipulations of symbols.

Earlier we wrote that in mathematics classrooms that follow an inquiry mathematics tradition explanations should describe actions on mathematical objects that are experientially real for the students. We refer the reader back to Episode 2 in an earlier part of this chapter as evidence that this sociomathematical norm was constituted in the differential equations class. In that episode both Dave and Bill talk about sinks, saddles, and sources as entities within their mathematical realities that they can identify and locate. In doing so, they not only act in accordance with this sociomathematical norm, they also demonstrate that their underlying beliefs about mathematical explanations are compatible with that norm.

In this differential equations class a more specific sociomathematical norm regarding acceptable explanation was constituted in the case of first order differential equations. In this case, to be acceptable, explanations had to be grounded in an interpretation of the rates of change as expressed by the differential equation(s). To clarify what we mean by this and to illustrate the constitution of sociomathematical norms, we include the following episode that occurred during the fourth week of the semester.

4.3.1. Episode 3

The dialogue is taken from a whole class discussion of the differential equation $dP/dt = 0.5P(1-P/8)(P/3 - 1)$ which models the rate of change in the population of fox squirrels in the Rocky Mountains. The task posed to the students was to figure out what interpretations might be given to the numerical values 0.5, 8, and 3. Students had discussed the problem in their small groups prior to the whole class discussion.

Instructor: Okay, so Jerry says that if the population gets above 8 they [the fox squirrels] are going to start dying. Tell us why you made that conclusion.

Jerry: Because some number greater than 8 over 8 is going to yield some number greater than one, which 1 minus something greater than 1 is going to give you a negative number and so something times a negative number is going to give you a negative number, so your slope is going to be negative.

Jerry's explanation is completely calculational in nature. At no time does he refer to what the quantities represent or to how they might be interpreted in the underlying scenario. We take this as evidence that Jerry believes descriptions of procedures such as this constitute adequate explanations. The instructor uses this as an opportunity to initiate a shift in the orientation of the discussion—away from a calculational orientation and towards a conceptual orientation (Thompson, Philipp, Thompson, & Boyd, 1994).

Instructor: So if P is bigger than 8, like you said, maybe 8 million or 8 thousand fox squirrels, then this term here is negative, like you said, right?

Jerry: Mmm hmm.

Instructor: And so what does that mean for us? That means what? If this term is negative, that doesn't tell us anything in itself in relation to the differential equation.

Jerry: The change is negative.

Instructor:	Okay, so this part here is negative, is this part negative or positive?
Sylvia:	Positive.
Instructor:	All right, if P is bigger than 8, certainly 8/3 - 1 is positive and so this is positive, and this [0.5P] is positive, so the rate of change, dP/dt, is negative. So that means dP/dt is negative, which means what?
Greg:	The population [is] reducing.
Instructor:	They're reducing, good. So the rate of change is negative that means the population, the number of fox squirrels is getting smaller. The population is decreasing. So the number of squirrels (i.e., P(t)) is decreasing.

In the preceding dialogue the instructor repeatedly asks for the meaning of the quantities involved, suggesting that none of the explanations given thus far have been adequate in addressing that issue. Eventually he explicitly tells the class his own interpretation when he says, "So the rate of change is negative *means* (emphasis added) the number of fox squirrels is getting smaller. The population is decreasing". In doing so, the instructor not only confirms Greg's response but also gives the students an opportunity to reorganize their beliefs about the criteria for meaningful explanations. As the discussion progressed, the instructor asked Dave how his group thought about the problem.

Dave:	Well, pretty much, kind of, the same as what Jerry was saying but just the opposite. In this case, it says the fertile adults have to be able to find other fertile adults to be able to increase. Well, if they don't, then the rate of change of that is going to be negative which makes everything else negative, so it's decreasing.

Dave's response indicates the effectiveness of the instructor's intervention. Although his response is not at all calculational (as Jerry's initial explanation was), Dave prefaces his response with, "pretty much the same as what Jerry was saying". In saying this, he indicates that he now understands that explanations are to be about rates of change and their significance within the scenario.

The above example illustrates what we mean by the sociomathematical norm that explanations had to be grounded in an interpretation of rates of change as expressed by first order differential equations. It also demonstrates that sociomathematical norms such as this are constituted in interaction. As with social norms, while the instructor typically initiates the negotiation of sociomathematical norms, other participants contribute to their constitution. In the above example, Dave's comment not only exemplified the norm, it contributed to its ongoing constitution. Further, we maintain that as students' actions are increasingly in accordance with the expectations for explanation, they demonstrate their evolving beliefs about what constitutes an acceptable mathematical explanation in this class.

5. CONCLUSION

The chapters in this book show that there are many different ways to investigate student beliefs about mathematics. Our purpose in this chapter has been to demonstrate that by coordinating sociological and psychological perspectives it is possible to explain how changes in beliefs might be initiated and fostered in mathematics classrooms. In particular, we have attempted to demonstrate how classroom social and sociomathematical norms and individual beliefs evolve together as a dynamic system. We have demonstrated that students' beliefs about their role and others' roles in the classroom and about the general nature of mathematical activity evolve in tandem with the social norms that students are expected to develop personally-meaningful solutions, to explain and justify their thinking, to listen to and attempt to make sense of the thinking of others, and to raise questions and challenges when they disagree or do not understand. Similarly, we have demonstrated that students' beliefs about what constitutes mathematical explanation evolved in tandem with the sociomathematical norm that, in general, explanations should signify actions on mathematical objects that are meaningful to the students and, in particular, in the differential equations class explanations had to be grounded in an interpretation of the rates of change as expressed by the differential equation(s). In doing so, we have shown that changes in beliefs and negotiation of classroom norms are inextricably linked.

By coordinating perspectives, we give primacy neither to the social nor the psychological. Rather, we maintain that each provides a backdrop against which to consider the other. Our purpose has been to clarify that as classroom norms are renegotiated, there is a concomitant evolution of individual beliefs. Verschaffel, Greer, and De Corte (1999) have noted that it is generally assumed that students' beliefs about mathematical activity develop "implicitly, gradually, and tacitly through being immersed in the culture of the mathematics classroom" (p. 142). We agree but would argue that one way to give explicit attention to student beliefs in the mathematics classroom is to be deliberate about initiating the negotiation of classroom norms.

6. ACKNOWLEDGEMENTS

The research reported in this chapter was supported in part by the National Science Foundation under grant No. REC-9875388. The opinions expressed do not necessarily reflect the views of the foundation.

7. NOTES

[1] The colleagues with whom we have worked over the past decade or more include Paul Cobb, Koeno Gravemeijer, Terry Wood, Grayson Wheatley and Diana Underwood. Paul Cobb and Erna Yackel developed the interpretive framework that forms the basis for this chapter.

[2] A detailed discussion of the nature of symbolic interactionism and its methodological position can be found in Blumer (1969).

[3] In 1973 Erlwanger published a study in which he investigated the beliefs of sixth-grade student, Benny. Benny was making good progress in mathematics using an approach to instruction that was based on individualized instructional technology. Erlwanger found upon talking with Benny that, despite his good progress, he had understood incorrectly some aspects of his work. In particular, he had many misconceptions about decimals and fractions and how to operate with them. In addition, Erlwanger found that Benny had developed learning habits and views about mathematics that would impede his progress in the future. Erlwanger concluded that the type of instruction Benny received "tends to develop in the pupil an inflexible rule-oriented attitude toward mathematics, in which rules that conflict with intuition are considered 'magical' and the quest for answers 'a wild goose chase'" (p.25).

[4] The classroom teaching experiment methodology referred to in this chapter was developed by Cobb, Wood, and Yackel (1991) to extend to the classroom setting the type of one-on-one teaching experiment

that has been used extensively by Steffe and colleagues to investigate children's mathematical activity and learning (Cobb & Steffe, 1983). For a detailed discussion of the classroom teaching experiment methodology see Cobb (1999) and Yackel (1995b).

[5] In using the labels "school mathematics tradition" and "inquiry tradition" we are following Richards (1991) who characterizes the "school mathematics tradition" as one in which students are treated as passive recipients of information and the "inquiry mathematics tradition" as one designed to teach students the language of mathematical literacy. Richards likens the discourse in the school mathematics tradition to "a type of 'number talk' that is driven by computation" (p. 16). By contrast, discourse in the inquiry tradition involves discussions in which individuals interact to attempt to explain and justify their mathematical activity to one another (cf. Thompson, Philipp, Thompson, & Boyd's discussion of calculational versus conceptual orientation, 1994).

[6] Voigt (1996) argues that of the various theoretical approaches to social interaction, the symbolic interactionist approach is particularly useful when studying children's learning in inquiry mathematics classrooms because it emphasizes the individual's sense-making processes as well as the social processes. Rather than attempting to deduce an individual's learning from social and cultural processes or vice versa, symbolic interactionism sees individuals as developing their personal understandings as they participate in negotiating classroom norms, including those that are specific to mathematics.

[7] Such a position is open to empirical verification. Our results from prior analyses confirmed that this is a viable position.

[8] Chris Rasmussen was the course instructor. Erna Yackel attended every class session. They, together with mathematics educator Karen King, formed the project team for the classroom teaching experiment.

[9] This small class size, which is much smaller than the typical differential equations class in an American University, made the class ideal for conducting a teaching experiment. It was possible for the members of the research team to have a rather intimate knowledge of each class member's conceptual understandings as the semester progressed. This information contributed significantly to the instructor's ability to develop appropriate instructional activities and strategies. Subsequently, Rasmussen has successfully used the instructional approach and activities first developed in this class with class sizes of thirty or more students.

[10] Realistic Mathematics Education instructional design theory, developed at the Freudenthal Institute, The Netherlands (Gravemeijer, 1994) informed the instructional design.

[11] The textbook emphasized graphical and other qualitative approaches along with the use of technology as means to solve problems involving differential equations. By contrast, traditional textbooks emphasize a variety of analytic methods to solve various types of differential equations.

[12] One of the ways we can judge the usefulness of theoretical ideas is the extent to which they are applicable in practice. In this case, we have repeatedly experienced in our own mathematics classrooms the rapidity with which we can initiate the renegotiation of social norms for classroom participation within the first one or two class sessions.

[13] In contrast to the previous section where we first documented students' beliefs and then analyzed the constitution of the corresponding social norms, in this section we discuss how students' specifically mathematical beliefs and the corresponding sociomathematical norms evolved together as a dynamic system. Therefore we will not separate the documentation of students' mathematical beliefs from the analysis of sociomathematical norms.

8. REFERENCES

Bauersfeld, H. (1988). Interaction, construction, and knowledge: Alternative perspectives for mathematics education. In T. Cooney & D. Grouws (Eds.), *Effective mathematics teaching* (pp. 27-46). Reston, VA: National Council of Teachers of Mathematics/Erlbaum.

Bauersfeld, H., Krummheuer, G., & Voigt, J. (1988). Interactional theory of learning and teaching mathematics and related microethnographical studies. In H. G. Steiner & A. Vermandel (Eds.), *Foundations and methodology of the discipline of mathematics education* (pp. 174-188). Antwerp: Proceedings of the Theory of Mathematics Education Conference.

Blumer, H. (1969). *Symbolic interactionism.* Englewood Cliffs, NJ: Prentice-Hall.

Cobb, P. (1985). Two children's anticipations, beliefs, and motivations. *Educational Studies in Mathematics, 16,* 111-125.

Cobb, P. (1986). Contexts, goals, beliefs, and learning mathematics. *For the Learning of Mathematics, 6*(2), 2-9.

Cobb, P. (1999). Conducting teaching experiments in collaboration with teachers. In A. E. Kelly & R. A. Lesh (Eds.), *Handbook of research design in mathematics and science education* (pp. 307-333). Mahwah, NJ: Lawrence Erlbaum Associates.

Cobb, P., & Steffe, L. P. (1983). The constructivist researcher as teacher and model builder. *Journal for Research in Mathematics Education, 14,* 83-94.

Cobb, P., Wood, T., & Yackel, E. (1991). Learning through problem solving: A constructivist approach to second grade mathematics. In E. von Glasersfeld (Ed.), *Constructivism in mathematics education* (pp. 157-176). Dordrecht: Kluwer.

Cobb, P., Wood, T., Yackel, E., & McNeal (1992). Characteristics of classroom mathematics traditions: An interactional analysis. *American Educational Research Journal, 29,* 573-604.

Cobb, P., & Yackel, E. (1996). Constructivist, emergent, and sociocultural perspectives in the context of developmental research. *Educational Psychologist, 31* (3/4), 175-190.

Cobb, P., Yackel, E., & Wood, T. (1989). Young children's emotional acts while doing mathematical problem solving. In D. B. McLeod & V. M. Adams (Eds.), *Affect and mathematical problem solving: A new perspective* (pp. 117-148). New York: Springer-Verlag.

Cobb, P., Yackel, E., & Wood, T. (1993). Learning mathematics: Multiple perspectives, theoretical orientation. In T. Wood, P. Cobb, E. Yackel, & D. Dillon (Eds.), *Rethinking elementary school mathematics: Insights and issues. Journal for Research in Mathematics Education Monograph Series, Number 6,* (pp. 21-32). Reston, VA: National Council of Teachers of Mathematics.

Erlwanger, S. H. (1973). Benny's concept of rules and answers in IPI mathematics. *Journal of Children's Mathematical Behavior, 1,* 7-26.

Gravemeijer, K. (1994). Educational development and developmental research in mathematics education. *Journal for Research in Mathematics Education, 25,* 443-471.

Mehan, H. (1979). *Learning lessons: Social organization in the classroom.* Cambridge, MA: Harvard University Press.

Richards, J. (1991). Mathematical discussions. In E. von Glasersfeld (Ed.), *Constructivism in mathematics education* (pp. 13-52). Dordrecht, The Netherlands: Kluwer.

Schoenfeld, A. H. (1983, April). *Theoretical and pragmatic issues in the design of mathematical "problem-solving".* Paper presented at the Annual Meeting of the American Educational Research Association, Montreal.

Smith, J. E. (1978). *Purpose and thought: The meaning of pragmatism.* Chicago: University of Chicago Press.

Thompson, A. G., Philipp, R. A., Thompson, P. W., & Boyd, B. (1994). Calculational and conceptual orientations in teaching mathematics. In D. Achiele & A. F. Coxford (Eds.), *Professional development of teachers of mathematics* (pp. 79-92). Reston, VA: National Council of Teachers of Mathematics.

Verschaffel, L, Greer, B., & De Corte, E. (1999). Pupils' beliefs about the role of real-world knowledge in mathematical modelling of school arithmetic word problems. In E. Pehkonen & G. Törner (Eds.), *Mathematical beliefs and their impact on teaching and learning of mathematics: Proceedings of the workshop in Oberwolfach* (pp. 138-145). Duisburg, Germany: Gerhard Mercator Universität.

Voigt, J. (1985). Patterns and routines in classroom interaction. *Recherches en Didactique des Mathématiques, 6,* 69-118.

Voigt, J. (1996). Negotiation of mathematical meaning in classroom practices: Social interaction and learning mathematics. In L. P. Steffe, P. Nesher, P, Cobb. G. A. Goldin, & B. Greer (Eds.), *Theories of mathematical learning* (pp. 21-50). Mahwah, NJ: Lawrence Erlbaum Associates.

Wood, T., Cobb, P., & Yackel, E. (1988, April). *The influence of change in teacher's beliefs about mathematics instruction on reading instruction.* Paper presented at the annual meeting of the American Educational Research Association, New Orleans.

Yackel, E. (1995a). Children's talk in inquiry mathematics classrooms. In P. Cobb & H. Bauersfeld (Eds.), *The emergence of mathematical meaning* (pp. 131-162). Hillsdale, NJ: Lawrence Erlbaum Associates.

Yackel, E. (1995b). *The classroom teaching experiment.* Unpublished manuscript, Department of Mathematics, Computer Science, and Statistics, Purdue University Calumet, IN.

Yackel, E., & Cobb, P. (1996). Sociomathematical norms, argumentation, and autonomy in mathematics. *Journal for Research in Mathematics Education, 27,* 458-477.

Yackel, E., Cobb, P., & Wood, T. (1991). Small-group interactions as a source of learning opportunities in second-grade mathematics. *Journal for Research in Mathematics Education, 22,* 390-408.

CHAPTER 19

PESSIA TSAMIR AND DINA TIROSH

INTUITIVE BELIEFS, FORMAL DEFINITIONS AND UNDEFINED OPERATIONS: CASES OF DIVISION BY ZERO

Abstract. In this chapter we describe a study in which we explore secondary school students' adherence to the perform-the-operation belief in the cases of division by zero. Our aims were: (1) to examine whether secondary school students identify expressions involving division by zero as undefined or tend to perform the division operation, (2) to study the justifications given for their approach, and (3) to analyze the effects of age (grade) on their responses. A substantial number of the participants argued, in line with the perform-the-operation belief, that division by zero results in a number. This intuitive belief was also evident in the justifications of students who correctly claimed that division by zero is undefined. Performance on division by zero tasks did not improve with age. Possible causes and educational implications of these findings are described and discussed.

1. INTRODUCTION

In several previous chapters of this volume (e.g., Furinghetti & Pehkonen; Goldin; Op't Eynde, de Corte, & Verschaffel; Törner) various definitions of beliefs and different facets of this construct were described and discussed. We have chosen to focus on intuitive beliefs, a specific type of cognitive beliefs which were defined and discussed by Fischbein (see, for instance, Fischbein, 1987). Intuitive beliefs are particular, immediate forms of cognition which refer to statements and decisions which exceed the observable facts. According to Fischbein, the characteristics of intuitive beliefs are: self evidence (persons perceive them as being true and in need of no further justification), intrinsic certainty (they are associated with a feeling of certitude, of intrinsic conviction), perseverance (intuitive beliefs are robust), coerciveness (the individual tends to reject alternative interpretations, those that would contradict his or her intuitions), extrapolativeness (intuitive cognitions have the capacity to extrapolate beyond an empirical support), and globality (intuitive beliefs are accepted as structured, meaningful, unitary representations, as opposed to logically acquired cognitions which are sequential and analytical). For instance, beliefs that do not appear to the individual as self evident are not considered intuitive. Instances of non-intuitive beliefs can be found in various cases, one of which relates to the formula of quadratic equations (see, for instance, Fischbein, Tirosh, & Melamed, 1981). The belief that the formula for solving quadratic

G. C. Leder, E. Pehkonen, & G. Törner (Eds.), Beliefs: A Hidden Variable in Mathematics Education? 331-344.

equations is correct is not considered an intuitive belief. Fischbein explained that even those who have no doubt about the validity of this formula (i.e., are certain that this formula is correct), do not regard it as self-evident and therefore do not question, or reject the possibility that one of the minuses in the formula is changed to plus (Fischbein, Tirosh, & Melamed, 1981, p. 492-493).

In numerous publications, Fischbein described and discussed students' common, intuitive beliefs about mathematical concepts and operations. He discussed, at length, the crucial impact of intensive, accumulating experiences on intuitive beliefs. Fischbein argued that "intuition summarizes experience, offers a compact, global representation of a group of data" (1987, p. 12). In respect to intuitive beliefs about mathematical operations, Fischbein (Fischbein, Deri, Nello, & Marino, 1985) argued that initial experience with these operations leads to the development of specific, intuitive beliefs about each of the four, basic arithmetic operations (e.g., "multiplication makes bigger"). In this chapter we argue that children's experiences with mathematical operations during their first years of schooling lead to the formation of intuitive beliefs not only about each particular mathematical operation but also about the general characteristics of mathematical operations. These experiences, which primarily consist of performing manipulations and arriving at numerical solutions, inevitably give rise to the intuitive belief that when faced with a mathematical problem, a mathematical operation should be performed and that, moreover, this must result in an answer. We call this the perform-the-operation intuitive belief.

Various documented behaviors of students can be interpreted to reflect the impact of this intuitive belief. Here we shall briefly mention two. The first relates to simplifying algebraic expressions. It has been widely reported that students tend to conjoin or complete algebraic expressions (e.g., 3+2x=5x or 3+2x=5). Researchers attribute this tendency to students' cognitive difficulty with "accepting lack of closure" (e.g., Booth, 1988; Collis, 1975; Davis, 1975). These researchers argued that students expect the behavior of algebraic expressions to be similar to that of arithmetic expressions, that is to say, they interpret symbols such as "+"as actions to be performed and expect the solutions to be single term solutions. Another case in which it was reported that students regarded expressions including operational signs as "incomplete" was described in a study that examined students' conceptions of complex numbers. Students were found to be reluctant to accept expressions such as 3+2i as "complete" numbers (see, for instance, Tirosh & Almog, 1989). In both these instances, students' responses could be interpreted as evolving from the intuitive, perform-the-operation belief. We would like to suggest that students' reported tendency to "solve" division by zero expressions could be regarded as another instance of the perform-the-operation belief.

Division by zero is usually the first undefined mathematical operation that students encounter in their school studies. Clearly, the mere existence of an undefined mathematical term violates the intuitive, perform-the-operation belief. Adherence to this belief might result in assigning numerical values to expressions involving division by zero. A tendency to perform the operation of division by zero, and to arrive at numeric results was indeed found among elementary and middle school students, prospective teachers and teachers alike (Ball, 1990; Blake &

Verhille, 1985; Grouws & Reys 1975; Reys, 1974; Reys & Grouws, 1975; Tsamir, 1996; Wheeler & Feghali, 1983).

Secondary school students encounter various undefined mathematical expressions in their studies and are expected to apply their knowledge about division by zero in various situations, such as identifying the domains of given rational functions and finding the excluded values of various rational equations and inequalities. It is reasonable to assume that high-school students' accumulating experience with undefined expressions would encourage them to regard division by zero expressions as undefined. On the other hand, intuitive beliefs are known to continue to affect students' responses to mathematical tasks even after pertinent, formal learning. This would lead us, in contrast with the above, to expect that high school students continue to perform the division by zero operation. It is, therefore, important to examine the extent to which these students perform the operation when presented with division by zero expressions. Such an inquiry could extend our understanding of the relationship between intuitive beliefs and formal definitions.

Our initial attempts to explore secondary school students' responses to tasks involving division by zero are described in previous papers (Tsamir, Sheffer, & Tirosh, 2000; Tsamir & Sheffer, 2000) and in this chapter. Our aims were: (1) to examine whether students in secondary schools tend to perform the operation in cases of division by zero, (2) to study the students' justifications for their responses, and (3) analyze the effects of age (grade) on students' responses.

2. METHOD

2.1. Participants

One hundred and fifty-three students from Grades 9, 10, and 11 in a secondary urban school in Israel participated in this study.

According to the *Israel's National Mathematics Curriculum* (INMC), students in Grade 4 should be learning that "adding (or subtracting) zero from a given number results in the same number, multiplying a number by zero results in zero, but division by zero is forbidden. An expression like $5 \div 0$ is meaningless and the expression $0 \div 0$ is also meaningless" (Ministry of Education, 1988, p. 60). The INMC does not specify how to teach this topic in Grade 4, yet popular elementary textbooks include the case of $\frac{a}{0}$ for $a \neq 0$, and avoid the special case of $0 \div 0$.

Teachers' guides recommend the use of practical models to illustrate why division by zero is undefined. An expression such as $6 \div 2$ is interpreted either as partitive division (e.g., "Six candies are divided evenly among two children, how many candies will each child get?") or in measurement terms ("I have six candies. I gave two candies to each child. How many children received candies?"). The books advise teachers to illustrate how all attempts to use these models demonstrate that an expression such as $6 \div 0$ leads to the conclusion that the expression is meaningless (that is, it is impossible to distribute six apples among no children, and no apples can be given to any number of children, so there is no one, definite suitable solution).

The teachers' guides recommend that this practical, "intuitive" introduction be followed by formal, mathematically-based explanations, namely, to show that division by zero is undefined because the related multiplication sentence has no solution (e.g., if $18 \div 0= \_$ then $\_ \cdot 0=18$, but $\_ \cdot 0=0$, therefore $18 \div 0$ has no solution).

Division by zero in the INMC is referred to again in Grade 8. Teachers are advised to explain first that $\frac{a}{0}$ for $a \neq 0$ is undefined because any definition of this operation can not coexist with the definition of division as the inverse of multiplication (e.g., if $\frac{4}{0}=c$ then $c \cdot 0 = 4$, but $c \cdot 0=0$ for every c, therefore $\frac{4}{0}$ is undefined). Teachers are then expected to address $0 \div 0$, while emphasizing that although this expression satisfies the equation $c \cdot 0=0$, the equation is satisfied by any number. Thus the single-value requirement of mathematical operations (i.e., the requirement that each operation results in only one number) is not fulfilled. Therefore, $0 \div 0$ is undefined. It is recommended that students then find the excluded values of various rational equations such as $\frac{x}{x(x-1)}=30$. In secondary schools, students are expected to apply their knowledge about division by zero in various contexts. Starting from Grade 8 on, students consider division by zero when solving algebraic equations and inequalities and when finding domains and ranges of given functions; in Grade 10 to 12, division by zero is discussed in Linear Planning and in Calculus.

2.2. Instruments

A written, paper-and-pencil instrument including 20 multiplication and division expressions was developed for this study. The students were asked to "read each expression. Solve it, if you can. If not, explain why it is impossible to solve it." These instructions suggest that some items might not result in a numerical answer.

The following five categories of expressions were included in this part of the instrument:

1. Four division expressions of the type "$a \div 0$ ($a \neq 0$)" (i.e., $12 \div 0$, $\frac{14}{0}$, $\frac{6}{3-3}$, $4 \div 0$).
2. Three division expressions of the type "$0 \div 0$" (i.e., $0 \div 0$, $\frac{4 \cdot 0+3 \cdot 0}{0}$, $\frac{0 \cdot (2+3)}{0}$).
3. Seven division expressions of the type "$a \div b$ ($a \neq 0$, $b \neq 0$)" (i.e., $\frac{4-0}{0-4}$, $\frac{16-4}{4}$, $9 \div 3$, $\frac{15}{3}$, $\frac{1+1}{0-1}$, $\frac{18}{6}$, $\frac{7-1}{62}$).
4. Four division expressions of the type "$0 \div a$ ($a \neq 0$)" (i.e., $0 \div 6$, $0 \div 16$, $\frac{0 \cdot 4}{3}$, $\frac{0}{10}$).
5. Two multiplication expressions of the type 'a·0" (i.e., $0 \cdot 0$, $4 \cdot 0$).

We were mainly interested in students' responses to the seven expressions involving division by zero (Categories 1 and 2), as these responses could point to adherence to the perform-the-operation intuitive belief. The other 13 expressions were designed to serve several additional purposes: (1) to reduce the possibility of receiving automatic, "undefined" responses by mixing undefined division expressions with defined division expressions; (2) to ensure that poor performance on division by zero expressions did not result from general incompetence in computing multiplication and division expressions, and (3) to record overgeneralizations of "undefined" responses to expressions involving zero. Previous studies have indicated that a substantial number of elementary and middle school students overgeneralized their undefined responses to all division expressions involving zero (see, for instance, Grouws & Reys, 1975). The expressions included in Categories 4 and 5 were found to evoke similar overgeneralized responses, if present, among secondary school students.

The division expressions in the questionnaire were written in two standard division notations: $a \div b$ and $\frac{a}{b}$. In each of the four categories involving division expressions, some expressions were presented in the first division notation, others in the second. Past studies have indicated that elementary and middle school students perform better with the $a \div b$ notation (Grouws & Reys, 1975). We were interested in examining whether different notations have a similar impact on the performance of secondary school students.

2.3. Procedure

The questionnaire was administered to the 153 participants during a mathematics class session of about 60 minutes. Responses to the questionnaires were typically elaborate, and thus provided substantial information about students' reasoning. Individual interviews were conducted with 30 students who either only stated that a given expression was undefined, or accompanied an "undefined" response with an illegible justification. The interviews were conducted about a week after we completed the analysis of students' written responses to the questionnaire. At the beginning of the interview, the interviewee was asked to explain his or her "undefined" response to the questionnaire. In cases where more than one of the written responses of the interviewee were not elaborated, he or she was asked to explain only one of these responses. Each interview lasted about 15 minutes. The interviews were audio-recorded and transcribed. (Pseudonyms have been used throughout this chapter.)

3. RESULTS

The main part of this section is devoted to examining the students' responses to division by zero expressions. Before reporting the results, four comments are in order:

1. Students performed similarly on the two different notations of the division expressions ($\frac{a}{b}$ and $a \div b$). Hence, these two types of expressions will be treated together.
2. Participants responded correctly to the division expressions that do not involve zero (Category 3: an average of 99% correct). Thus, inadequate performance on

division involving zero expressions can not be explained as resulting from general incompetence in solving division expressions.

3. All the students correctly solved the multiplication expressions involving zero (Category 5).
4. Most students (about 86% on average) knew that $0 \div a = 0$, for $a \neq 0$ (Category 4). Thus, overgeneralization of "undefined" responses was not frequent among secondary school students (although the typical incorrect response was "undefined").

We will now describe students' reactions to division by zero expressions. First, we analyze students' written responses to the seven expressions involving division by zero. Then, we examine the justifications made by students who claimed that "division by zero is undefined", highlighting those that imply an impact of the perform-the-operation belief.

3.1. Responses to Division by Zero Expressions

The data shown on Tables 1 and 2 indicate that in all seven division by zero tasks, the majority of the participants correctly claimed that division by zero is undefined. From Table 1 it can be seen that the most common incorrect response to the four expressions concerning division of a non-zero number by zero, was zero. This was true for all grade levels. Other incorrect responses were the dividend and ∞. The manner in which the students wrote the "infinity" solutions (e.g., $12 \div 0 = \infty$), the fact that they provided no justifications for their responses, and their explanations during the interviews (e.g., "Infinity is the largest number", "Infinity is larger than any other number") suggest that they regarded ∞ as a specific, numeric answer. The only incorrect response to the three division of zero-by-zero expressions in all grade levels was zero (see Table 2). In the seven division by zero tasks, the differences among the responses of students in different grades were not statistically significant.

3.1.1. Justifications to "Division by Zero is Undefined" Responses

We identified three types of justifications to the "division by zero is undefined" response: It is impossible to divide by zero; division by zero is sometimes defined and sometimes undefined, and, division by zero results in a "problematic number". We shall describe each of these responses. Specific attention will be given to the extent to which the responses testify to the impact of the intuitive, perform-the-operation belief.

3.1.2. It is Impossible to Divide by Zero.

In this category we include mathematically-based, practically-based and rule-based justifications to "undefined" responses to division by zero expressions. The mathematically-based justification for the claim that division of a non-zero number by zero is undefined incorporates a notion of division as the inverse of multiplication. Rina (Grade 9) provides a justification referring to this notion:

It is impossible to perform 6÷0 because division is defined as the inverse of multiplication. Thus, if 6÷0 is the number a, then a·0=6, but a·0=0 therefore division by zero is undefined.

Table 1. Percentages of responses to a÷0 (a≠0) expressions by grade and level of mathematics achievement (in %)

		Grades			Total
		9	10	11	
	n=	39	58	56	153
12÷0	**Undefined**	**77**	**72**	**71**	**74**
	0	15	10	21	15
	12	5	10	3	6
	∞	0	3	5	3
$\frac{14}{0}$	**Undefined**	**86**	**72**	**70**	**76**
	0	14	13	19	16
	∞	0	0	3	1
$\frac{6}{3-3}$	**Undefined**	**77**	**67**	**69**	**70**
	0	18	10	12	13
	6	0	10	5	6
	∞	0	3	3	2
4÷0	**Undefined**	**82**	**76**	**76**	**77**
	0	15	13	19	16
	∞	0	0	3	1

Table 2. Percentages of responses to 0÷0 expressions by grade and level of mathematics achievement (in %)

		Grades			Total
		9	10	11	
	n=	39	58	56	153
0 ÷ 0	**Undefined**	**62**	**53**	**61**	**56**
	0	38	47	39	44
$\frac{4\cdot 0+3\cdot 0}{0}$	**Undefined** 0	**64** 36	**60** 40	**66** 34	**60** 40
	$\frac{0\cdot(2+3)}{0}$	**64** 36	**63** 37	**69** 31	**63** 37

The mathematically-based justification for the claim that 0 ÷ 0 is undefined referred to the notion that division is an operation; as such, it should fulfill the single-value requirement. Alon (Grade 10) was among the students who provided such a justification:

> $0 \div 0$ is undefined because if $0 \div 0 = x$, then $x \cdot 0 = 0$. But this is true for every number. Division is an operation and therefore the answer to $0 \div 0$ must be one number. Therefore it is undefined.

Rina and Alon argued that division by zero is undefined by hypothetically performing the operation and then realizing that any numerical solution would violate either the definition of division as the inverse of multiplication or the single answer requirement. They thus concluded that this division by zero could not be performed.

The practically-based justifications applied the notion of division as sharing and zero as nothing. Typical responses were given by Sara (Grade 9) and Amit (Grade 10):

Sara: I'll refer to division, to $8 \div 2$. This means 8 apples are divided between two kids, and that each kid gets 4 apples. Then, $8 \div 1$ means that only one kid takes the 8 apples. So, $8 \div 0$ is impossible. It is impossible to do that; we are not performing an operation, and therefore it is undefined.

Amit: $0 \div 0$ is undefined, because it is impossible to perform such a division. It is meaningless to divide zero apples among no children.

Sara and Amit pointed out that it is impossible to divide eight apples, or no apples, among no children. They referred to the impossibility of practically performing the operation.

Rule-based justifications to both the claim that division of a non-zero number by zero is undefined and to the claim that $0 \div 0$ is undefined revealed students' views that division by zero is undefined because of an existent rule stating this principle. We interviewed two students (Betty, Grade 9 and Vered, Grade 11) who provided a response of this type:

Betty: My teacher explicitly emphasized that there is such a rule in mathematics, that division by zero is undefined.
I: Why is this the rule?
Betty: That's the way it is. It's like an axiom.
I: But why?
Betty: The question is not 'Why is this the rule?' You just have to know the rule. Clever mathematicians make rules and we should memorize them. The problems we want to solve by applying the rules are what we have to understand.

In this interview, Betty expressed her view that clever mathematicians established the rules to be the way they are. She argued that these rules should be memorized and that there is no need to understand why these are the rules. The other student, Vered, voiced a more extreme position, claiming that there is no point in trying to understand why mathematical rules are set the way they are as some of them are illogical. During an interview she explained herself:

> It is not allowed to divide by zero. In mathematics we have rules, and we operate according to them. These rules often do not seem reasonable. For instance, it is illogical

that minus times minus is plus. When studying mathematics, we have to obey the rules and to work with them. There is no point at all in looking for explanations. One just has to accept them.

3.1.3. Division by Zero is Sometimes Defined and Sometimes Undefined.
Another justification, provided in interviews by students who had merely responded: "undefined" in the questionnaire, was that division by zero is undefined in certain domains (e.g., arithmetic, algebra), while it is defined, as infinity, in others (e.g., calculus). Ruth (Grade 11) was one of these students:

Ruth:	For certain topics you use the word 'undefined' when talking about division by zero. You do that in arithmetic, or even in algebra, but in calculus you don't.
I:	What is division by zero in calculus?
Ruth:	It tends to infinity
I:	Can you explain a bit more?
Ruth:	Yes. If I take 12 and divide it first by 1, and then by 1/2, by 1/4 and so on, I get larger and larger numbers. So, division by zero tends to infinity.
I:	But in your written answer you wrote that it is undefined.
Ruth:	Yes. It is 'locally undefined'. It is undefined in arithmetic and in algebra, but defined in calculus.
I:	Can you explain yourself?
Ruth:	In calculus division by zero is infinity. I know that it seems problematic, that it appears not to make any sense, but division by zero can be defined and also undefined.
I:	How would you answer my original question now?
Ruth:	Division by zero is locally undefined.

In this and in other interviews, interviewees expressed the idea, in various ways, that division by zero is "locally undefined". When the interviewer probed further into their "in calculus, division by zero tends to infinity" or "division by zero is infinity" responses, it was revealed that the majority of these students performed a series of division operations, starting from 1 and choosing smaller and smaller divisors (e.g., 1, 1/2, 1/4,...). They explained that division by smaller and smaller numbers tends to infinity and therefore division by zero is infinity. In these justifications, there was clear evidence that the "tending to infinity" notion was based on performing the division operation, and that division by zero was assumed to equal infinity.

3.1.4. Division by Zero Results in a "Problematic Number".
Two types of justifications are included under this heading. The first, "division by zero is undefined because it is infinity, and infinity is undefined" was given for division of non-zero numbers by zero. The second, "0 ÷ 0 is undefined because it is "zero-undefined" was given to 0 ÷ 0 expressions. In an attempt to deepen our understanding of these written responses, we interviewed several students who provided an "infinity-undefined" and "zero-undefined" responses. Excerpts from three of these interviews, with Avner, (Grade 10), Ravit, (Grade 10), and Orit, (Grade 11) are given below:

I:	In the questionnaire, you wrote: '12÷0 is undefined because it is infinity, and infinity is undefined.' Can you explain your answer?

Avner: What I meant was that if you divide $12 \div 1 = 12$, $12 \div \frac{1}{10} = 120$, $12 : \frac{1}{100} = 1200$, and so on, the numbers get larger and larger, and at zero they reach infinity.

I: Is infinity a number?

Avner: Yes, but it is undefined, like Π. We know how to write it, but we don't know its exact location on the number line ...

Ravit provided the following explanation to the same question:

Ravit: In physics, when we talked about lenses, we learned that infinity could be any number. The same happens here: When you get closer to zero, you get different larger and larger numbers.

I: Can you say a bit more about that?

Ravit: Yes. If I take 5 and divide it first by 2, and then by 1, and then by 1/2, by 1/4 and so on, I get bigger and bigger numbers. So, division by zero is infinity.

I: Is infinity a number?

Ravit: Yes. But it is not one specific number, like 4 or 7. It could be any number. Sometimes infinity is 4, sometimes it is 7, and sometimes it is a very large number.

Avner and Ravit both related to successive division processes. For both students, division by zero could be performed and the result would be a number. This number (∞) was undefined either because its exact location on the number-line was unknown, or because its value was not fixed. Hence, for these students, responding that division by zero is undefined did not contradict the perform-the-operation intuitive belief, because they performed the operation and attached "the undefined number ∞".

A somewhat similar tendency, exhibited by students who claimed that zero divided by zero is undefined, was based on a perception of zero as an undefined number. Following is a fragment taken from this interview with Orit:

I: Orit, you wrote that the expression [pointing at $0 \div 0$] is 'undefined' and you also wrote 'it is zero – undefined'. Can you explain this to me?

Orit: Yes. It is impossible to divide by zero. But zero divided by a number is zero. Therefore, when we have $0 \div 0$ the answer is zero-undefined.

I: I'm not sure I follow you. Can you explain yourself in a different way?

Orit (thinking): Zero is a strange number, and many strange things happen with it. Last week in class we learned that $0!=1$. This is strange, but zero is there. Also, zero is neither negative nor positive. So, for $0 \div 0$, the answer is zero. But in mathematics, we say that it is undefined to show that it is impossible to divide by zero.

Interviews with other students who provided the same response (i.e., "zero - undefined") concluded similarly. Most of these interviewees were convinced that $0 \div 0$ is the number zero, but because dividing by zero is problematic, it seemed essential to them to indicate that the answer is "zero-undefined". Evidently, these students performed the division-of-zero-by-zero operation, and obtained the solution zero,

4. DISCUSSION AND CONCLUSIONS

It is widely documented that intuitive beliefs about mathematical operations affect students' thinking processes. Such intuitive beliefs are often incompatible with formal mathematical definitions and theorems. In these situations, intuitive beliefs tend to determine students' responses to mathematical tasks (e.g., Fischbein, Deri, Nello, & Marino, 1985; Fischbein, 1987; Greer, 1992; Hart, 1981; Tirosh & Graeber, 1991). This study aimed at exploring two main issues regarding a particular intuitive belief, namely, what we termed the perform-the-operation belief: (1) Do secondary school students tend to perform the division by zero operation and to assign numerical values to division by zero expressions? And (2) What are the effects of age (grade) on students' responses to such expressions?

The findings indicate that about a third of the participating students performed the division of a non-zero number by zero and reached zero, the dividend or "the number infinity". The claim that $0 \div 0 = 0$ was more frequent: about 40% of the students came up with this response. Our data also show that the performance of secondary school students on division by zero tasks did not improve with age (grade).

It may correctly be argued that our results show that most participants in this study knew that division by zero is undefined. This may then lead to the conclusion that students who encounter various instances of undefined operations, and who apply their formal knowledge related to the non-definition of division by zero in various situations, which contradict the perform-the-operation belief, can resist the impact of this intuitive belief. This conclusion might, however, be too hasty. During interviews, students who gave "infinity-undefined" and "zero-undefined" responses indicated that they performed the division by zero operation, arriving at either infinity or zero. They further explained that infinity and zero are problematic, and therefore undefined. Hence, although these students performed the division by zero operation, they argued that the resulting number is undefined. Such somewhat unexpected responses may result from conscious or unconscious attempts on the part of the student to reconcile the apparent contradiction between the perform-the-operation intuitive belief and the formal, mathematical decision not to define division by zero.

The response "division by zero is locally undefined" may also reflect an attempt to reconcile the apparent contradiction between the perform-the-operation intuitive belief and the non-definition of division by zero. Students who applied the "locally-undefined" notion claimed that it is impossible to divide by zero in certain mathematical domains but that such an operation can be performed in others. Clearly, these students had performed the division by zero operation, at least in one mathematical domain. Once again, we learned that percentages of correct and incorrect responses to given tasks tell only part of the tale, and may often lead researchers to inadequate conclusions. Impact of the perform the operation belief was evident among students who argued that division by zero is undefined.

Another major finding of the study is that secondary school students' performance on division by zero tasks did not improve with age (grade level). Students in higher grades have more experience with mathematical definitions and theorems; they have been exposed to more cases of undefined expressions and applied their knowledge that division by zero is undefined in more situations. Thus, one would assume that in comparison to students in lower grades, they are better equipped to control for the effects of the perform-the-operation intuitive belief and to rely only on formal definitions when responding to division by zero tasks. This, however, was not the case. Our data indicate that in the case of division by zero, as in many other instances in which formal, mathematical definitions contradict intuitively-based beliefs, the frequencies of correct responses are relatively stable across ages (e.g., Fischbein, Tirosh, & Hess, 1979; Fischbein, 1987).

Such a situation naturally raises the following question: How can learners overcome the effect of the perform-the-operation intuitive belief and accept that some mathematical expressions are undefined? We should admit at the outset that this task is indeed complicated. In his writings, Fischbein argued that "I see the solution only in resorting to metacognitive procedures. Students should be helped to consider consciously all the implications of the formal definitions taught and to try to apply them to particular cases, especially those known by the teacher to produce certain difficulties" (1994, p. 106). Similar ideas have been expressed by other researchers (e.g., Garofalo & Lester, 1985; Schoenfeld, 1985, 1988; Vinner, 1991).

With respect to the cases of division by zero it is important, in class, to relate explicitly to the perform-the-operation intuitive belief, to discuss its possible sources, and to demonstrate its impacts on our reasoning processes. Other common intuitive beliefs about mathematical operations, e.g., that addition and multiplication "make bigger", or that division "makes smaller" (e.g., Fischbein, Deri, Nello, & Marino, 1985) could be addressed as well. This approach can lead to a more comprehensive discussion about the impact of intuitive beliefs on our thinking processes, the differences between intuitive beliefs and formal definitions, the need to analyze and control these intuitive beliefs and the necessity of leaning on formal, mathematical definitions. Together with Fischbein (1987, 1999) and Vinner (1991) we would argue that a major aim of mathematics education is to develop students' ability to analyze their intuitive conceptions and keep them under control, as well as to assist our students in building new intuitions, consistent with the mathematical definitions.

Another issue that should not be overlooked relates to secondary school students' views of mathematics. Rule-based views of mathematics were evident in responses such as: "There is a mathematical rule stating that division by zero is undefined and there is no need to understand why this rule had been devised just so", or, "Mathematical rules often do not appear to be reasonable". Such views contrast sharply with the view of mathematics that various reform documents encourage, namely, that mathematics is a challenging and continuously expanding discipline whose logic can be grasped (Australian Education Council, 1991; Ministry of Education, 1988; National Council of Teachers of Mathematics [NCTM], 1989, 2000). These documents emphasize the importance of challenging students' non-productive, rule-based views of mathematics, and of encouraging them to appreciate mathematics as a human-made discipline. Moreover, there is increasing evidence that the beliefs that students hold about mathematics affect their understanding of mathematical concepts and their performance on mathematical tasks (see, for instance, Garofalo, 1989; Mcleod, 1992; Schoenfeld, 1985, 1988; Tirosh & Graeber, 1994).

In a recently published study, Szydik (2000) reported that students with external sources of conviction for the determination of mathematical truth (i.e., appeal to authority) gave more incoherent or inappropriate definitions of limit, held more misconceptions of limit as bound or unreachable, and were less able to justify limit calculation than those with internal sources of conviction (i.e., appealing to empirical evidence, logic and consistency). The interviews that we conducted with students who exhibited rule-based views of mathematics suggest that they tended to rely on external sources of conviction ("My teacher explicitly emphasized that.. . .", "Rules are made by clever mathematicians"). Thus, it seems worthwhile to encourage students to explore some general issues related to the decision not to obey the perform-the-operation intuitive belief in the case of division by zero. This approach should also be adopted with respect to other issues regarding undefined expressions (e.g., Why are division by zero expressions doomed to be undefined? What other operations are bound to be undefined, and why? Is there a general rule by which mathematical operations are either defined or undefined?) as well as to those issues pertaining to definition of mathematical operations (e.g., How do mathematicians make decisions about the definitions of mathematics operations? What are the main properties of mathematical operations?). This would encourage students to appreciate more fully mathematics as a human-made discipline.

In the introduction, we stated that this chapter focuses on the perform-the-operation intuitive belief about mathematical operations. We studied the impact of this intuitive belief on secondary school students' responses to, and justifications for, division by zero tasks. This intuitive belief appears to originate in arithmetic. At first glance, it might be classified as a subject matter belief, according to Törner's hierarchies of beliefs (Törner, this volume). However, as stated in the introduction to this chapter, possible impacts of this intuitive belief are reflected in students' responses to tasks embedded in various mathematical domains (e.g., algebra). We also found that students' responses and justifications for division by zero tasks were related to their views of specific mathematical concepts (e.g., the number zero, infinity) and to their global beliefs of mathematics. We would like to end this chapter with a call for further research on the nature of intuitive beliefs, and the connections between these beliefs and other types of both cognitive and affective beliefs.

5. REFERENCES

Australian Education Council (1991). *A National Statement on Mathematics for Australian Schools.* Melbourne, Australia: Curriculum Corporation.

Ball, D. (1990). Prospective elementary and secondary teachers' understanding of division. *Journal for Research in Mathematics Education, 21*(2), 132-144.

Blake, R., & Verhille, C. (1985). The story of 0. *For the Learning of Mathematics, 5*(35), 35-47.

Booth, L. (1998). Children's difficulties in beginning algebra. In A. Coxford (Ed.), *The Ideas of Algebra: K-12* (1988 yearbook) (pp. 20-32). Reston, VA: National Council of Teachers of Mathematics.

Collis, K. (1975). *The Development of formal reasoning*. Newcastle, Australia: University of Newcastle.

Davis, R. (1975). Cognitive process involved in solving simple algebraic equation. *Journal of Children's Mathematical Behavior, 1*, 7-35.

Fischbein, E. (1987). *Intuition in science and mathematics.* Dordrecht, Holland: Reidel.

Fischbein, E. (1994). Tacit models. In D. Tirosh (Ed.), *Implicit and explicit knowledge: An educational approach* (pp. 96-110). Norwood, New Jersey: Ablex.

Fischbein, E. (1999). Intuitions and schemata in mathematical reasoning. *Educational Studies in Mathematics, 38*, 11-50.

Fischbein, E., Deri, M., Nello, M., & Marino, M. (1985). The role of implicit models in solving verbal problems in multiplication and division. *Journal for Research in Mathematics Education, 16*, 3-17.

Fischbein, E., Tirosh, D. & Hess, P. (1979). The intuition of infinity. *Educational Studies in Mathematics, 10*, 3-40.

Fischbein, E., Tirosh. D., & Melamed, U. (1981). Is it possible to measure the intuitive acceptance of a mathematical statement? *Educational Studies in Mathematics, 12*, 491-512.

Garofalo, J. (1989). Beliefs and their Influence on mathematical performance. *Mathematics Teacher, 82*, 502-505.

Garofalo, J., & Lester, F. (1985). Metacognition, cognitive monitoring, and mathematical performance. *Journal for Research in Mathematics Education, 16*, 163-176.

Greer, B. (1992). Multiplication and division as models of situations. In D. A. Grouws (Ed.), *Handbook of research on mathematics teaching and learning* (pp. 276-295). New York: Macmillan.

Grouws, D., & Reys, R. (1975). Division involving zero: an experimental study and its implications. *Arithmetic Teacher, 22*, 74-80.

Hart, K. (1981). *Children's understanding of mathematics: 11-16.* London: John Murray.

Mcleod, D. (1992). Research on affect in mathematics education: A reconceptualiztion. In D. A. Grouws (Ed.), *Handbook of research on mathematics teaching and learning* (pp. 575-596). New York: Macmillian.

Ministry of Education. (1988) *The National Mathematics Curriculum.* Jerusalem, Israel: Ministry of Education. (Hebrew).

National Council of Teachers of Mathematics. (1989). *Curriculum and evaluation standards for school mathematics.* Reston, VA: Author.

National Council of Teachers of Mathematics. (2000). *Principles and standards for school mathematics.* Reston, VA: Author.

Reys, R. (1974). Division by zero: An area of needed research. *Arithmetic Teacher,21*, 153-157.

Reys, R., & Grouws, D. (1975). Division involving zero: some revealing thoughts from interviewing children. *School Science and Mathematics, 78*, 593-605.

Schoenfeld, A. (1985). *Mathematical problem solving.* New York, NY: Academic Press.

Schoenfeld, A. (1988). "When good teaching leads to bad results: the disasters of "well taught" mathematics classes." *Educational Psychologist, 23*, 145-166.

Szydik, J. (2000). Mathematical beliefs and conceptual understanding of the limit of a function. *Journal for Research in Mathematics Education, 31*, 258-276.

Tirosh, D., & Almog, N. (1989). Conceptual adjustments in progressing from real to complex numbers. In G. Vergnaud, J. Rogalski, and M. Artigue (Eds.), *Proceedings of the 13th International Conference on the Psychology of Mathematics Education* Vol. 3 (pp. 221-227). Paris, France.

Tirosh, D., & Graeber, A. (1991). The effect of problem type and common misconceptions on preservice elementary teachers' thinking about division. *School Science and Mathematics, 91*, 157-163.

Tirosh, D., & Graeber, A. (1994). Implicit and explicit knowledge: The case of multiplication and division. In D. Tirosh (Ed.), *Implicit and explicit knowledge: An educational approach* (pp. 111-130). Norwood, New Jersey: Ablex.

Tsamir, P. (1996) *Teaching prospective teachers about division by zero.* Unpublished manuscript, Tel-Aviv University. Tel-Aviv. (Hebrew).

Tsamir, P., Sheffer, R., & Tirosh, D. (2000). Intuitions and undefined operations: The case of division by zero. *Focus on Learning Problems in Mathematics*, 22(1), 1-16.

Tsamir, P., & Shefffer, R. (2000). Concrete and formal arguments: The case of division by zero. *Mathematics Education Research Journal. 12*, (2), 92-106.

Vinner, S. (1991). The role of definitions in teaching and learning of mathematics. In D. Tall (Ed.), *Advanced mathematical thinking* (pp. 65-78). Dordrecht, Holland: Kluwer.

Wheeler, M., & Feghali, I. (1983). Much ado about nothing: preservice elementary school teachers' concept of zero. *Journal for Research in Mathematics Education, 14*(3), 147-155.

CHAPTER 20

FRANK K. LESTER, JR.

IMPLICATIONS OF RESEARCH ON STUDENTS' BELIEFS FOR CLASSROOM PRACTICE

Abstract: This chapter offers a critical analysis of each of the five reports related to students' beliefs about mathematics, highlighting the most salient aspects of each report. In addition to assessing each report, an argument is presented for the importance of studying students' beliefs about the nature of mathematics, how it is learned, and how the subject is taught.

It is fitting that I begin this brief paper with a disclaimer; namely, that I am not currently engaged in research on students' beliefs about anything, let alone their beliefs about the nature of mathematics, the teaching of mathematics, or the learning of mathematics (although I have contributed to this area in previous years). In fact, although I regard research in this area as quite important, I am now an amateur when it comes to talking about beliefs and how they are related to mathematics learning and teaching. Imagine, then, my surprise when I was invited to write a synthesis of the set of research reports dealing with students' beliefs about mathematics. In spite of my amateur status as a beliefs researcher, I am an experienced researcher and a past editor of a leading research journal and I have given considerable thought over the years to issues associated with the role beliefs play in mathematics learning and teaching. So, presumably the editors of this volume thought it would be interest to those involved in research on beliefs in what I have to say about the research represented by this set of reports.

In preparation for synthesizing these reports, I read what various individuals whose work on beliefs I know well have to say about beliefs. Specifically, I read syntheses of the beliefs research prepared by McLeod (1992, 1994) and Pehkonen and Törner (1998), and I re-read a paper I co-authored with two colleagues some years ago (Lester, Garofalo, & Kroll, 1989). So, although my background reading has a decidedly American flavor, it has been enhanced by the addition of a Finish perspective.

My remarks are organized into three sections. In the first section, I briefly discuss a fundamental problem inherent in the study of people's beliefs and I offer some suggestions for solving it. In the second section, I provide a short analysis of each of the five reports, highlighting what I consider the most salient aspects of each report. In the final section, I offer some thoughts about why we should care about

G. C. Leder, E. Pehkonen, & G. Törner (Eds.), Beliefs: A Hidden Variable in Mathematics Education? 345-353.

the beliefs our students hold about the nature of mathematics, how it is learned, and how the subject is taught.

1. A FUNDAMENTAL PROBLEM FACING BELIEFS RESEARCHERS

In the history of mathematics education research, research on students' beliefs is relatively new; it has been only during the past 15 to 20 years that beliefs research has come to be viewed as an essential ingredient in mathematics education research programs (McLeod, 1992). Indeed, in a comprehensive review of American and Canadian research in mathematics education reporting on research conducted primarily in the 1960s and 1970s, the word "beliefs" does not even appear in the index for this nearly 500-page volume (Shumway, 1980). As beliefs research has grown in importance, so too has the realization that the study of beliefs is extremely problematic and complex. A central difficulty is that the fundamental assumption undergirding much of this research rests on a shaky logical foundation. Specifically, a basic assumption is that beliefs influence peoples'—both students' and teachers'—thinking and actions. However, it is also often assumed that beliefs lie hidden and so can be studied only by inferring them from how people think and act. For researchers to claim that students behave in a particular manner because of their beliefs and then infer the students' beliefs from how they behave involves circular reasoning[1]. The reasoning goes something like this:

Question:	"How do you know that students' beliefs influence how they do mathematics?"
Answer:	"Because in our study students did mathematics in a certain way."
Question:	"But how do you know that the students' beliefs contributed to this behavior?
Answer:	"Because they would not have behaved this way if they did not hold these beliefs."

There are at least two ways to solve this problem. One solution is to insist that studies of beliefs involve very careful conceptual and methodological analyses. Such analyses will help to uncover reasoning fallacies. Another solution is to develop research methods to uncover the beliefs directly, rather than to infer them from students' actions. Of course this latter solution will result only from attention to the first solution. I was pleased to find that the authors of these reports were able to avoid the logical difficulty to which I referred above.

2. OVERVIEW OF THE STUDIES

Study 1. Do students' beliefs influence their interest in and motivation to learn mathematics? (Kloosterman)

Continuing a tradition within the American research community that takes for granted the overlapping, interactive nature of cognitive and affective aspects of mathematical behavior, Kloosterman reports on a study in which he documents the importance of including a range of types of beliefs when attempting to study students' interest in doing mathematics and motivation to learn it. He has been interested in the roles beliefs and motivation play in mathematics learning for several years. Thus, it is natural that he has now turned his attention to the factors that influence students' motivation to learn mathematics. Kloosterman argues that the study of beliefs is important because beliefs "can have a substantial impact on their interest in mathematics, their enjoyment of mathematics, and their motivation in mathematics" (Kloosterman, this volume). A particular strength of his work is his insistence upon establishing a sound conceptual framework based on contemporary psychological theories—attribution theory, self-worth theory, self-efficacy theory, and goal-orientation theory. Another strength is that, after several years of working on the development of Likert scales to measure students beliefs, attitudes, and motivations, he has come to recognize the limitations of this approach. His interview instrument is constructed in a way that allows him to avoid the circular reasoning fallacy because it asks students *directly* about their beliefs (about themselves as doers of mathematics, about what it takes to do well in mathematics, and about mathematics teaching), rather than trying to infer them from what students do. Unfortunately, because beliefs typically are elicited in a context, it is too bad that his instrument did not ask the students questions with respect to specific mathematics task. As a result, he is unable to make any claims about the influence students' beliefs has on their motivation to do mathematics. Notwithstanding these concerns, in my view, Kloosterman has made a good start by taking pains to situate his work in current psychological theories.

Study 2. Why don't students worry about whether their solutions to word problems make sense? (Greer, Verschaffel, & De Corte)

The research reported in this chapter continues a systematic, long-term program of exemplary research on elementary and middle grades students' solutions to word problems (also referred to as "story problems") begun by Verschaffel and DeCorte about 15 years ago. The addition of Greer makes this team formidable indeed! A particularly appealing aspect of their research is their realization that contextual factors affect students' beliefs about mathematics and how it is learned and that students' particular beliefs are part of a system of beliefs about the nature of mathematics and how it relates to their world. For example, in one of my early research studies, I found that many third graders believed that all story problems could be solved by applying the operations suggested by the key words present in the story (Lester & Garofalo, 1982). For some students this belief was so strong that they used this key word strategy even when it was clearly inappropriate. These students did not bother to attempt to understand relationships expressed in problems, to monitor their actions, or to assess the reasonableness of their results, because they saw no need to do so. Perhaps the most disturbing finding was that some of the students stated that their teachers had taught them to look for key words in story problems. Furthermore, an examination of the students' mathematics textbook revealed that all of the story problems in their textbook could be solved correctly using this strategy. Here we have an example of two contextual factors—the teacher and instructional materials—forming a quite harmful belief.

Another attractive feature of this study is the desire of Greer and his colleagues to track down the source of the students' belief that there is no need to make sense of story problems. They decided to look for this source outside the purely cognitive realm and, in particular, to look at the role teachers play in the development of students' tendency to disregard sense making when they work to solve word problems whose solutions necessitate realistic considerations. They correctly point out that there is no direct evidence that teachers' beliefs are responsible for students' inclination to overlook or disregard their real-world knowledge when they attempt to solve word problems, but they also correctly note that it makes sense to wonder if students beliefs can be changed through instruction.

A final feature of their report, and perhaps the most controversial, is their "radical proposal whereby word problems become reconceptualized as exercises in mathematical modelling" (Greer, Verschaffel & DeCorte, this volume). In my view, this proposal represents a natural evolution in how we think about problem solving and its relation to other forms of mathematical activity. In a recently prepared paper, Lester and Kehle (in press) have made this very same proposal by suggesting that for research purposes problem solving should be considered from a mathematical-modeling perspective. Thus, I will follow with considerable interest the further work in this direction of these researchers.

Study 3. Do students' beliefs influence their ability to connect real-world and school mathematics? (Presmeg)

In this report, Presmeg discusses two research projects that examined the ontological beliefs about mathematics of students and the possibility of changes in such beliefs that resulted from a variety of mathematical experiences provided in classes she taught. She offers evidence that she says supports her claim that students' beliefs about the nature of mathematics enable or constrain the process of constructing connections between home or cultural activities and school or college mathematics topics. The "evidence" she provides is in the form of personal reports of teachers' and students' changes in their beliefs. But, how much confidence can we have in personal verbal reports? The problem is that it is difficult to determine whether the teachers' and students' beliefs actually changed as a result of the experiences or if they simply told the interviewers what they thought the interviewers wanted to hear. Moreover, we know nothing about the teachers' and students' degree or strength of belief (Goldman, 1999). That is to say, we have not learned whether the newly-held beliefs are held strongly enough to influence the teachers' and students' behavior. I mention this concern because beliefs are notoriously resistant to change, even in the face of overwhelming evidence to the contrary. So, the skeptic in me wonders: Did the subjects in Presmeg's studies really change their core beliefs about mathematics or did they merely learn that they should hold certain beliefs?

Of special note is Presmeg's use of "semiotic chaining" to help students create connections between school/college and non-school mathematics topics. The use of semiotic analysis as a methodological tool in educational research has grown tremendously in popularity in recent years, but this report is the first instance I have read of using semiotic chaining as an instructional device in mathematics classrooms. I find this notion intriguing because semiotics is primarily about meaning making, and an individual's beliefs are, after all, the results of her or his attempts to make sense of the world. Semiosis is the cognitive process of structuring our experiences and making sense of the world. As Lemke (1997) puts it:

> Semiosis is selective contextualization; it is making something meaningful by seeing it as a part of some wholes rather than others, as being an alternative to some options rather than others, as being in some particular relation to some things rather than others. (p. 39-40)

Although Presmeg provides too little specific information about the way she helped students in her classes construct chains of signifiers to shed light on the new meanings they saw in mathematics as practiced in and out of school, I think this technique has tremendous potential as a research tool in the study of the formation of beliefs.

Study 4. How do classrooms influence the development of beliefs? (Yackel & Rasmussen)

Over the past dozen years or so, Yackel and her various collaborators have developed an interpretive framework for analyzing classrooms that coordinates both psychological and sociological perspectives. Fundamental to this framework is the assumption that students' beliefs are essentially cognitive in nature; that is, beliefs are students' understandings of things. The authors claim that by coordinating the psychological (i.e., student beliefs) and sociological (i.e., classroom social and sociomathematical norms) perspectives it is possible to develop ways to explain how changes in beliefs might be formed and nurtured in mathematics classrooms. They illustrate this thesis with the case of a university mathematics class on differential equations. An important conclusion they make is that if teachers wish deliberately to affect their students' beliefs about mathematics, they must give direct attention to the negotiation of classroom norms.

There is no doubt that beliefs come from somewhere. A person does not simply wake up one morning and say "I believe in God" without something or some experiences having led her or him to this belief. Similarly, students do not spontaneously believe that they should make an effort to make sense of (i.e., understand) the mathematics they are learning. Such a belief develops because the students have experienced something that has led them to conclude that sense making is important (in this classroom). Furthermore, it is natural to conclude that at least some of these experiences have taken place in their mathematics classrooms and it is clear that the current state of the students' beliefs about mathematical activity in the classroom constrains the nature of the interactions that occur. Yackel and Rasmussen's report is a compelling example of how classroom norms influence the development of beliefs (at least beliefs about what sorts of behaviors are expected in the classroom in this study). It also challenges us to address a fundamental question: Why should we care about the beliefs students have about what is expected of them in the classroom?

Study 5. How do students' intuitive beliefs about mathematical operations affect their thinking processes? (Tsamir & Tirosh)

The authors of this report note that students' "intuitive" beliefs about mathematical operations affect their thinking processes. They define "intuitive beliefs" to be "particular, immediate forms of cognition that refer to statements and decisions which exceed the observable facts" (Tsamir & Tirosh, this volume). My own interpretation of this definition is to think of intuitive beliefs as "personally-held beliefs" that may be empirically warranted, but often on the basis of an inadequate set of observations. So, an intuitive belief may or may not be "true" and may or may not be justified. Common examples of erroneous intuitive beliefs from research into the beliefs of elementary and middle grades students include: (a) "You can't take a larger number from a smaller one," (b) "Multiplication always results in a larger number," and (c) "[In a story problem] 'left' means 'take away'."

I comment on the meaning of intuitive belief because I think Tsamir and Tirosh's distinction is quite similar to the distinction made between "concept image" and "concept definition" first introduced by Vinner and Hershkowitz (1980). Research on the differences between students' concept images and the formal concept definition has proved to be very fruitful in informing us about the development of students' mathematics concepts. I wonder if similar benefits would accrue from studying the relationship between students' intuitive beliefs and formal definitions. My hunch is that they would, so I hope these authors will continue to pursue this approach to the study of beliefs.

A final observation I wish to make is to comment on the fact that the performance of students in Tsamir and Tirosh's study did not improve with age (grade level), even though the older students had considerably more experience with mathematics. This result merely supports the oft-cited result that many beliefs, even bad or erroneous ones, are extremely resistant to change. To me the resistance of the older students to changing their belief about division by zero might be due to the fact that to say that division by zero is undefined simply does not *make sense* to them. I hope Tsamir and Tirosh will think about what sort of instructional experiences would be needed to help students make sense of this elusive notion.

3. HOW DOES THE STUDENTS' BELIEFS RESEARCH INFORM CLASSROOM PRACTICE?

Recently, Dylan Wiliam and I have argued that mathematics education research should focus on the pursuit of knowledge that causes real, lasting changes not only in the way people think about learning and teaching, but also in how they act (in the classroom, in making policy, in designing curriculum, etc.) (Lester & Wiliam,

2002). That is, we suggest that research should aim to cause people to sit up and take notice; to do something with the results. In what follows, I will attempt to clarify what in the students' beliefs research might cause us to take action.

In my view, for most of the history of research in mathematics education—at least in the United States—our research has been what Kenneth Ruthven has called "internally-focused" rather than "externally focused" (K. Ruthven, personal e-mail communication, December 3, 1998). By internally focused he means that concerns with matters of epistemology and methodology have been predominant. Were our research more externally-focused, we would be much more concerned about the credibility of our inquiry with practitioners and its capacity to make a productive contribution to both policy and practice. Or, as I put it above, our research would cause teachers (and policy makers, curriculum developers, etc.) to sit up, take notice, and do something with the results. With this thought in mind, I asked myself if the set of research reports on students' beliefs in this volume has the potential for changing mathematics teachers' practice. My answer to this question is as follows.

The first, and perhaps most important, contribution of these reports is that they add clarity to our understanding of the role of beliefs in learning and doing mathematics, especially classroom mathematics. For several years I was perplexed about the distinction between beliefs and knowledge. For me these discussions did little to help me understand why I should care about students' beliefs. These papers, in particular Yackel and Rasmussen's, have made me begin to think that to make a sharp distinction between beliefs and knowledge is unhelpful and probably wrong headed. Now I think of a belief as a special form of knowledge—namely, personal, *internal knowledge*. This form of knowledge can be contrasted with *external knowledge*, which is knowledge resulting from the consensus of some community of practice. The formal, mathematical knowledge typically learned in school is, then, external knowledge because it represents a body of information agreed upon by the community of mathematicians. Some of what an individual believes about mathematics happens also to be external knowledge, some is not. But so what? It is the internal knowledge of an individual that matters because it is this kind of knowledge that directs her or his actions and subsequent learning. So, it is vital that we as teachers pay attention to students' beliefs. Beyond this central contribution, each of the reports has something special to offer to teachers.

Kloosterman's study helps us realize that as teachers we need to pay considerable attention to the beliefs students have about the nature of mathematics because these beliefs can have a tremendous impact on their interest in, enjoyment of, and motivation to do mathematics. In particular, we should ask ourselves: "What influence do the activities I engage my students in and the sort of emphases I place on certain aspects of mathematics have on developing beliefs about what doing mathematics involves? For example, if I stress only procedural aspects of mathematics, what beliefs am I fostering about the nature of mathematics?"

Greer, Verschaffel and DeCorte's study presents the very encouraging news that it is possible to change students' beliefs about what is involved in solving word problems. But, what is perhaps even more important is that their research offers documentation for the view that beliefs about mathematics word problems constitute a part of a more general, complex set of beliefs about the nature of mathematics and how it relates to the world both inside and outside of the classroom. This view is consistent with the one posited by Yackel and Rasmussen in their report. In fact, taken together, the two papers present a coherent, compelling message: Students' beliefs about mathematics and mathematics learning can be developed and changed by teachers, if teachers consider both sociological and psychological perspectives in the way they organize their classrooms, establish social norms, and select tasks for students to engage with.

I am not as confident about the potential benefit to teachers of Presmeg's study. As I noted above in my discussion of her report, her work is useful inasmuch as it describes a new tool for researchers interested in studying the formation of beliefs. But, although I think semiotic chaining may prove to be quite helpful as an instructional strategy in mathematics classrooms, at present her research provides no direct implications for classroom practice. We will have to wait for future research to tell us if any direct implications exist.

The research reported by Tsamir and Tirosh offers further evidence that what students believe about mathematical ideas may be quite different from what their teachers expect them to believe and, furthermore, attempts to change these beliefs can be very difficult. The mismatch between a student's internal knowledge (i.e., beliefs) and the external knowledge of the mathematics community can be quite pronounced, just as a student's concept image can be very different from the concept definition. Teachers who are aware of such a mismatch will be alert to the importance of providing a wide range of ways to think about the concepts and procedures they ask their students to learn.

4. FINAL OBSERVATIONS

At the beginning of this essay I indicated that I consider myself an amateur in the area of research on students' beliefs. Now, at the end of the essay, I must admit to still being an amateur, but one who feels much better "tuned in" to what this body of literature—represented by the five reports I have discussed—has to tell us. Two observations have come to mind as I have mulled over what researchers in this field need to be concerned about.

With regard to my first observation, I am reminded of the title of a PME [*The International Group for the Psychology of Mathematics Education*]address made some years ago by Kath Hart, who was then president of PME. In her title, she wondered: "Do I know what I believe? Do I believe what I know?" My version of this question is "Do students know what they believe?" I wonder! I am skeptical about the credibility and reliability of the data presented by these five reports because I doubt that they accurately indicate what the students really believe. Put another way, how much confidence should I have that these authors have really tapped into the students' core beliefs about the nature of mathematics and how it is learned? I am simply not sure that core beliefs can be accessed via interviews (see Kloosterman, Presmeg, and Tsamir & Tirosh) or written self reports (see Presmeg and Yackel & Rasmussen) because interview and self-report data are notoriously unreliable. Furthermore, I do not think most students really think much about what they believe about mathematics and as a result are not very aware of their beliefs. So, although I think these researchers are leading the way in efforts to develop good methodological tools for studying students' beliefs, a considerable amount of work remains to be done.

My second observation is more positive. With the exception of Tsamir and Tirosh, the authors of the other reports clearly acknowledge the tremendous influence socio-cultural context has on the formation of beliefs. Moreover, as Greer et al. point out, this context extends beyond the narrow confines of the classroom to include the total school environment, the educational system, and society in general. Although this realization makes the study of beliefs more complicated, it promises to lead to the development of a body of knowledge about the role of beliefs in mathematics learning that can cause real, lasting changes not only in the way people think about learning and teaching, but also in how they act.

5. NOTES

[1] I am grateful to an anonymous reviewer for calling attention to the circularity in the reasoning of some beliefs researchers.

6. REFERENCES

Goldman, A. I. (1999). *Knowledge in a social world.* New York: Oxford University Press.

Lemke, J. L. (1997). Cognition, context, and learning: A social semiotic perspective. In D. Kirshner & A. Whitson (Eds.), *Situated cognition: Social, semiotic, and psychological perspectives* (pp. 37 - 55). Hillsdale, NJ: Erlbaum.

Lester, F. K., & Garofalo, J. (1982, April). *Metacogntive aspects of elementary school students' performance on arithmetic tasks.* Paper presented at the annual meeting of the American Educational Research Association, New York.

Lester, F. K., Garofalo, J., & Kroll, D. L. (1989). Self-confidence, interest, beliefs and metacognition: Key influences on problem-solving behavior. In D. B. McLeod, & V. Adams (Eds.) *Affect and mathematical problem solving: A new perspective* (pp. 75 - 88). New York: Springer-Verlag.

Lester, F. K., & Kehle, P. (in press). From problem solving to modeling: The evolution of thinking about research on complex mathematical activity. In R. Lesh & H. Doerr (Eds.), *Mathematical models and modeling.* Mahweh, NJ: Lawrence Erlbaum.

Lester, F. K., & Wiliam, D. (2002). On the purpose of mathematics education research: Making productive contributions to policy and practice. In L. English (Ed.), *International handbook of research in mathematics education.* Mahweh, NJ: Erlbaum.

McLeod, D. B. (1992). Research on affect in mathematics education: A reconceptualization. In D. A. Grouws (Ed.), *Handbook of research on mathematics teaching and learning* (pp. 575 - 596). New York: Macmillan.

McLeod, D. B. (1994). Research on affect and mathematics learning in the JRME: 1970 to the present. *Journal for Research in Mathematics Education, 25*(6). (25th anniversary special issue).

Pehkonen, E., & Törner, G. (Eds.) (1998). *The state-of-art in mathematics-related belief research: Results of the MAVI activities.* University of Helsinki, Department of Teacher Education Monograph Series.

Shumway, R. J. (Ed.) (1980). *Research in mathematics education.* Reston, VA: National Council of Teachers of Mathematics.

Vinner, S., & Hershkowitz, R. (1983). Concept images and common cognitive paths in the development of some simple geometrical concepts. *Proceedings of the Fourth International Conference for the Psychology of Mathematics Education* (pp. 177 - 184). Berkeley, CA.

INDEX

A

abstraction............295, 303, 305, 307
academic choice............104
acceptable explanation.........322, 324, 325, 326, 327
achievement motivation............120
activation............81
activities............182, 183, 187, 192
activity, reflexive............205
affect.... 14, 15, 19, 23, 27, 33, 38, 39, 59, 60, 61, 62, 63, 64, 116, 211
affective............16
  ~ domain............59, 115, 211
  ~ pathways............61, 69
  ~ system............212
analysis
  conceptual............346
  methodological............346
  semiotic............349
analytical tools............200
assessment............16, 252, 257, 268
Atlanta Math Project............170
attitude........61, 68, 95, 118, 213, 248
  ~ toward mathematics............261
  measurement............102
attribute............187, 189, 190
attribution............248, 263
  ~ theory............347
authority............144

B

behavior............178, 179, 180, 181
belief
  ~ objects............78
  ~ structure.........59, 60, 64, 68, 69, 70, 178, 179, 180, 181, 187, 189, 191, 192
  ~ system............26, 27, 29, 30, 36, 38, 39, 59, 64, 67, 68, 85, 179, 191
  intuitive............331, 332, 341
  primary.... 187, 189, 190, 191, 192
beliefs.........101, 162, 177, 178, 179, 180, 181, 182, 183, 184, 187, 188, 189, 191, 192, 212, 313, 314, 316, 317, 318
beliefs about
  ~ learning............204
  ~ mathematics... 15, 18, 19, 20, 21, 22, 23, 25, 28, 29, 30, 31, 32, 35, 36, 38
  ~ mathematics curriculum.......156
  ~ mathematics learning............22
  ~ mathematics teaching......18, 19, 22, 31
  ~ the self............17, 29, 30, 33, 34
  ~ the social context............19, 20, 21, 22, 34
beliefs and knowledge............351
beliefs
  ~ research............345, 346
  change of..........153, 154, 282-283, 313, 314, 316, 328
  contradictory............28, 32
  definitions............73, 95
  domain-specific............87
  empirical relevance of ~............74
  epistemological............18, 74
  espoused............135
  inferential............179, 187, 189,

G. C. Leder, E. Pehkonen, & G. Törner (Eds.), Beliefs: A Hidden Variable in Mathematics Education? 355-362.

190, 191, 192
intuitive....350
mathematical....73, 313
measurement....95
motivational. 14, 17, 19, 26, 33, 34
normative....64
of cognitive psychologists....287-288
ontological....294, 298, 302, 305, 310, 311, 348
personally-held....350
quasi-logical character of ~....81
quasi-psychological character of ~....81
specifically mathematical 313, 315, 324, 329
subject-matter....87
teachers' about curriculum....149, 156
warrants for....60
breadth of a belief object....78

C
calculational orientation....326, 327,329
cases....198
causal attributions....117
central....179, 180, 188, 192
certitude....80
chains of signifiers....293
challenge....262, 266
change....128, 177, 178, 179, 180, 181, 185, 190, 192
changes in practice....294
characterization....40
children's thinking....206
circular reasoning....346, 347
class context....13, 28, 29, 30, 34, 35, 36
classroom
~ norms....349
interactions....323, 329
participation structure....314, 320, 322, 323
social norms....35
cognition....59, 60, 61, 62, 69, 116
cognitive
~ components....212
~ development....211
~ strategies....15
collaboration....170, 172
communication....59, 61, 179, 185, 186, 187, 189, 190, 191
community of practice....25, 196, 200, 207
competence....238, 239, 240
conative....14, 15, 23
concept
~ definition....350, 352
~ image....77, 350
conceptions....14, 16, 19, 23, 26, 28, 30, 41,75, 212
conceptual
~ orientation....326, 329
~ tools....198
conflict....182, 184, 189
pedagogical....189, 190, 191
connection(s)....183, 185, 186, 187, 189, 192, 294, 295, 298, 302, 310
consciousness....80
constitution....143
~ of norms....327
constructivism, constructivist....133, 134, 135, 234, 238, 239
content set....78
context....128, 132
socio-cultural....353
contextual factors....131, 347, 348

control 213
core beliefs 348, 352
course portfolios 317, 318, 319
cultural practices 293, 295, 296,
301, 302, 309, 310
culture 240, 242
curriculum
~ initiatives 149
~ materials 150, 151, 152,
.... 155, 156, 158
reform-oriented 154
teachers' treatment of 157

D
debate.................................. 134, 139
decentering........................... 134, 136
definition............... 13, 16, 17, 25, 26,
30, 31, 35, 115, 118
derived 204
developing
~ new beliefs 155
~ the identity 205
development
~ of an identity 206
~ of identity............................205
~ of teachers............................150
didactical contract.................281,284
differences, individual 276, 284
differential equations ... 313, 316, 317,
320, 321, 322, 325, 326, 327, 328
disaster studies 281
discourse 201
discourse, discursive practice 235,
.... 236, 237, 238, 240
discursive use 206
disposition...................... 75, 130, 141
~ mathematical 14, 15, 36
division 334
~ by zero 331, 332, 333,
336, 338, 340

E
efficacy 211
~-beliefs 211
effort 248, 249, 250, 251,
.... 253, 255, 256, 266, 267
electronic journals......... 317, 318, 325
Elise 181, 182, 183,
188, 189, 190, 191
emotion 14, 15, 19, 28, 33, 59,
61, 62, 68, 69, 116, 212
emotional 212
enjoyment of mathematics 254, 255
equity 295
ethics 61, 66, 67, 68
ethnicity 121
evaluation maps 81
everyday practices 293, 295,
302, 303, 310
evolution 61
excitement...............................61, 62
expectations 314, 315, 316, 318,
319, 320, 322, 323, 324
~ of teachers............ 252, 257, 268
experience 181, 184, 185, 186
187, 188, 190, 192
Experience Sampling Form (ESF)......
.... 105, 106, 109, 111
Experience Sampling Method (ESM).
.... 103, 104, 105, 108, 111, 120
experiential 185, 186, 192
experientially-real mathematical object
....316

F
factors, sociocultural 286-287
fear 60, 62, 63
~ of mathematics.......................63

feelings 212
focus of interactions.......................204
formal 334
forms of participation 196
four-component-definition.............77
framework............. 13, 16, 17, 18, 23, 28, 29, 30, 31, 36
framing 237, 238, 239
frustration......................61, 63, 69, 70
fun...............................183, 187, 189

G
gender 115, 284, 285, 286
generalization...............294, 295, 301, 302, 305, 307, 310
global affect 61
goal orientation 248, 251, 255, 256, 267
growth........... 177, 179, 180, 181, 191

I
identity........................179, 188, 192, 235, 236, 240, 241
ideology 75
image 75
infinity 336, 339, 340
~-undefined...............................342
inquiry..........................135, 137, 143
~ mathematics 149
~ mathematics tradition ..314, 316, 320, 326, 329
~-based instruction..................151
integrity, mathematical 62
interaction patterns........315, 318, 324
interactive constitution.........320, 322, 323, 325
interpretive framework 313, 315, 316, 328
intimacy, mathematical..................62

J
joy..62, 63
justification 331, 335, 336, 340
mathematically-based 337
practically-based......................337
rule-based................................339

K
key word strategy..........................347
knowledge................... 13, 14, 15, 16, 17, 18, 22, 23, 24, 25, 26, 27, 28, 29, 30, 31, 32, 36, 38, 39, 74, 130, 131
~ developed as learners...........155
~ structure 237
external 351, 352
internal............................351, 352
pedagogical content 198

L
learning
~ environments 206, 208
~ from classroom experiences increases as teachers 153
~ in practice 200
~ in practice at university.........207
~ opportunities 152
~ to teach at the university.......207
Likert scales 347
lived experiences 295, 296, 297
local affect 63

M
Mark 181, 184, 185, 186, 188, 189, 190, 191
materials, reform-oriented 152
mathematical

~ ability 67, 68, 69
~ explanations 315, 326
mathematical modeling 272-273, 282-283, 285, 348
perspective 348
mathematically-based 334
mathematics 59, 60, 61, 62, 63, 64, 65, 66, 67, 68, 69, 70
~ methods course 206
~ teaching, segments of 198
folk perception of 288
nature of 252, 257, 259, 260, 262, 263, 293, 294, 295, 296, 302, 303, 309, 310, 311
relation to reality 288
teaching of ~ 211
usefulness of ~ 249, 250, 251, 254, 266
meanings 293, 294, 295, 299
~ of the artifacts 206
measures, physiological 99
membership degree function(s) 80
memorization 252, 255, 260, 262, 263, 267
mental manipulatives 272, 288
meta-affect 59, 62, 63, 122
metacognitive, ~ knowledge 13, 14, 15
metaphor 188, 190
model 13, 16, 17, 19, 21, 22, 23, 25, 26, 28, 30, 31, 33, 38
modeling 164, 170, 172
morals 61, 66, 67, 68
motivation 14, 15, 16, 18, 19, 23, 28, 33, 38, 211, 247, 249, 250, 251, 252, 253, 255, 257, 260, 261, 263, 346, 347, 351
multiplication 334
mutual engagement 200, 206

N

NCTM 100
negotiation 318, 320, 322, 323, 327, 328, 329
~ of meanings 196, 197, 208
non-motivated 24
normative patterns of interaction 24
norms
social 313, 315, 316, 318, 320, 321, 322, 323, 324, 325, 327, 328, 329, 352
sociomathematical 34, 282, 284 313, 315, 316, 324, 326, 327, 329, 349

O

obligations 314, 316, 320, 324
observations 99, 111
operation 331, 332
performing 333
ownership 300, 301

P

participation .. 195, 206, 208, 240, 241
~ in mathematics 105
partial 198
peripheral form 203
patterns 301, 303, 304, 305, 306, 307, 308
pedagogy, pedagogical . 132, 133, 134
perception 75
performance 237, 238, 239
peripheral 196
~ member 200
personal meaningfulness 300
personally-meaningful solutions 316, 324

perspectives 177, 179, 180, 186, 190, 192
psychological 349
situated learning 196
sociological 349
philosophy 75
plan/teach/debrief cycles 165, 171
play 181, 183, 187, 188
position 234, 236, 237, 238, 240
practice 130, 233, 234, 235, 236, 240, 241, 242
dimensions of ~........................199
elements of ~...........................199
reflective 205
repertoire of the ~ 203
preservice mathematics program ..220
problem solving 13, 14, 15, 16, 17, 18, 22, 23, 24, 26, 28, 30, 31, 32, 33, 34, 35, 36, 37, 38, 39, 117, 139, 143, 182, 183, 186, 187
proof 258, 259
psychological
constructivism 314
perspective 313, 327
pupils' ways of thinking 198

R

rate of change........320, 326, 327, 328
real world 185
realistic mathematics education ...295, 310 329
real-life experiences......................295
reasoning, pedagogical 198, 206
reflection......................132, 137, 171, 172, 187, 189, 301
Reflective Teaching Model (RTM) 163
reflexivity.....................315, 316, 317
reform 128, 132, 252
~ mathematics curricula...........263
reification.....................195, 206, 208
reifying 198
relativistic 132
renegotiation . 320, 321, 323, 328, 329
representation modes 199
research
~ methods 120
~ report 129
rule-based 337

S

scale
Likert ~ 98
semantic differential ~ 98
Thurstone's interval ~ 98
scaling, Gutman 98
schema 25, 27, 28
school
~ climate 220
~ mathematics tradition 314, 319, 329
secondary ~ 247, 253
self
~-beliefs 213
~-confidence 117, 250, 251, 255, 267
~-efficacy 117, 248
~-identity, mathematical 62
~-regulation 122
~-worth 248
semiosis 302, 349
semiotic chaining .. 295, 302, 349, 352
semiotics 302, 349
sense 128, 131
~-making, suspension of .. 274, 281
shared affect...................................61
sign 302, 306
signified 302, 306, 309

signifier................302, 306, 309, 310
social class...................121, 284, 286
social
~ context.................................121
~ interaction...........314, 315, 316, 317, 322, 324, 329
~ practice.................................208
sociocultural.........................236, 240
sociological perspective..............313, 314, 327
sociology..............236, 239, 240, 242
story.....................................181, 188
strategy 183, 187, 189, 190
student teaching............................207
students' beliefs.....13, 14, 15, 16, 17, 18, 19, 21, 22, 23, 24, 28, 30, 31, 32, 33, 34, 36, 37, 345, 348, 349, 351
students' beliefs about
~ mathematics.........13, 19, 29, 31
~ mathematics learning........19, 21
~ mathematics teaching 21
~ the self.......................13, 33, 34
~ the social context........13, 19, 36
students
~' mathematics-related beliefs..15, 16, 18, 26, 29, 30
mature age..........95, 103, 105, 110
study habits in mathematics.........252, 256, 268
symbolic interactionism.......313, 314, 328, 329
symbolism...................................302
systematization 301, 310

T

taken-as-shared............................316
task difficulty...............................262
teacher
~ thinking.................................177
learning of ~.............................195
teachers.......................................152
~' beliefs................128, 150, 151, 152, 278.281, 348
development......................151, 158
~' beliefs about curriculum......157
~' beliefs and knowledge.........200
~' thinking.......................180, 181
teaching problems, practical.........199
techniques
projective..................................99
repertory grid............................99
textbook(s)...................150, 156, 158
traditional mathematics...........151
texts, traditional...........................152
theory
goal-orientation........................347
psychological...........................347
self-efficacy 347
self-worth................................347
theoretical..33, 234, 236, 237, 239
tools, repertoire of the....................07
Treviso Arithmetic..........71, 272, 273
truth...31
tune goals......................................99

U

undefined...............31, 333, 335, 339
understanding(s), understand.........28, 129, 132, 134, 135, 136
unit of analysis.....................240, 241

V

validity.............59, 61, 64, 65, 66, 67
values..........60, 61, 63, 66, 67, 68, 69
viability..65
visual imagery.............................294

W

word problem........................347, 351
~ game 273-274, 276-278, 284
~ schema..................................273
algebraic ~s..............................287
history of ~s 271-272
nature of ~s 271-273
non-routine ~s..................275, 282
purpose of ~s............................272
world view..............................75, 121

Z

zero-undefined..............................342

Mathematics Education Library

Managing Editor: A.J. Bishop, Melbourne, Australia

1. H. Freudenthal: *Didactical Phenomenology of Mathematical Structures.* 1983
ISBN 90-277-1535-1; Pb 90-277-2261-7
2. B. Christiansen, A. G. Howson and M. Otte (eds.): *Perspectives on Mathematics Education.* Papers submitted by Members of the Bacomet Group. 1986.
ISBN 90-277-1929-2; Pb 90-277-2118-1
3. A. Treffers: *Three Dimensions.* A Model of Goal and Theory Description in Mathematics Instruction The Wiskobas Project. 1987 ISBN 90-277-2165-3
4. S. Mellin-Olsen: *The Politics of Mathematics Education.* 1987
ISBN 90-277-2350-8
5. E. Fischbein: *Intuition in Science and Mathematics.* An Educational Approach. 1987
ISBN 90-277-2506-3
6. A.J. Bishop: *Mathematical Enculturation.* A Cultural Perspective on Mathematics Education. 1988 ISBN 90-277-2646-9; Pb (1991) 0-7923-1270-8
7. E. von Glasersfeld (ed.): *Radical Constructivism in Mathematics Education.* 1991
ISBN 0-7923-1257-0
8. L. Streefland: *Fractions in Realistic Mathematics Education.* A Paradigm of Developmental Research. 1991 ISBN 0-7923-1282-1
9. H. Freudenthal: *Revisiting Mathematics Education.* China Lectures. 1991
ISBN 0-7923-1299-6
10. A.J. Bishop, S. Mellin-Olsen and J. van Dormolen (eds.): *Mathematical Knowledge: Its Growth Through Teaching.* 1991 ISBN 0-7923-1344-5
11. D. Tall (ed.): *Advanced Mathematical Thinking.* 1991 ISBN 0-7923-1456-5
12. R. Kapadia and M. Borovcnik (eds.): *Chance Encounters: Probability in Education.* 1991 ISBN 0-7923-1474-3
13. R. Biehler, R.W. Scholz, R. Sträßer and B. Winkelmann (eds.): *Didactics of Mathematics as a Scientific Discipline.* 1994 ISBN 0-7923-2613-X
14. S. Lerman (ed.): *Cultural Perspectives on the Mathematics Classroom.* 1994
ISBN 0-7923-2931-7
15. O. Skovsmose: *Towards a Philosophy of Critical Mathematics Education.* 1994
ISBN 0-7923-2932-5
16. H. Mansfield, N.A. Pateman and N. Bednarz (eds.): *Mathematics for Tomorrow's Young Children.* International Perspectives on Curriculum. 1996
ISBN 0-7923-3998-3
17. R. Noss and C. Hoyles: *Windows on Mathematical Meanings.* Learning Cultures and Computers. 1996 ISBN 0-7923-4073-6; Pb 0-7923-4074-4

Mathematics Education Library

18. N. Bednarz, C. Kieran and L. Lee (eds.): *Approaches to Algebra.* Perspectives for Research and Teaching. 1996 ISBN 0-7923-4145-7; Pb ISBN 0-7923-4168-6
19. G. Brousseau: *Theory of Didactical Situations in Mathematics.* Didactique des Mathématiques 19701990. Edited and translated by N. Balacheff, M. Cooper, R. Sutherland and V. Warfield. 1997 ISBN 0-7923-4526-6
20. T. Brown: *Mathematics Education and Language.* Interpreting Hermeneutics and Post-Structuralism. 1997 ISBN 0-7923-4554-1
Second Revised Edition. 2001 Pb ISBN 0-7923-6969-6
21. D. Coben, J. O'Donoghue and G.E. FitzSimons (eds.): *Perspectives on Adults Learning Mathematics.* Research and Practice. 2000 ISBN 0-7923-6415-5
22. R. Sutherland, T. Rojano, A. Bell and R. Lins (eds.): *Perspectives on School Algebra.* 2000 ISBN 0-7923-6462-7
23. J.-L. Dorier (ed.): *On the Teaching of Linear Algebra.* 2000 ISBN 0-7923-6539-9
24. A. Bessot and J. Ridgway (eds.): *Education for Mathematics in the Workplace.* 2000 ISBN 0-7923-6663-8
25. D. Clarke (ed.): *Perspectives on Practice and Meaning in Mathematics and Science Classrooms.* 2001 ISBN 0-7923-6938-6; Pb ISBN 0-7923-6939-4
26. J. Adler: *Teaching Mathematics in Multilingual Classrooms.* 2001 ISBN 0-7923-7079-1; Pb ISBN 0-7923-7080-5
27. G. de Abreu, A.J. Bishop and N.C. Presmeg (eds.): *Transitions Between Contexts of Mathematical Practices.* 2001 ISBN 0-7923-7185-2
28. G.E. FitzSimons: *What Counts as Mathematics?* Technologies of Power in Adult and Vocational Education. 2002 ISBN 1-4020-0668-3
29. H. Alrø and O. Skovsmose: *Dialogue and Learning in Mathematics Education.* Intention, Reflection, Critique. 2002 ISBN 1-4020-0998-4
30. K. Gravemeijer, R. Lehrer, B. van Oers and L. Verschaffel (eds.): *Symbolizing, Modeling and Tool Use in Mathematics Education.* 2002 ISBN 1-4020-1032-X
31. G.C. Leder, E. Pehkonen and G. Törner (eds.): *Beliefs: A Hidden Variable in Mathematics Education?* 2002 ISBN 1-4020-1057-5; Pb ISBN 1-4020-1058-3

KLUWER ACADEMIC PUBLISHERS – DORDRECHT / BOSTON / LONDON

GPSR Compliance

The European Union's (EU) General Product Safety Regulation (GPSR) is a set of rules that requires consumer products to be safe and our obligations to ensure this.

If you have any concerns about our products, you can contact us on ProductSafety@springernature.com

In case Publisher is established outside the EU, the EU authorized representative is:

Springer Nature Customer Service Center GmbH
Europaplatz 3
69115 Heidelberg, Germany

Batch number: 10614741

Printed by Printforce, the Netherlands